D0142171

Encyclopedia of Geography Terms, Themes, and Concepts

Encyclopedia of Geography Terms, Themes, and Concepts

Reuel R. Hanks

Contributing Author
Stephen J. Stadler

Santa Barbara, California • Denver, Colorado • Oxford, England

Library of Congress Cataloging-in-Publication Data

Hanks, Reuel R.
Encyclopedia of geography terms, themes, and concepts / Reuel R. Hanks ; contributing author, Stephen J. Stadler.
p. cm.
Includes bibliographical references and index.
ISBN 978–1–59884–294–4 (hardcopy : alk. paper) — ISBN 978–1–59884–295–1 (ebook)
1. Geography—Encyclopedias. I. Stadler, Stephen John, 1951– II. Title.
G63.H38 2011
910.3—dc23 2011022895

ISBN: 978–1–59884–294–4
EISBN: 978–1–59884–295–1

15 14 13 12 11 1 2 3 4 5

This book is also available on the World Wide Web as an eBook.
Visit www.abc-clio.com for details.

ABC-CLIO, LLC
130 Cremona Drive, P.O. Box 1911
Santa Barbara, California 93116-1911

This book is printed on acid-free paper ∞

Manufactured in the United States of America

Contents

List of Entries

List of Related Entries

Human Geography Entries

Physical Geography Entries

Acknowledgments

This volume is the product of a great number of people, all of whom contributed to it either directly or indirectly. The first who must be acknowledged is my late father, Cliff Hanks, who many years ago bought my first subscription to *National Geographic* magazine and ignited my interest in other places, peoples, and cultures. I have never stopped learning about geography since receiving that first issue. I am in the debt of many scholars and professors at the University of Missouri and the University of Kansas, where I received several degrees in the discipline of geography. The department at Missouri gave me my first opportunity to teach geography, and much to my astonishment (and that of some of my early students, I am sure), I have been doing just that for the past 25 years. Professor Jesse Wheeler, who taught geography to thousands of students at the University of Missouri for several decades, played an important role in further stimulating my interest in the world, as did my advisor there, Professor Trent Kostbade. My advisor at the University of Kansas, Professor Leslie Dienes, gave generously of his time and expertise, and also taught me a great deal about economic geography and the vast stretches of Eurasia. There are many others who must remain unnamed, but who helped me in innumerable ways. I am of course grateful to Steve Stadler for his contributions to this work and sharing his knowledge of physical geography. I worked with several editors at ABC-CLIO over the course of writing: Lynn Jurgensen, David Paige, Kim Kennedy-White, and Robin Tutt all provided useful and timely feedback, as well as a nearly limitless reserve of patience. Finally, my wife Oydin, mother Frances, and daughter Kami all tolerated my long hours and piles of books as I worked through the manuscript. I thank them most of all.

Introduction

Geography, like all academic disciplines, has its own nomenclature and terminology. Those learning about the field for the first time may be bewildered by the many specialized terms and concepts that geographers use, a feeling frequently compounded by the broad interests that geographers pursue and the many subfields that make up the study of geography as a whole. But it is the very diversity of geography that gives the discipline depth and strength, and allows us to think in integrative and complex ways about the myriad spatial processes we encounter on a daily basis. Almost everyone carries a "mental map" that allows us to organize the world into an order that we recognize and expect—we are by nature all geographers. Geography is the science of location, and the entire world is the laboratory of the geographer. To appreciate the wide body of knowledge that geography offers, a mastery of the language of the field is essential.

Such an appreciation is vital in a globalizing world. Those who have a weak knowledge and understanding of the world around them are destined to be at a disadvantage, both professionally and personally. Geography provides the fundament for such understanding. For the past several decades, studies have shown that American students, and the general public, lack a solid grounding in geography. When tested about their knowledge of other peoples and countries, for example, Americans typically rank near the bottom when compared to most Europeans. Some have explained this as simply the result of North America's relative isolation, the fact that the country borders on only two other states, one of which is primarily English-speaking and shares a similar culture, and other factors, most of which, ironically enough, are geographic. Yet these excuses are no longer acceptable, or even relevant. Physical distances, in the age of modern transportation systems, have become a secondary consideration, and contact with the remainder of the world is no longer dependent on actual travel, although visiting other locations is certainly an excellent way to learn about them. In the era of the Internet and globalization, we can no longer think in terms of the world being "out there somewhere." It is here and now, often right outside our doors, perhaps even in our homes. The things we eat and drink, the vehicles we use for transportation, even the clothing we wear all come to us from "out there." A failure to appreciate this diminishes us both intellectually and culturally.

There are, of course, other reasons that Americans are not as aware of geography as they should be. For a considerable time, several decades at least, geography as an independent subject almost disappeared from the curricula of American schools. Exactly why and how this happened is a matter of some discussion and debate among geographers, but the effect was to fold the study of geography into the study of history, civics, or some other discipline, where in fact it often was ignored, or at least minimized in importance. This coincided with a decline in the field at the university level starting in the 1940s and extending into the 1980s, when colleges and universities, some of which had traditionally supported quite vigorous programs in geography, collapsed those programs and departments into other related disciplines.

Fortunately, this trend abated in the late 1980s, and since that time, due to the efforts of the professional organizations in the field, scholars and teachers at all levels and many legislators and policymakers, geography has experienced an academic resurrection of sorts. Occasionally today an unenlightened administrator will question the need for geography in his or her curricula, but these incidents have become relatively rare. However, the need for information and understanding of the way the world is organized and functions is so compelling to most that it is no longer a matter of whether geography should be studied, but rather a question of how to motivate students to learn as much geography as possible, and why this is important to their future success.

The discipline of geography is distinguished by its emphasis on observation and analysis in a spatial context. Any phenomenon that is expressed through the dimension of space, therefore, may be and probably has been examined by a geographer. Yet there are new areas of study emerging every day for scholars, making the study of geography quite broad and inclusive. The result is that geographers typically have close ties to many other fields of inquiry, not only in the social sciences, but in the humanities and natural sciences as well. In addition, many geographers specialize in one or more of the techniques that are used when analyzing geographical data or information. In recent years these have become quite sophisticated, using computer data bases containing layers of spatial information to conduct quite complex studies of multiple phenomena. In general, geographers may be placed in one of two categories: physical geography or human geography. Physical geographers concentrate on "natural" patterns—the distribution of vegetation, landforms, climate types, etc., while human geographers are concerned with systems and distributions created by human activity. The latter may include religions and languages, economic systems, the spatial expression of political power, and a virtual limitless number of others. Some geographers work in both areas. Those scholars who study cultural ecology or environmental geography, for example, frequently include in their work an analysis of the interaction between the human and natural worlds. All studies in geography attempt to answer, either directly or obliquely, two basic questions: where, and why?

This handbook is a modest effort to aid in the study of a fascinating and crucial approach to the world—the approach of geography. My colleague at Oklahoma State University, Dr. Stephen J. Stadler, wrote the entries in this volume focused on physical geography, and I covered the material that deals with human geography. Such an endeavor presents a number of challenges. First, one must select those terms and concepts deemed most relevant, important, and necessary to write about. There will inevitably be concepts and vocabulary left out that might, and perhaps should, be included. Second, there is the task of presenting and explaining a complex lexicon of geography using a quite limited number of words. Condensing the information that in many instances might fill several pages of a textbook was a significant and difficult task indeed. My hope is that we have managed to convey the essence of these ideas, while presenting them as concisely as possible. Our purpose here is to present those terms and concepts most likely to be encountered by students and teachers of geography, as well as by the layman who seeks to enlarge his or her understanding of the world, and the spatial relationships that give rise to its diversity and wonder. Most of all, the aim of this volume is to inspire many to learn more about geography and read further about those ideas that brought them to this book in the first place. The selected bibliography at the end of the work provides a listing of both classic and contemporary works that may be consulted by readers who seek a deeper discussion than what we are able to offer in these pages.

Reuel R. Hanks

A

Agglomeration

Agglomeration may have two general meanings to geographers. To urban geographers, it is a term that designates a large urban concentration. To economic geographers, the word refers to the tendency of producers in a given industry to cluster together in a given space.

In urban geography, agglomeration is a broad term that is used to identify a large, extended area of urban development. It is often used as a synonym for similar terms, such as conurbation, metropolitan area, or metropole. There is no established definition for how large an urban concentration must be to be classified as an agglomeration, and scholars of urban geography tend to use the term in a general sense, frequently employing it in a case where two or more significant urban clusters have fused to form a larger urban space. Agglomeration may even occur across political boundaries, as in the cases of Detroit (United States) and Hamilton, Ontario (Canada); Brownsville, Texas and Nuevo Laredo (Mexico); and Spokane (United States) and Vancouver (Canada). While these units actually occupy different sovereign spaces, they nevertheless function as a single urban unit in many respects, especially in the case of economic linkages between U.S.–Canadian agglomerations.

For economic geographers, agglomeration signifies the tendency of units of economic production to group together in the same location. This clustering provides many potential economic advantages, including achieving economies of scale, utilization of a common transport structure, lower shipping and transport costs between firms making specialized products, concentration and transfer of capital and labor, and increased communication among various units. Typically agglomeration occurs near or in large metropolitan areas, facilitated by the large pools of capital, labor, and consumers located there. Agglomeration increases the advantages brought on by so-called network effects, which often result in lower operating costs from increased competition among suppliers, a larger and more diverse pool of potential employees, and attracting a larger number of consumers to a central location. The latter may be simply illustrated by the grouping of gasoline service stations around a central intersection in a town. Although each station is in direct competition with the others at this location, the stations cluster due to the large number of potential customers who frequent the location.

Agglomeration is a phenomenon that is frequently associated with the production of sophisticated, high-value goods that require a technically skilled labor supply and the input of multiple components in the assembly of the final product. The manufacture of automobiles is a process typically marked by agglomeration. Because the production of automobiles requires a large amount of high-quality steel, automobile plants, at least in the early days of the industry, were often located near iron and steel manufacturing facilities, or at least close to railroads or water transport that could be used to bring steel to the plant at relatively low cost. The automotive industry also requires the production of a large number of specialized products, such as tires, automotive glass, etc., and firms supplying these commodities naturally congregate in the same geographic space as their major clients, which in turn frequently attract yet additional businesses who service these manufacturers, leading to yet additional agglomeration.

Agribusiness

There are two connotations for the term "agribusiness." In a more general sense, the word is used to identify the integrated and diverse components of modern crop production, including the actual grower, but also many others: silo operators; fertilizer, seed, and farm implement salespeople; agricultural technicians; and others involved in agricultural production. When used this way, the word is inclusive of all aspects of modern agricultural economics, including the marketing and advertising of agricultural commodities, and is not focused simply on production techniques and methods. In a more specific sense, the term may be used interchangeably with the phrases "industrial farming," "corporate farming," or "factory farming." In this context, "agribusiness" means the application of mass production techniques to farming along with the advent of large agricultural units, typically under corporate ownership. Production in this system relies on achieving economies of scale, either through the acquisition and combination of farmland into larger parcels (for crop production), or through the concentration of animals into high-density facilities, where feeding, breeding, and processing costs are minimized (for livestock production).

Agribusiness in the sense of factory farming is a widespread phenomenon in the agricultural ecology of the industrialized world and may be seen as a continuation of the **agricultural revolution** initiated in Great Britain in the early 1700s. The most obvious spatial manifestations of agribusiness are extensive farms, and animal husbandry methods utilizing high-density units like poultry houses, hog parlors, or other so-called Confined Animal Feeding Operations, or CAFOs, a designation assigned in the United States by the U.S. Department of Agriculture. Since the 1940s, shifts in many areas of agriculture have occurred, resulting in larger, more intensively utilized production units. For example, in 1970, the average dairy farm in the United States had 19 cows, but by 2007 that number had risen to 128 cows per farm. In some regions, particularly in western states, the increase was much larger than the national average. The overall size of farms has increased substantially as well. The size of the average farm in the United States increased from about 150 acres in the 1930s to more than 400 acres in 2002, although the total number of farms in the United States has decreased steadily since the 1940s. Nevertheless, non-family farms, that is, those corporately owned, account for only about two percent of all farms in the country, although the corporate farms produce almost 14 percent of total agricultural output by value. Moreover, corporate farming is particularly important in the production of meat and animal products. A handful of corporations slaughter the great majority of livestock in the United States, and in the case of beef, account for more than 80 percent of the output.

The emergence of agribusiness may be traced to the beginning of the Industrial Revolution. It was only a short time before the mass-production techniques of the factory began to be applied to the countryside. This was especially true in the case of mechanization, with a steady progression of machines replacing both man and animal power in the fields after 1820. The advent of steam-powered equipment gave even greater momentum to the process of mechanization, since such machines could do the work of many men, or several draft animals, faster and more efficiently. This had the effect of drastically reducing the agricultural labor force, as well as intensifying competition, as the relative prices of agricultural commodities fell because of mass production. By the 1930s affordable commercial fertilizers had become widely available in the United States, a development that substantially increased yields. A decade later, advances in genetic science and botanical engineering, still a major component of agribusiness, paid enormous dividends in total production, especially the hybrids generated as a result of the Green Revolution. Food production in many parts of the world skyrocketed, as the new technology was married in many cases with advanced business strategies and an expanding market. After the application of High Yield Varieties (HYVs) of wheat and rice in India in the 1960s, the country has avoided major food shortages for the past 50 years, in spite of a rapidly growing population.

Agribusiness today is characterized by technological innovations and managerial methods that have vastly increased production. An example of the former is the technique of hydroponics, a technology that enables higher-value vegetables to be grown intensively in large greenhouses rather than in agricultural fields. There are multiple advantages to producing vegetables in this manner. First, the greenhouse presents a closed environment that allows for easier control of insect pests, limits the incidence of plant disease, and avoids the impact of adverse weather on the plants. Soil is not necessary for this type of production, as the plants are grown in an inert media saturated with a solution of water and plant food that provides all the nutrients necessary for the plant to thrive. In some cases individual plants are often fed via a system of drip irrigation, which greatly reduces the amount of water required, another major advantage in the process. Artificial lighting in these facilities can supplement natural sunlight, and in the United States, yields may be 6–8 times what a comparable field would produce. A related approach is utilized in aquaculture, in which both plant and animal products are produced in a confined aquatic space, usually a pond or large tanks. A large percentage of fish and shellfish are now produced this way rather than being harvested from the ocean or fresh-water bodies.

Animal products have also seen enormous increases in yields due to agribusiness advances. Much of the world's meat and eggs is produced using factory farming techniques, which cluster animals in high-density enclosures. These facilities allow for intensive farming and maximum output, and advocates of these production strategies argue that they are not only the most efficient mode of production, but that they provide humane conditions for livestock. An example of the increased efficiency may be seen in the fact that over the last 40 years, the number of hog farms in the United States has declined precipitously, but production of pork meat products has greatly increased, as hog farming has shifted to factory farming. There is no question that such approaches to meat production have allowed the ever-rising demand for animal products to be met, but the new production techniques have brought new challenges and problems as well. The large scale of factory farming operations and the high density of animal confinement results in environmental challenges, as large amounts of waste, noise, and other undesirable side effects must be addressed and properly contained. Moreover, ethical questions concerning the welfare of animals have become a major issue in North America and Europe, with many animal rights advocates calling for the curtailment or complete abandonment of some procedures, such as the removal of the beaks of poultry and the use of so-called "sow stalls" for breeding pigs. Some countries in Europe and a few U.S. states have limited or banned the use of some confinement techniques, although defenders of factory farming suggest that these procedures are necessary to prevent animals from injuring themselves or other livestock. Factory farmers also contend that producing animal products on a scale that will meet future demand requires the continuation of confinement and mass-production strategies.

Concern over agribusiness production techniques has led to a number of alternatives. Critics question the **sustainable development** of some agribusiness production techniques. The rise of *organic farming* is in many cases in direct response to concerns over the mass production strategies pursued by large farming corporations, especially the use of large amounts of fertilizers, pesticides, and herbicides. Organically produced vegetables and fruits have become a multi-billion dollar industry, and legislation governing what may be labeled "organically grown" has been passed in many states. The condition of livestock in CAFOs has elicited responses from animal rights' activists, who argue that such confinement and the techniques employed in the production of animal commodities is inhumane and unnecessary. The use of gestation crates in pork production, for example, is viewed as cruel by activists, and indeed this technique has been outlawed in some countries and U.S. states. Because of the controversy, some companies have pledged to eliminate the practice in the near future. Likewise, mass production of eggs and chickens by using large-scale chicken houses has been answered by the *free range* movement, which advocates allowing animals to roam freely outside rather than being confined. Free-range poultry accounts for up to a quarter of the total produced in some countries, but there are disadvantages to this method as well. Free-range poultry units must be much larger in area to allow the animals to roam and forage and are economically less efficient than methods based on confinement. Other opponents of large-scale agribusiness promote the return to "multi-species" farming, where animals are not confined and are allowed to roam and mingle on pastureland; but while such alternatives may be viewed as more humane, they typically are far less productive and result in higher prices. With global **population** and demand for food steadily rising, some commentators question whether the alternative approaches in production to large agribusiness methods can adequately and cost-effectively satisfy the world's need for food.

Agricultural Regions. *See Cultivation Regions.*

Agricultural Revolution

"Agricultural revolution" is a term that has broad meaning to geographers and historians, and may refer to a number of events and processes. Some scholars use the phrase in reference to the Green Revolution (see sidebar) of the latter 20th century, while some historians use "agricultural revolution" to identify the Neolithic period when plants were first domesticated and societies developed systems of irrigation and settled cultivation. Various cultures and regions may be said to have undergone such a "revolution" at some point in their history. The most common application of the term is to describe the fundamental shift in agricultural theory and practice that took place in Great Britain from the 1700s to the early 1900s, resulting in marked increases in overall production by volume, as well as sharp increases in productivity per agricultural laborer and per unit land. A somewhat similar process occurred in the United States in the late 1800s and early 20th century.

The agricultural revolution in Great Britain was an extended process that stretched across several centuries. By 1700, scientific innovations in agricultural techniques were being more commonly applied on British farmsteads, resulting in higher yields and more effective use of farmland. A famous advocate of the application of new, scientific means of farming was Jethro Tull, who was one of the first scholars to study agricultural systems of production in a comparative way. Tull was familiar with farming techniques used outside of Great Britain and emphasized the control of weeds, soil preparation, and proper planting techniques. He also was an early advocate of mechanization in agriculture, and invented a seed drill that gradually replaced the hand-sowing of seeds for a number of crops in England, although the new technology was not widely adopted until several decades after his death. Tull also made important modifications to the horse-drawn plough, which improved the furrowing of the soil, and in subsequent decades his successors in agricultural innovation continued modifying the device, resulting in the application of cast iron ploughs based on Chinese designs that were superior to those used in Great Britain at the time, in a classic example of **cultural diffusion**. The seed drill and improved ploughs were still dependent on draft animals, mostly horses, however, which limited their efficiency and made them more physically demanding to operate.

Changes in land tenure during the 1700s were also a key component of the agricultural revolution. One of the most profound and controversial was the process of enclosure, which was designed to consolidate smaller units of agricultural land into larger, contiguous parcels that could be worked more efficiently. In addition, land that had been held in common, especially grazing pastures, was enclosed and awarded private title, reducing the availability of open range to those who owned little or no land, and who relied on the common pasturelands to feed their livestock. The open field system widely followed in central England gave way to land tenure using severalty, meaning private title to the land and its use. Although the enclosure process in Great Britain can be dated to the Middle Ages, by the mid-1700s the concept was supported by the English Parliament, which beginning in the 1740s passed legislation accelerating enclosure. Parliament codified many individual laws on enclosure with the General Enclosure Act in 1801, which left only a small percentage of the agricultural land in common holding. Enclosure dramatically changed the geography of agriculture in England, and in the long term the movement led to higher agricultural productivity. In the short term, however, it removed many smaller farmers from the countryside, a large number of whom migrated to the cities to find work in the newly emerging factories of the Industrial Revolution.

The next phase of the agricultural revolution in Great Britain may be dated from the early 1800s, with the widespread introduction of machines, first powered by draft animals and later using steam power. As was the case with enclosure, increased mechanization that replaced farm labor was extremely unpopular, especially with the large number of landless agricultural workers, a group made quite numerous by the previous decades of enclosure and other changes in the agricultural system. In some instances farm workers reacted violently to the application of machinery to agriculture, as in the case of the Swing Riots of the early 1830s, in which thousands of laborers attacked and destroyed threshing machines that had come into use across England. Such revolts had little effect on the introduction of machinery, however, and by the 1870s steam-powered mechanical ploughs and other devices were transforming British agriculture. Mechanization had several long-term effects, including a steady rise in productivity, a huge loss of jobs in the agricultural sector, and a subsequent shift of labor to the urban areas, where it was quickly absorbed by rapidly expanding industry. This trend would continue through the first decades of the 20th century and was mirrored by similar changes in North America and elsewhere in Europe.

In the years between 1700 and 1900 the agricultural landscape of Great Britain was completely transformed, and the age of modern agricultural production emerged. The benefits of this process are evident: after the early 1700s, food shortages due to crop failures in England and Scotland were extremely rare, and outright famine was essentially unknown in the region, with the obvious exception of the Irish Potato Famine of the mid-19th century. Techniques such as crop rotation, fertilization, contour plowing, and soil conservation, along with the increasing application of powered machinery to production, resulted in astonishing gains in both the productivity per unit of agricultural land and of the average British farmer. These spectacular gains were not achieved without significant social and economic disruption, however, as large numbers of rural workers had to leave the countryside to find employment in the alien environment of Britain's growing industrial cities. Indeed, without the revolution in agriculture begun several decades earlier, the Industrial Revolution in England likely would have taken quite a different form.

Some would argue that the agricultural revolution is a continuous process, not only in Great Britain, but around the world. In many lesser-developed countries, the mechanization of agricultural production and shifts in land tenure seen in Europe and North America over the last two centuries are only now taking place. For these regions, the revolution is only beginning. In the developed world, evolution of the agricultural structure continues apace as well. This most recent stage of the revolution continues many of the trends of previous centuries, with profound changes in the geographic structure of production, land ownership, and innovative techniques designed to enhance production. The advent of agribusiness, especially the growing influence of corporate farming or so-called industrial agriculture, along with changes in livestock breeding and husbandry may be taken as indicators that the "revolution" has yet to run its course. Just as in the past, these changes are not without controversy, and are not universally accepted as beneficial. Some may even point to a "counter-revolution," as reflected in the popularity of organic farming, rising emphasis on sustainable farming practices, problems with losses of farmland to urban sprawl, and concerns with the decline of the family farm, particularly in the United States. It is likely that the forces of modernization and globalization will only intensify the revolutionary nature of contemporary agriculture.

Air Masses

An air mass is a huge, identifiable body of air possessing a relative homogeneity of **temperature** and moisture characteristics within compared to the air surrounding it. Most air masses have dimensions of hundreds of thousands of square kilometers and are sometimes pushed thousands of kilometers from their source regions. Boundaries between air masses are known as **fronts** and the various types of fronts are described elsewhere.

Although air is a continuous fluid, its properties are such that it frequently organizes into air masses having differences substantial enough to cause significant weather changes as they pass. When air of the lower troposphere passes over **Earth's** surface, it exchanges energy and moisture characteristics with the surface. Air slowly moving large horizontal distances starts to take on the characteristics of the surface, be it tropical ocean or polar tundra. Some areas are well known as air mass source regions. Source regions occur in all latitudes and are characterized by non-mountainous terrain and frequent dominance of high pressure. High pressure is associated with light winds and this allows air to take on the surface temperature and moisture characteristics over the course of several days. As air masses move out of their source regions, they are modified. It is not rare for Arctic air to leave Siberia and travel over central North America to the Gulf Coast. While its original temperature might be –40°C, it might moderate to temperatures slightly below freezing. This is a cold shock to the Gulf Coast but a testimony to air mass modification.

One well-appreciated instance of air mass modification is lake effect snow. In early winter the Great Lakes of the United States and Canada are not yet frozen over. Polar and Arctic air masses stream over the lakes. These cold air masses do not contain much moisture but have high relative humidities. As the air passes over a few tens of kilometers of lake surface, the air mass gains water vapor and saturates. Downwind of the lakes, the arrival of the air over land initiates a small bit of lifting that cools the air lower than its dewpoint temperature. Condensation and the precipitation processes begin and make copious amounts of snow without the presence of a winter storm.

A simple air mass classification scheme considers the surface and latitude over which air passes. For instance, consider the differences in air types that can be generated over the great Antarctic ice sheets versus the tropical reaches of the Indian Ocean. Air masses having their origins in these places will provide vastly different weather as they progress over a location.

Maritime Tropical air (mT) is a product of tropical and subtropical oceans. It is warm at any time of the year and associated with a high amount of **humidity** and latent heat. Its air is a prolific bringer of precipitation into the middle latitudes as it is transported away from its tropical origins. This air mass is usually unstable and it has numerous summer thunderstorms. Ironically, it is this air mass that provides most of the moisture falling as snow in the middle latitudes. The maritime Tropical air is drawn into **middle latitude cyclones** and lifted and mixed with polar air.

Continental tropical air (cT) is a hot, dry air mass associated with the subtropical deserts of the world. It produces the hottest temperatures on the planet while not containing enough water vapor to produce significant precipitation. Record summertime temperatures are associated with the circulation of this air into the middle latitudes. Although this air mass is unstable in the first few hundred meters above the surface because of extreme heating of the surface, thought of as a whole, continental Tropical air has great stability because of the subsidence of air into the subtropical high.

Continental Polar air (cP) has source regions on the continents of the upper middle latitudes and, with the maritime Polar air mass, continental Polar is behind most of the cold fronts passing through the middle latitudes. This air is cold, dry, and usually stable. Because of these characteristics, continental Polar air is not associated with significant precipitation.

Maritime polar air (mP) has its source regions over the oceans of the upper middle latitudes. It is moister and milder than continental Polar air. Landfalling middle latitude cyclones fueled by maritime Polar air can bring prolific precipitation to mountainous coasts of the upper middle latitudes.

Arctic air (A) is the coldest air of the Northern Hemisphere. It is very cold and stable in the winter and not associated with significant precipitation. Because of the land/sea configuration around the Arctic Ocean, Arctic air originates in northern Canada, Alaska, and Siberia in the winter.

Antarctic air (AA) is the coldest air mass on Earth. In fact, it is much colder than the Arctic air of the Northern Hemisphere. It is generated on the Antarctic continent, which is covered by ice sheets having surfaces frequently in excess of two kilometers above sea level. Besides being cold, the air is very stable so that organized low pressure systems and their consequent precipitation are rare. Most middle latitude **locations** have considerably greater snowfall totals than Antarctica because of the inability of Antarctic air masses to hold much moisture.

Equatorial air (E) is an extremely humid air mass found in the lowest latitudes. Maritime and continental varieties are not usually designated because both types of sources produce similar temperatures and humidities. The temperatures tend not to be as hot as continental Tropical and maritime Tropical air masses because clouds and high humidities inhibit surface heating. Equatorial air masses are present year round near the Equator with incursions of other air masses being extremely rare. Equatorial air becomes unstable by the middle of the afternoon and produces daily air mass thunderstorms for which the deep tropics are so well known.

The classification scheme can also include a third letter, a lower case k or w. A k in the air mass designation indicates the air mass is colder than the surface over which it passes while a w denotes the air mass is warmer than the surface. Examples of this would be cPk air as it passes over North America in the summer. The continental surface is warm, the air is cool, and this sets up instability in the lower atmosphere. Conversely, a winter mPw air mass can be warmer than snow-covered North America and tend to be stable.

Climatologically, air masses have preferred areas of occurrence that enlarge, contract, and shift latitudes with seasons. Seasonally, the zones of dominance shift. For instance, the maritime Tropical source region strengthens, enlarges, and edges into the lower middle latitudes during the summer season. Also, there are geographic variations associated with the positioning of continents. Continental Tropical air vanishes from North America in winter because of the small size of the continental surface at those latitudes. In North Africa this air mass persists year round.

Some locations are associated with monotony of weather caused by the firm entrenchment of air masses. If one alights in Belém, Brazil in the heart of tropical Amazonia, she/he is struck with the sameness of weather day after day as dictated by the presence of Equatorial air. In the middle latitudes in particular, there are places that are neither air mass origin zones nor dominated by one air mass

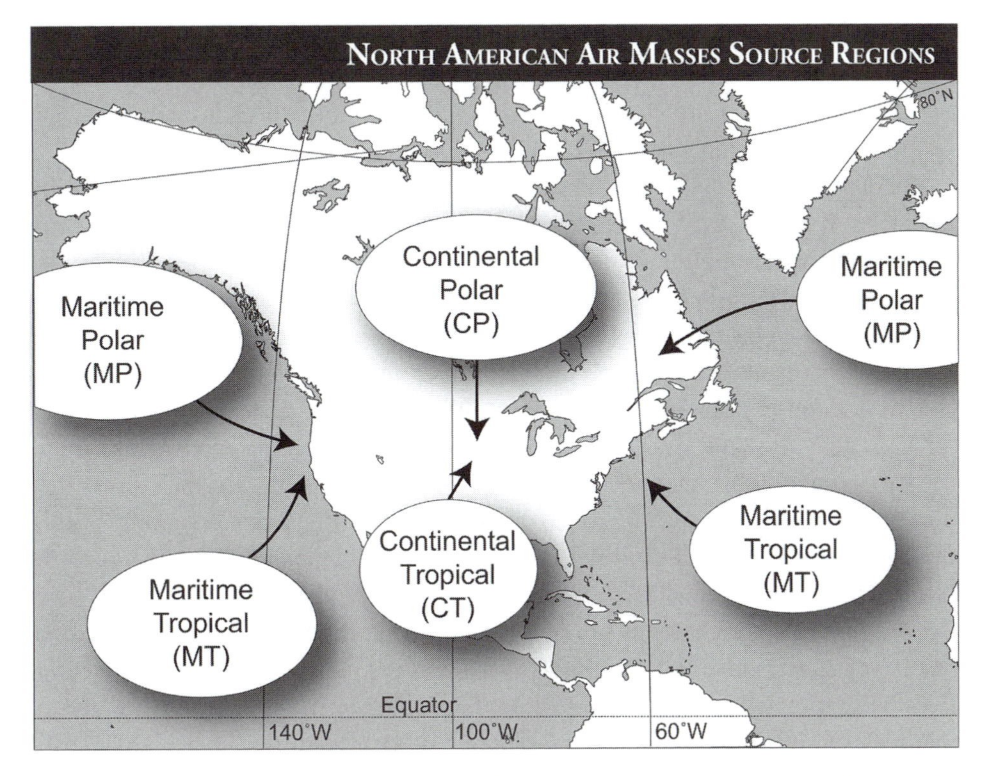

The interaction of major air masses over North America provides the continent with diverse and sometimes violent weather. (ABC-CLIO)

through the year. In these places, there is a hearty daily variation of the weather, especially in winter as the polar front jet stream forces the clash of tropical and polar air masses.

Altitudinal Zonation

Applied primarily in Latin America, this term refers to the differentiation of distinct environmental regions characterized by specialized agricultural production, based on elevation. Each elevation layer or "zone" is marked by a unique, or nearly unique, set of agricultural products suited to the growing conditions at that particular altitude. This structure is most pronounced in countries in the Caribbean, Central America, and tropical South America where the local topography is characterized by coastal lowlands that rise abruptly into mountainous regions in the interior. The various zones of production are typically referred to by the Spanish terms *tierra caliente* (hot land), *tierra templada* (temperate land), *tierra fria* (cold land), and, if the elevation reaches above 12,000 feet, *teirra helada* (frozen land). Some scholars identify yet another zone in Andean South America, the *tierra nevada*, which stretches from roughly 15,000 feet to higher elevations, and which is generally unoccupied by humans.

The *tierra caliente* is located between sea level and approximately 3,000 feet in elevation. It experiences warm, humid temperatures year round, and frequently receives abundant rainfall, in some places exceeding 60 inches a year. Naturally occurring vegetation in this region is typically tropical rainforest, or in somewhat drier areas, tropical savanna. **Population** densities here are typically high where the land is being cultivated. The climate provides an ideal setting for the commercial production of tropical specialty crops, many of which are exported northward to the large North American market. Bananas, sugar cane, tobacco, various tropical fruits, including mangoes, papaya, plantains, and others are raised on large plantations, as well as various staple crops such as wetland rice and corn. Within the zone there is some altitudinal layering as well: bananas, rice, and sugar are raised in the lower elevations, while tobacco, corn, and sometimes coffee are planted on the higher reaches of the slopes. In areas where the soils are less productive or there is insufficient rainfall to support the plantation crops, commercial livestock, primarily cattle and pigs, are raised. The *tierra caliente* is a region of intensive cultivation in most instances, and for some countries, especially Caribbean islands that lie at lower elevations and that depend largely on the production of a single crop (sugar cane in Haiti, for example), represents the only commercial agricultural region.

A different set of crops characterizes the *tierra templada*. This zone occupies roughly the elevations between 2,500 feet and 6,000 feet. It is considered temperate because the average temperatures are usually much cooler than those of the *tierra caliente*, and there is typically less rainfall at these altitudes. As in the *tierra caliente*, human population densities are relatively high, due to the more comfortable climate, and the intensive cultivation in many places of the central specialty crop of the *templada*, coffee. Coffee may be cultivated only in the higher reaches of the *tierra caliente*, due to the coffee plant's sensitivity to high temperatures and humidity, but the *tierra templada* offers an ideal climate for coffee production. Moreover, the well-drained, volcanic soils found in many sections of the *tierra templada* make the location perfect for coffee. Coffee cultivation is supported by the enormous demand of the North American and European markets, and coffee bean production is the single most important component of the agricultural economy for many Latin American countries. Brazil and Colombia together account for more than 40 percent of global production in an average year. Tobacco is a secondary cash crop produced in the *tierra templada*, and corn is also grown there, mostly for local consumption. A relatively new product of this zone is cut flowers, also focused toward the North American export market. Latin American growers supply much of the U.S. floral market during the winter months.

The region known as the *tierra fria* is found approximately from 6,000 feet to 12,000 feet, and represents the highest zone of significant human habitation. Population densities here are much lower than in the two lower zones, due to the lack of large-scale commercial agriculture, as well as the frigid temperatures, encountered especially at the upper reaches of the *fria*. Frosts are common in this **region**, and the growing season is much shorter than at lower elevations. The natural vegetation at this level consists mostly of hardy evergreens at the lower elevations, and brush, grasses, and stunted trees at the top of the *tierra fria*. The growing season at the lower end of the *fria* is frequently long enough to allow for the cultivation of wheat or barley, and highland pastures at this elevation often support sheep, or indigenous livestock such as llamas or alpacas. In the Andean countries of South America, indigenous peoples make up a high percentage of the population of the *tierra fria*.

Lying between 12,000 and 15,000 feet, the *tierra helada* represents a harsh environment of sparse settlement and restricted activity. This is the highest level of permanent human occupation, with the majority of villages clustered along the lower margin of the zone. Few crops can be grown at these extreme elevations, and animal husbandry is limited almost exclusively to hardy indigenous animals adapted to the cold temperatures and thin atmosphere, like the llama, alpaca and vicuna, although sheep can be raised in some locations in the lower part of the *helada*. Crops are restricted to those that may be grown underground such as tubers, or in some instances fast-maturing varieties of grains at the lower end of the zone. Only subsistence agriculture is found at this elevation, as the climate is too cold to support any commercial crops.

The highest zone is the *tierra nevada*, found only in the Andean region of South America. Located above 15,000 feet, the *nevada* is an environment dominated by snow pack and glaciers, and does not support any permanent human settlement. Agriculture is nonexistent, and even animals adapted to high altitude do not survive for extended periods in the extreme conditions of the *tierra nevada*. Although of no direct economic importance to the countries of South America, the *nevada* plays a vital role as a supplier of water to the rivers and streams of the Andean region, providing a crucial resource to agricultural production down slope.

Anticyclones

First conceived by Sir Francis Galton in the 1860s, anticyclones are, in some senses, the opposite of cyclones. Where cyclones are associated with the world's great storms, anticyclones are the cause of bright, dry weather. An anticyclone is a center of high pressure and is the pressure analog to a topographic hill. On weather maps, the highest pressure in an anticyclone is labeled with an "H" and it is common for publicly presented maps to dispose of the isobars to concentrate on the "H." There is no quantitative rule as to how high, absolutely or relatively, the pressure must be for an anticyclone to exist. It is sufficient that there are closed isobars that engender a circulation. Although anticyclones may be circular, they can take on all sorts of irregular shapes.

Air near **Earth's** surface rearranges itself from the center of high pressure to the lower pressure on the fringes, but not directly. The Coriolis Effect and force of friction modify surface air from moving in the direction of the pressure gradient force. In the Northern Hemisphere, the flow is from the high to low pressure side of the isobars at an acute angle so as to set up a clockwise circulation. In the Southern Hemisphere, the opposite Coriolis Effect causes counterclockwise circulation. The air that leaves the anticyclone center lowers the center's pressure and helps air to descend from the upper troposphere.

In the middle and upper levels of the troposphere, there is no friction, so anticyclonic winds are the result of pressure gradient and Coriolis Effect together. At altitude, anticyclonic winds are parallel to the isobars and flow clockwise.

Anticyclones are stable as compared with their unstable cyclone counterparts. That is, anticyclonic systems invariably resist the rise of air from Earth's surface so as to inhibit cloud development and production of **precipitation**. Precipitation can occur when the air is stable, but the precipitation will tend to be light because of the relative lack of uplift and the lack of moisture.

The strongest incarnations of anticyclones are those that build up in the Arctic, Antarctic, and Subarctic air mass source regions in the winter time and become Arctic, Antarctic, and Polar **air masses**. Over a course of weeks these air masses are generated and portions of the air masses break loose under the influence of the polar front jet stream to visit the middle latitudes and, occasionally, subtropical latitudes. Unlike cyclones that are associated with the mixing of two or more air masses, anticyclones are composed of single air masses. Their leading edges are cold fronts that usher in changes in wind direction, lower humidity, and colder temperatures. However, behind cold fronts are loosely spaced isobars of anticyclones making for light and variable winds. The calm and stable air near anticyclonic centers is quite dry, usually cloudless, and optimal for the nighttime loss of longwave radiation to space. The result is the coldest air experienced in the middle latitudes. Indeed, record cold middle latitude temperatures are usually associated with anticyclones. It might be thought that anticyclones are types of blizzards, but this is not so; the cold air at the front edge of the anticyclone helps to cause lift that might be associated with a blizzard by interacting with middle latitude cyclones. The cores of anticyclones are locations where precipitation is relatively unlikely.

The winter configurations of **middle latitude cyclones** and anticyclone are ordinarily geographically related. Cyclones are strung along the waves of the polar front (see **Middle latitude cyclones**). Following to the north and west of the cyclones are anticyclones that have edges circulating air into the cyclones and the fronts. Wintertime anticyclones can obtain immense proportions and are usually hundreds of thousands of square kilometers in size; at times they can stretch from Canada to Mexico. The pressures in winter anticyclones can sometimes pass 1050 mb and their centers travel at rates of about 50 kph.

The Siberian Express is, perhaps, the most nefarious incarnation of a middle latitude anticyclone. In the winter, the coldest air of the Northern Hemisphere—including the north polar region—is usually over Siberia. A large ridge in the polar front jet stream will move this air from eastern Siberia to the eastern United States in a matter of a few days. This fast-moving anticyclonic incursion is sometimes termed a cold wave and is responsible for several days of cold, if not low-temperature, record-setting weather. The "Arctic Express" is a similar event when the anticyclonic air has origins in northwest Canada or northern Alaska; this "express" also barrels along but it does not tend to be as cold.

Winter anticyclones pass over warmer and warmer surfaces as they leave their source regions. Over a course of days, the anticyclone will moderate from its cold, dry origins. For instance, an anticyclone that started at –55°C in Siberia may modify up to 0°C by the time it reaches the Gulf of Mexico. This is not bitterly cold weather as when it was in Siberia, but is much colder than the winter averages over the Gulf.

Summer anticyclones are also noted for bright, dry weather with minimal winds. In summer, the central pressure in a strong anticyclone may only be around 1,000 mb. In the middle latitudes these set up when the polar front jet stream is far poleward and there are no strong winds directly aloft and anticyclones languish and can become stationary. Such a pattern becomes hotter and drier with time and is strongly associated with heat waves and droughts.

In the subtropical regions are the semi-permanent anticyclones. These are known as the subtropical highs within the global circulation. Air in the Hadley circulation leaves the equatorial region at high altitudes in the troposphere and sinks as it moves poleward. This descending air is profoundly important to world climates because it causes the annual and seasonal dryness experienced in tropical and subtropical latitudes. Even though the near-surface air is very hot, its propensity to rise is countered by the strong descent of air from aloft. When the subtropical high is overhead of a **region** there is usually a paucity of precipitation. Indeed, the presence of the Sahara, Australian deserts, the Atacama, and the Kalahari share the common characteristic of the year-round presence of subtropical highs. Adjoining regions are not as dry because the subtropical highs are not entrenched throughout the year.

Over the oceans, the subtropical highs are particularly pronounced and are sometimes referred to as semi-permanent. They exert major control in the average wind directions and ocean currents. Over the subtropical ocean basins, winds and currents are clockwise in the Northern Hemisphere and counterclockwise in the Southern Hemisphere.

Ironically, the presence of light, diverging anticyclonic flow in the upper troposphere can have a profound effect on the formation of **hurricanes**. Hurricanes are cyclonic storms built from the warm sea upward. To develop, the vertical change of wind speed and direction with height (wind shear) must be minimal. Jet streams and other fast winds aloft literally break the top off of a developing hurricane. The winds of upper anticyclones provide minimal wind shear and the divergence provides outward evacuation of the air rising from the surface and helps to deepen the surface low and increase hurricane wind speeds.

Areal Differentiation

Areal differentiation represents one of the classic philosophical approaches to geographic inquiry. Some of the earliest geographical scholars, including Strabo and Ibn Khaldun, sought to describe and catalog variations in the places and cultures they encountered, or were informed about by others. A central concept of areal differentiation is that the surface of the earth may be divided into **regions**, which may be distinguished and categorized using various spatial criteria. Thus, areal differentiation provides the theoretical foundation for regional geography, by conceptualizing space as consisting of identifiable units that may be distinguished from one another on the basis of a set of phenomena or criteria. For much of the historical development of geography as a science, this was the approach followed. Geographers partitioned the world as they encountered it into sections based on differences they codified, formulating these distinctions in a descriptive narrative designed to provide a sense of place. During the **quantitative revolution**, this view was derided by its fiercest critics as simplistic, static, and sterile, but in recent decades an emphasis on areal differentiation has reappeared in some subdisciplines of geography. For example, areal differentiation has offered a basis for new directions in human geography, especially postmodern analyses that focus on the social and cultural processes that construct a sense of place for a given location; and studies in the geography of economic development, which seek to

Alexander von Humboldt (1769–1859)

Humboldt was the scion of a wealthy Prussian family. He was university educated with a wide range of interests. Of greatest importance was his economic freedom to travel bringing measurement instrumentation with him. His careful measurements and splendid write-ups are considered precedent setting in terms of modern, scientific geography. He traveled to Europe, Russia, Central America, and South America and was a mainstay on expeditions of the early 1800s on which his diverse experiences included exploring the Orinoco River, climbing Andean volcanoes, and measuring temperature and velocity in the Peru Current. He then lived in Paris where he published 30 volumes (1805–1834) about his studies in the Americas. His works covered many geographic themes including landforms, plant geography, and climate, and he was celebrated as one of the leading intellectuals of the age. In 1827 he moved to Berlin and became chamberlain to the Prussian king where he continued his pursuits with such projects as the first temperature map of the world. His last work, *Kosmos*, integrated his ideas about the world in a series of five volumes and was immediately considered to be the best world scientific work ever written. Alexander's brother, Wilhelm, founded the University of Berlin in 1810 and it was renamed Humboldt-Universität to honor both men.

conceptualize the forces and factors that result in an uneven geography of economic and political activity, as well as the spatial variation between places in terms of the opportunities and conditions that results from such variation.

The most detailed modern explanation and vigorous defense of the perspective of areal differentiation was provided by Richard Hartshorne in his influential monograph, *The Nature of Geography*, published in 1939. In this work Hartshorne argued that geography is based on a "chorographic point of view," which distinguishes it from "systematic sciences," and that geography "seeks to acquire a complete knowledge of the areal differentiation of the world. . . ." Hartshorne closely correlates the study of geography with that of history, arguing that both must derive their basis from the integration of other scholarly sciences and philosophical approaches, and that both are so-called "naïve sciences," meaning that they rely on describing phenomena as they actually exist in the world. Many of Hartshorne's critics suggest that he is too dependent on the view of the "German School," especially Carl Ritter, Friedrich Ratzel, and Alexander von Humboldt in his description of the evolution of geography, but his work remains a classic and is standard reading for students of the discipline.

In the second half of the 20th century the perspective of areal differentiation was challenged by new methodologies and theories. The foremost of these was the evolution of **locational analysis** as a method of examining spatial phenomena, and the larger contextual approach of landscape analysis. Proponents of these new approaches held that areal differentiation was overly descriptive, lacked an integrative perspective, and limited geography to a narrow philosophical foundation that ignored the development of emerging statistical and mathematical methodologies. Nevertheless, areal differentiation continues to inform the scholarship of many geographers who study the distribution and interaction of human systems, and the perspective remains a cornerstone of geographical analysis and theory.

Atmosphere

Geographers are quite interested in the atmosphere because it interacts with continental and oceanic surfaces, is variable by latitude, and variable over time. The spatial characteristic of the atmosphere impart important possibilities and constraints to the human mosaic. Where is it too cold to grow crops or where must irrigation water be used? What sorts of materials (e.g., trees or grasses) do indigenous peoples incorporate in their constructions? Surely, the atmosphere is the most vital resource of our planet.

Of all the planets surrounding the sun, **Earth** is the only one that has an atmosphere containing enough oxygen and the right temperatures to support plentiful life. Our atmosphere acts as a blanket shielding us from the harsh radiative environment of space while vigorously circulating to prevent Earth from becoming impossibly hot or cold in any latitude. Crucially, the circulation also brings with it the moisture of the **hydrologic cycle** needed to maintain life on continental surfaces.

Usually, we think of "atmosphere" as synonymous with "air" but that is true only within a very few kilometers of sea level where air has the energy and moisture characteristics capable of supporting life. In truth, the atmosphere varies considerably with altitude. The atmosphere is usually taken to extend from the surface to about 10,000 km. However, the top of our atmosphere is not a definite boundary and is considered atmosphere merely because it is not as totally devoid of mass as is space. Most of the atmosphere is near sea level and this is because of the sharp drop in gravity with altitude above sea level. Along the ascent to the top of Mt. Everest (Sagamartha) at 8.8 km above sea level, atmospheric pressure drops to somewhat over 300 mb, a third of the average pressure at sea level; this is a condition that cannot sustain human life. Above these altitudes the pressures are even lower and by the altitude of 50 km, the 1 mb of pressure compares favorably to pressures produced in laboratory vacuum chambers. Thus, the life-sustaining properties of our atmosphere exist only within a very few kilometers of sea level.

Air is not a chemical compound but a mechanical mixture of gases that is kept mixed by solar heating in the lower atmosphere. There are many gases in the atmosphere, but only a few main ones. Nitrogen is the dominant gas of the atmosphere, comprising about 78 percent of air without water vapor. Breathable oxygen is next at almost 21 percent and is the byproduct of photosynthesis in green plants (see **Biogeochemical cycles**). The third gaseous element is argon with somewhat over 0.9 percent of the atmospheric volume. Argon is seldom heard of because unlike the more plentiful gases, it is chemically inert. After argon, come gases such as neon, helium, methane, krypton, hydrogen and others. They are present in such small amounts that none of them makes up more than a couple of thousandths of a percent of the atmosphere. The atmospheric amounts of these gases have remained quite steady over time, and they will change meaningfully only over time spans of millions of years.

Another set of gases are significantly variable over times that can range from minutes to years. They are water vapor, carbon dioxide, carbon monoxide, ozone (O3), sulfur dioxide, and nitrogen dioxide. Water vapor is, by far, the most variable over time and place, and it can range from nearly 0 percent of air's volume over the Antarctic continent in winter to almost 4 percent in the first few meters over a tropical ocean. Water vapor can be extremely variable by location and, for instance, its presence can increase or decrease by several orders of magnitude with frontal passages in the middle latitudes.

Air has always been impure and carries with it many small solids and liquids corporately called particulates. These can be salts from the breaking of ocean waves, dust from the deserts, smoke from burning vegetation, and volcanic ash among other substances. The particulates have tiny masses and can be buoyant in the winds and air currents if they are small enough. Pollutants are human additions of mass to the atmosphere. They include the same particulates and gases provided to the atmosphere by natural means and also many chemical substances not found in nature. Pollutant concentrations tend to be highest in cities where human activities are at a maximum. There are potential interferences with biogeochemical cycles as human **population** increases along with the variety of its activities.

Atmospheric carbon dioxide and ozone have already been affected by the presence of humans. Since the middle of the 1880s, carbon dioxide has increased by a third. Carbon dioxide receives much publicity but is now and always destined to be a rather small part of the atmosphere. It is approaching 400 parts per million due to human activities; however, its small relative volume belies its ability to absorb and reradiate thermal infrared energy as part of the greenhouse effect. Ozone is the substance that absorbs most solar ultraviolet radiation before it reaches Earth's surface. In that sense, it keeps us alive, and it been shown to be decreasing over the south polar latitudes because of the human release of chlorofluorocarbons (CFCs) at other latitudes. These releases have been in the form of CFC-based coolants, spray can propellants, and fire retardants. Inert near sea level, CFCs react with ozone in the presence of ultraviolet energy to the effect that some of the ozone has broken down. Indeed, a hole in the Antarctic ozone is now a regular feature of our atmosphere.

The atmosphere is quite varied with altitude. There are usually four temperature-based layers denoted. The lowest layer is the troposphere. Averaging the first 18 km of the atmosphere, the troposphere is a region that is heated by radiation from Earth's surface and so declines in temperature with altitude. The average planetary temperature is 15°C near the surface and this decreases with altitude to −65°C at sharply defined top of the troposphere (the tropopause). The tropopause acts like a lid so that daily weather—storms, clouds, and appreciable amounts of humidity—are limited to the troposphere.The stratosphere is the second temperature-defined layer and extends from 18 km to 50 km. This layer has no surface-like weather and the air, although relatively well mixed, exists at tiny fractions of its surface pressure.

The stratosphere contains most all of the world's natural ozone, which nature creates and destroys at these elevations. The absorption of ultraviolet energy by ozone makes some of it break apart while reradiating the energy in thermal infrared wavelengths. This release heats the stratosphere so that the temperature signature of the stratosphere is one of increasing temperature with altitude. At the top of the stratosphere the temperature is slightly above 0°C. Above the stratosphere the atmosphere is not well mixed and there are altitudes that are rich in oxygen or in helium, etc. The amounts of the gases are so minuscule that these altitudes are not usually thought of as having air. The mesosphere extends up to 80 km where the temperature declines to lower than −80°C where the atmosphere is at its coldest. Interestingly, the next layer above the mesosphere is the hottest part of the atmosphere. **Solar energy** interacting with the scant molecules and ions at these high altitudes cause dramatic heating. Average temperatures are well over 1,000°C and can approach 2,000°C when the sun is active. The huge temperature swings are not felt near the planetary surface and not well understood as how they relate to weather and climate. It is clear, however, that day-to-day weather changes are not a response to the activity of the sun. The ionosphere is a non-temperature-based region of the atmosphere extending from 60 to 400 km. It is noteworthy because of the occurrence of electrically charged molecules and atoms (ions). Ultraviolet and x-ray energy emanating from the sun peel electrons off some of the matter; this is called ionization. Although ionization affects only 1 percent of the gases, they become strong electrical conductors. The ionization thus affects the propagation of radio waves such that the transmissions can be greatly enhanced or blocked.

Atmospheric Stability

The great ocean of air that is the **atmosphere** is far from quiescent. Place-to place heating differences make air flow horizontally (see **winds and pressure systems**) and also vertically. The vertical rise of air has particular significance because it is the rise of air that is associated with **clouds** and **precipitation**. How strong is the rise of air? The answer is that it depends on the stability of the atmosphere and that stability differs by location, season, and time of day. If lift is gentle or air is descending there is not likely to be significant precipitation. It is the strong ascent of unstable air that is associated with all of the world's great storminess. The basis of understanding of atmospheric stability is rooted in the ready ability of air to compress and decompress. From physics, the Gas Law (Equation of State) holds that temperature, pressure, and density are interrelated. As air is heated and cooled, its density can change dramatically. In this regard, air is usually considered in small, discrete pieces known as air parcels. Air parcels are useful constructs with which to consider the thermodynamic properties of air. A few meters across, an air parcel has consistency of **temperature**, pressure, and density within and is compared to the air around it (known as the environmental air).

As air parcels rise they decompress; that is, the same mass takes up more volume and the density lowers. This decompression decreases the number of molecular collisions in the air parcel and causes a corresponding drop in temperature. Importantly, the temperature drop is the result of the spatial rearrangement of the molecules and *not* a loss of energy to the surrounding environmental air. This is known as the adiabatic process and is a profound influence on the temperature the rising and sinking air. Rising air parcels that are not saturated cool at the unsaturated (dry) adiabatic lapse rate of 10°C for every kilometer of ascent. In saturated air, the decrease of temperature averages 6°C for every kilometer (the saturated or wet adiabatic lapse rate). This retardation of the saturated adiabatic lapse rate is the result of the conversion of latent heat to sensible heat during the cooling of the air parcel. Descending air parcels warm at the unsaturated adiabatic lapse rate, whether or not they begin the descent saturated: descending air warms so that it soon becomes unsaturated. Air parcel temperatures can change dramatically in a few minutes because of adiabatic processes; similar temperature change because of gain or loss of radiant energy would take hours to occur. Consider an air parcel rising from the surface. The air parcel is lifted to altitude X and its temperature decreases at the appropriate adiabatic lapse rate. The air parcel at altitude X is less dense and warmer than the environmental air, so the air parcel will *rise spontaneously*, and this air is absolutely unstable. An air parcel already at altitude X has the same density as the environmental air, will remain at altitude X, and it has neutral stability. If an air parcel is denser and cooler than the environmental air at altitude X, the air parcel will *sink spontaneously*, and this air is termed absolutely stable. On average, the troposphere is absolutely stable because the average environmental lapse rate is less than the saturated and unsaturated adiabatic lapse rates. How, then, does the troposphere sometimes destabilize? The answer is straightforward: the environmental lapse rate is quite variable over short amounts of time. The atmosphere destabilizes as the environmental lapse rate increases to be greater than the adiabatic lapse rates. The environmental lapse rate might have been increased by warming near Earth's surface or by cooling of the air aloft. Warming of the lower air is accomplished by air moving over a warm surface, and daytime heating by the influx of solar energy, or advection of warm air. Air aloft can be cooled by air and clouds radiating energy to space or by cold air advection resulting from disturbances in the jet stream. Moreover, both the warming of the lower air and cooling of air aloft can be simultaneous and result in rapid destabilization. Over the southern Great Plains of the United States maritime tropical air flows at low level and the polar front jet stream flows at the top of the troposphere. The combination makes for thousands of meters of instability and some of the largest thunderstorms on Earth. The presence of plentiful water vapor helps makes the air more unstable. Although air can be unstable with modest amounts of vapor, greater amounts of vapor make the air more unstable to greater altitude. The key is in the saturated adiabatic lapse rate. Moister air parcels become saturated with less lift than dryer parcels. The retardation of the cooling of a saturated parcel via the saturated environmental lapse rate keeps the rising air parcel warmer, less dense, and unstable to heights greater than possible in unsaturated air.

Clouds provide visible depictions of stability conditions. Clouds can be present when the air is stable, but are generally taller when the air is unstable. Clouds of vertical development such as cumulus, altocumulus, and cirrocumulus are indicators of instability at various elevations. Cumulonimbus clouds sometimes reach all the way through the troposphere and bring lightning, thunder, rain, hail, and high winds because of massive instability in a moist environment. The atmosphere's lifting mechanisms (see **Precipitation**) are made much more potent by instability. If an air parcel is nudged upward by one of the lifting mechanisms, it will sink or remain at the lifted altitude if the air is stable or neutrally stable. However, if unstable air is lifted it rises spontaneously above the initial lift and is much more likely to cause significant cloudiness and precipitation.

There is a strong geography to the stability conditions of the atmosphere. They are largely dependent on the global circulation. The intertropical convergence and the polar front are zones where instability and, hence, storminess are common. Conversely, regions dominated by the polar highs and subtropical highs are quite stable and are associated with lack of strong ascent and sparse precipitation.

Instability has diurnal and seasonal fluctuations. Because the atmosphere is heated from the ground upward during daylight, a sunny day frequently produces surface-based convection. Convection represents an unstable condition and the top of the convection depends on the characteristics of the environmental air. At night, the cooling atmosphere usually becomes much more stable. Therefore, surface-based instability cycles with diurnal heating. Seasonally, locations can have significant variations in surface heating and upper air conditions so that the atmospheric profiles are more prone to instability in some seasons than others. In most climates, winters have less instability than summers. Also, instability can markedly increase in **monsoon** regions as the flow regime shifts to vapor-rich maritime tropical air arriving over a continent.

Inversions are conditions in which the temperature of the environmental air increases with altitude as compared to the average decline typical of the troposphere. The air parcel in an inversion cannot rise spontaneously. So, inversions are associated with stable air. Low-level inversions are common overnight via radiational cooling during regionally quiet weather and higher inversion layers relate to the descent of air from aloft, especially associated with the subtropical highs. Whatever their root causes, inversions act as lids on weather, preventing significant upward mixing. Stagnant air and high pollution levels are associated with locations that have frequent inversions.

B

Balkanization

The breakup or division of a region or **nation-state** into smaller political, ethnic, or religious units, often who share strong mutual animosities. The term derives from the frequent disruption and modification of **boundaries** in the **shatterbelt** of southeastern Europe. It is an expression of **territoriality** and is often a violent process. Balkanization of a **region** may be brought on by **devolution** and **centrifugal forces**, and has been a common phenomenon in the aftermath of **imperialism**. This is because the colonial powers ignored for the most part the multiple expressions of **cultural identity** in the administrative regions they created and did not promote the concept of a unified identity among the groups living in those regions. Indeed, in many cases the formation of such an identity was actively discouraged by colonial policy. Trends toward balkanization have been evident in Eurasia and Africa in recent decades, as states that gained independence in the post–World War II era struggled to solidify both their territorial integrity and their national identity.

The potential consequences of balkanization may be seen in the term's origin. "Balkanization" was first popularized as a geopolitical concept in the late 19th century. At the Congress of Berlin in 1878, the major European powers established a series of independent or semi-independent states in the region of the Balkan Peninsula, all carved from lands previously controlled by the Ottoman Empire. These included the Kingdoms of Serbia, Romania, and Montenegro, as well as a Bulgarian state that was semi-independent. For the next several decades, the region was beset by almost constant political conflict and border changes. The states either fought one another in limited regional conflicts, as in the Serbo-Bulgarian War in 1885; banded together to fight the Ottoman state in the form of the Balkan League in the first Balkan War (1912); or fought a region-wide war among themselves over territory, as in the second Balkan War (1913). Ultimately, the tensions in the region and the resultant alliances would serve as the spark for World War I, only a year after the second Balkan War concluded. Balkanization came to signify the disintegration of any region into smaller states that are plagued by seemingly intractable territorial and ethnic rivalries. Ironically, after almost 70 years as a unitary state, the balkanization of the Yugoslav state in the 1990s presented a modern reminder of both the origin and the consequences of the term.

The process of balkanization has resonance in current international conflicts and in the War on Terrorism. Some observers have suggested that Pakistan, a strategic partner of the United States and a key player in the campaign to control Islamic radicalism, is a prime candidate for balkanization because of the country's numerous ethnic groups and the large Pushtun population it shares with Afghanistan. Indeed, historically one of the principal factors driving the process of balkanization has been the geographic separation of an ethnic group by an international border. The Pushtun community was divided by the Durand Line, an international border imposed in the late 19th century that resulted in approximately two-thirds of the Pushtuns residing in Pakistan, and one-third living across the boundary in southern Afghanistan. Furthermore, the Afghan state is composed of multiple ethnic groups, many of whom share close linguistic and historical ties with cousins lying just across one of the country's borders. This is the case with the Tajiks, Uzbeks, Turkmen, and Baluchis, in addition to the Pushtun community. A successful effort to unify the Pushtuns or any of the other minorities could result in the balkanization not only of Pakistan, but of Afghanistan as well.

Yet another example is Georgia, a country that has experienced the process of balkanization since independence in 1991. Although the country covers only about 27,000 square miles (slightly larger than West Virginia), it contains two breakaway regions that have fought the government in Tbilisi since 1995. Abkhazia in the northwest and South Ossetia in the north-central part of the country both border on Russia, and both have received strong support from that nation. The Ossetians represent yet another example of an ethnic community splintered by an international boundary. The Georgian government no longer exercises political control in either region, and both now govern their respective territory as semi-independent states. In addition, the southwestern corner of this tiny country, a region called Ajaria, has also indicated signs of centrifugal tendency since the early 1990s, although unlike in Abkhazia and South Ossetia, not via violent separatism. The de facto balkanization of the Georgian state has greatly increased international tensions between Georgia and Russia in recent years, leading to a short but destructive war between them in 2008. Here as elsewhere, the process of balkanization involves not simply the division of a country into smaller, ethnically based units, but spills over into a larger regional context, creating the potential for much wider conflict.

On the other hand, balkanization has been suggested by some policymakers as a reasonable response to overcoming internal animosities in some states. In the case of Iraq, some American leaders suggested a policy that would promote at least limited balkanization of the country. Senator Joseph Biden, who later became Vice President of the United States, offered such a plan in 2006 that would divide Iraq into three ethno-religious zones: a Kurdish sector in the north, a Sunni region in the center, and a Shiite zone in the southern reaches of the country. The Kurds, an ethnic group without a nation-state, are divided between four adjacent countries in the region: Iraq, Turkey, Iran, and Syria. Iraq's Shiite community, concentrated in the southern reaches of the country, shares strong religious affinities with neighboring Iran. Biden's proposal was not accepted by either the American or Iraqi governments, however, and critics of the plan pointed out that the call for an autonomous Kurdish territory might possibly lead to the destabilization and balkanization of some of Iraq's neighbors. Once initiated, balkanization is a process that has historically often gathered more momentum than expected and proven difficult to control or direct.

Bid-Rent Theory

A theory of land valuation based on the geographic distance from an established point of maximum value represented by the Central Business District, or CBD. The theory is generally applied to urban environments but is adapted from the **von Thunen model** of rural land valuation. The urban geographer credited with developing bid-rent theory is William Alonso. The theory proposes an urban model in which a city occupies an unbroken plain, with no topographical barriers. The quality, cost, and availability of transport throughout the urban area are uniform, and at the heart of the urban space is a CBD that provides the maximum economic opportunity in the city and contains all the jobs in the urban area. It also assumes an atmosphere of perfect economic competition, where no one party has an advantage over others. Within this scenario, the following circumstances of land valuation and land use will occur. First, the land lying within the CBD and immediately adjacent to it will have the highest value, or bid rent. This is because land in the CBD represents access to the largest pool of consumers for businesses, and the shortest distance to work for residential users. Demand therefore for land within and close to the CBD will be quite high, and this will be reflected in correspondingly greater land costs. Manufacturing and retail businesses will be more willing to pay these higher costs for location closer to the CBD than residential users, who gain economic advantage only via lower transportation costs by locating in the CBD.

The result is a series of zones or rings surrounding the CBD in which land values decline with distance from the CBD. Each zone is marked by a cluster of similar users whose specific distance from the CBD is determined by willingness to pay higher land rents. The innermost zone will be dominated by high-volume, high-value retailers who are willing to pay the maximum in land costs, due to the higher number of consumers and greater volume of trade made possible by the central-most location. The outermost ring will be occupied mostly by residential users, who accrue only limited economic benefit by being centrally located, in the form of lower transportation costs. In between will be zones dominated by lower-volume retailing, wholesale suppliers, manufacturers, and other users. The desire of each set of users to pay higher land costs as a function of distance from the CBD can be graphically illustrated via a bid-rent curve for each group. The curve for those willing to pay the highest bid rents in the CBD is quite steep, while the curve for those users who will only pay lower bid rent, located in the outer zone, is almost flat. In bid-rent theory, any residential users living closer to the CBD must cluster in high-density units (i.e., high-rise apartment buildings) to offset higher bid rents. The resulting pattern of urban land use is a set of circular zones, emanating from the CBD. This pattern of **urbanization** is called the Concentric Zone Model, and is one of several generalized models of urban morphology.

Biogeochemical Cycles

Along with the circulation of the **atmosphere** and **oceans** there are energy and mass circulating through and interacting with the various life subsystems of **Earth**. The corporate name for these translocations is biogeochemical cycles. The term is apt in that it connotes the interplay of life and its chemical environment on, over and under Earth's landscapes. In toto, the magnitudes of these flows are unimaginably vast and complicated but their components have been studied to the extent that the basics are well known. The cycles provide needed supplies of mass energy because these commodities are finite on our planet. In short, the biogeochemical cycles bathe the planet in ways that provide for sustained life. Although it can be shown that humans have had major, unintentional impacts of some of these cycles, it is evident that there is much resilience in the way they function.

The energy cycle is usually the first to be noted because it is the fuel by which all the other cycles act. A huge amount of solar energy is incident on the Earth system. Instantaneously, the amount is about 174 petawatts (10^{15} watts), which is billions of times the rate of electric energy generated by human devices. As **solar energy** passes through the atmosphere there are all manner of pathways as described elsewhere.

The key energy for life is supplied by photosynthesis. Although incredibly complex with many nuances yet to be understood, photosynthesis is responsible for energizing the large bulk of biomass on the planet and, ultimately, is the source of energy on which the human organism depends. Photosynthesis occurs in plants and uses carbon dioxide and water commonly available in the environment. Not all solar energy received is used. Perhaps 0.6 percent of solar energy incident on Earth's surface interacts with the above ingredients in plant cells through the auspices of the green pigment chlorophyll. The radiant solar energy is converted into chemical form as a series of carbohydrates. This storage allows plants a steady supply of energy for respiration, the processes that keep the plant alive. Of incredible importance to our planet is the oxygen released as a byproduct; the bulk of our atmospheric oxygen supply was produced by photosynthesis.

Plants manufacturing carbohydrates from solar energy are known as primary producers. Once fixed into chemical form, solar energy is available for further use by other organisms. Some of the primary producers are consumed by animals that use the carbohydrates to sustain their life processes and these animals are known as primary consumers. Of course, some animals eat other animals thereby gaining the energy the primary consumers gained from plants; in this case the animals are known as secondary consumers. Plants and animals are intertwined by energy and mass pathways known as food chains. This is, perhaps, oversimplification in that most ecosystems are complex webs of recycled energy and matter. The relationships do not represent perfect usage of energy. At every step, about 90 percent of the energy is lost. Ironically, any top predator like an eagle or a shark is several steps away from the capture of solar energy by plants, and is dependent on the existence of huge amounts of biomass to survive. Humans have survived because of their ability to either eat plants directly or eat animals that have eaten plants.

Geographers are interested in global net primary productivity (NPP). It is the measure of photosynthesis minus respiration and is commonly given as kilograms of carbon per square meter per year. NPP is geographically varied by climate and landscape environments. Tropical rainforests have the greatest net primary productivities, on the order of 2.2 $kg/m^2/yr$, whereas tundra manages only 0.14 $kg/m^2/yr$. Although some upwelling areas of the oceans are quite productive, there are large stretches that are unproductive. For example, the stable waters in the subtropical highs might have NPPs of .002 $kg/m^2/yr$. Moreover, these numbers are only for the top waters into which solar energy can penetrate.

The **hydrologic cycle** is the most massive of the cycles on the planet with about 400,000 cubic kilometers leaving the oceans each year. Water is a part of all Earth's life so that its circulation around the planet is of major importance to the sustenance of life.

Oxygen is plentiful in the atmosphere and is well known for its use by the animal life on our planet. Like other components of our atmosphere, oxygen has evolved over time. In Earth's earliest atmospheres, the two- and three-atom oxygen so common now could not exist in quantity because of the propensity of unmitigated solar energy to break apart these molecules. Two things happened to increase the oxygen supply. They were (1) the formation of the ozone (O_3) layer that screens most of the ultraviolet energy form the lower atmosphere, and (2) the buildup of breathable oxygen (O_2) as a chemical byproduct of photosynthesis that was allowed once the ozone layer was established. Additionally, oxygen enters the atmosphere as part of water vapor during evapotranspiration and via the weathering of rocks (oxygen is the most prominent component of rock at the planetary surface). Oxygen leaves the atmosphere through the precipitation of water and inhalation by animals. At present, Earth's atmospheric oxygen supply is stable and there are no worries that it will become scarce in the foreseeable future.

The nitrogen cycle is interesting in the way it provides useful nitrogen into plants and animals. Nitrogen forms the bulk of the atmospheric gases and is, yet, not usually useful to life until it is fixed into one of a series of compounds called nitrates. Plants can directly take up these compounds to supply life processes. Only a tiny fraction of gaseous nitrogen is fixed. On land, most nitrates are produced by bacteria in the root nodules of certain plants and in the soil itself. Exotically, cosmic rays and lightning both produce small amounts of nitrates. Also, minor amounts are fixed by marine life. Fixed nitrogen can return to the atmosphere by denitrification caused by bacteria. Humans manufacture and concentrate nitrates as fertilizer for agricultural purposes. Runoff from fields and other fertilized surfaces has had unintentional, deleterious effects on life in streams and small water bodies because increased nitrate concentrations decrease dissolved oxygen.

The carbon cycle is significant to life and has had major interference from human activities. Carbon, hydrogen, and oxygen atoms in combination are building blocks for myriad organic molecules. Carbon enters plants via photosynthesis and animals via the consumption of plants. Carbon-based carbohydrates include chemical energy derived from solar radiation. As animals consume plants this carbon is passed through the web of life. Animals exhale this carbon dioxide. Organism death and decay result in the release of carbon to the atmosphere. Inorganic sources such as the outgassing of volcanoes and the weathering of certain rocks also provide carbon back to the atmosphere. Every burning process, natural or human-related, combines oxygen and carbon that becomes carbon dioxide.

Of great concern are the agricultural, industrial, and transportation processes now releasing carbon dioxide into the atmosphere in amounts that surpass natural sources. Since the start of the Industrial Revolution in the middle 1800s, the carbon dioxide content of the atmosphere has risen by over a third. Although intake by photosynthesis and solution into oceans buffer the planetary atmosphere from containing all of this carbon dioxide, it is unknown how much capacity these "sinks" possess. Such an increase in carbon dioxide is not toxic nor will it decrease breathable oxygen in the atmosphere but carbon dioxide is one of the primary greenhouse gases such that humans may be causing unintended impacts on Earth's energy budget.

In a short article about the cycles it is impossible to convey a full sense of how interconnected they are. A change in one cycle can markedly affect the functioning of another. There are a number of other biogeochemical cycles that have not been mentioned here (for example, calcium, zinc, sulfur, and phosphorus). Although vital at various times and places, they do not have the overall volumes and energy contents of the major cycles illustrated above; life can be threatened if these other cycles are somehow interrupted.

Biomes

The biome is a concept that allows spatial organization of vegetation and animals at the planetary scale. More specifically, it accounts for the integration of climate-plant-animal relationships over large areas. Biomes are the world's major life communities classified by the predominant vegetation, often the most prevalent feature on the natural landscape. The biome concept implies that plants and animals have adapted to each biome in ways specific to the biome. Each biome has species that the geographer views as emblematic of that biome. For instance, the solitary baobab tree in the midst of the savanna grasses and the grove of scrubby trees near the Mediterranean Sea evoke regional relationships of vegetation, animals, and people.

The zone of transition between two biomes is an *ecotone*. Although biomes usually grade gradually from one to another, small-scale biome maps such as that shown below give the misimpression that there are sharp boundaries. Horizontally, biomes usually grade into each other. For instance, the Sahel—the semiarid zone south of the Sahara—contains the transition between the desert and the tropical savanna biomes. In the southern Sahel there is greater biomass with grasses and shrubs found more plentifully than in the northern Sahel, where there is sparse biomass emblematic of the Sahara. Yet, when pressed to find an exact boundary, it becomes apparent that the edges of biomes blend together as do the climates that cause them. Contrastingly, in mountainous areas there are often several sharply defined vertical biomes visible.

The biomes that can be observed today have not existed throughout **Earth** history, even relatively recent history (see accompanying map). Over very long periods evolution and continental drift have impacted the extents of biomes. Climate changes have shifted ecotones over scales of hundreds and thousands of years. In North America the Pleistocene ice age saw biomes shifted hundreds of kilometers south of their current positions because of the presence of the continental ice sheets. In places, the boreal forest reached all the way to the Gulf of Mexico.

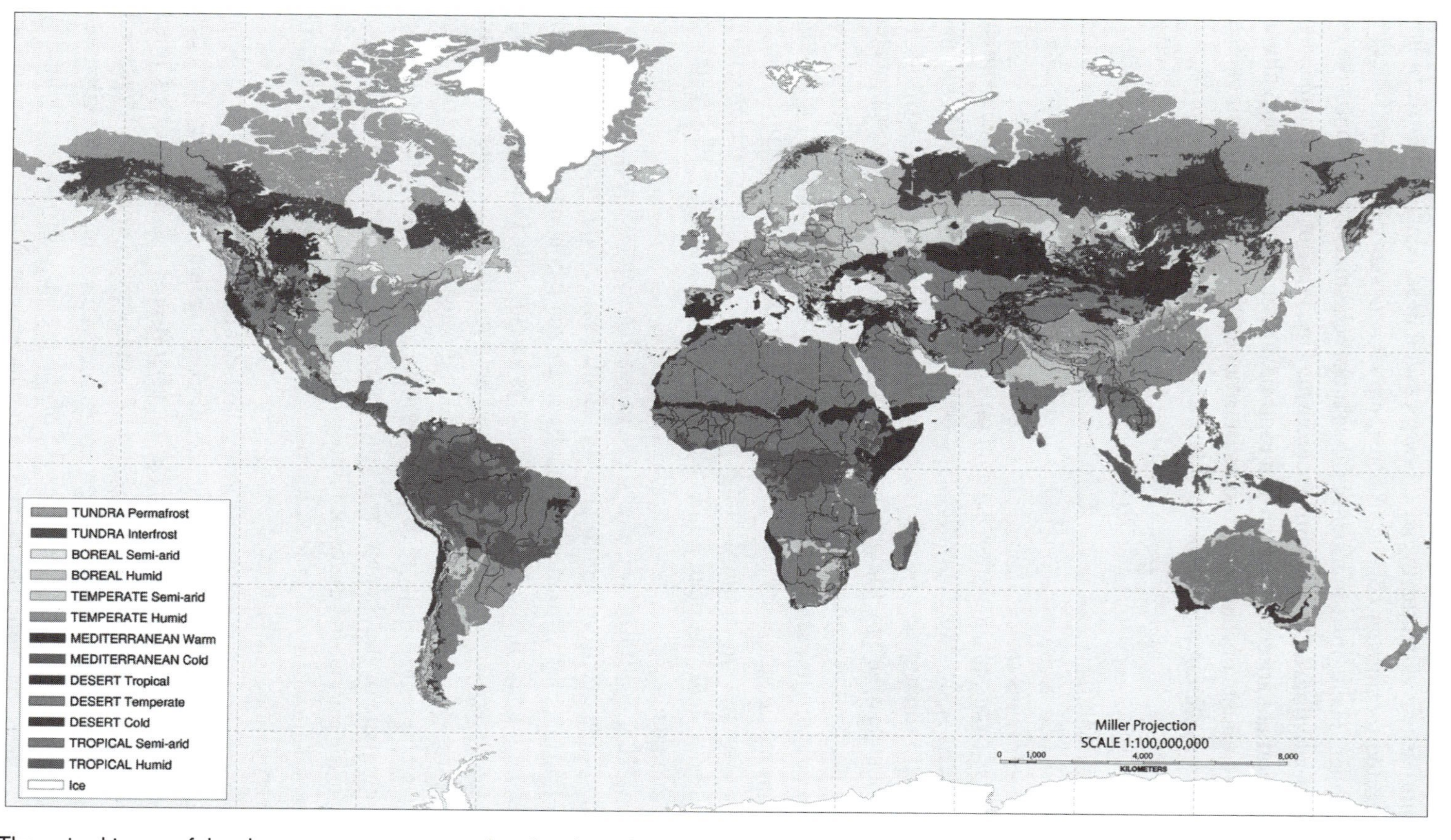

The major biomes of the planet represent a geography of ecological zones, represented by unique combinations of flora and fauna. (U.S. Department of Agriculture)

Of great importance in the understanding of present-day biomes is that humans have now heavily influenced the mix of species over the planet. Natural parts of biomes are quite rare in the populated parts of Earth. Illinois had many acres of prairie but corn and soybean farming has reduced this to a very few hectares. In many **regions** of Earth deforestation has changed the landscape over to crops. **Urbanization** is an unrelenting fact of social geography so that much land is paved over and/or revegetated with human-placed species such as lawn grasses.

Similar to climates and soil classifications, there are several biome classifications with various "flavors" and, thus, different numbers of biomes and various complexities. For this *Handbook* we will use a simplified classification while appreciating that there are other ways with greatly increased complexity to spatially sort life on Earth.

Biomes are divided into aquatic and land-based varieties. Aquatic biomes cover most of Earth. Three marine (saltwater) biomes are usually recognized. The first is **ocean**. The world's oceans are deep as well as extensive, so oceans represent the most diverse biome. Plants and animals range from microscopic zoo- and phytoplankton to whales that can measure over 30 m in length.

The ocean biome is usually divided into four zones: abyssal, benthic, pelagic, and intertidal. The abyssal portion is the part of the ocean deeper than 4,000 m yet contains some plant and animal species in the lightless, highly pressurized, cold (2°C to 3°C) water. Here, the biomass is much less than other oceanic biomes but can be locally rich where methane, sulfides, or thermal energy seep out of the sea bottom.

The benthic zone is generally between 1,000 and 4,000 m depth and ranges from 4°C to 8°C around the world. This is a lightless zone with pressures exceeding 200X those in the air above. Life is a bit more diverse than in the abyssal zone but the relative lack of plankton limits life.

Above the benthic zone is the pelagic zone, sometimes called the open ocean. This is the first 1,000 m or so in which sunlight penetration directly fuels the food chain anchored by plankton. This zone, particularly the first few hundred meters, is directly affected by climate and exchanges mass and energy with the atmosphere. Lack of nutrients causes many pelagic areas to be unproductive in terms of biomass and biodiversity. Selected areas, such as off the western coast of South America, are subject to upwelling, which brings benthic nutrients to the pelagic waters and are hugely productive.

The intertidal zone is the area between the lowest and highest tides. Generally, the lower intertidal locations have a greater biodiversity than the higher locations because the higher locations are less often covered by water. Rocky coasts tend to have greater biodiversity than sandy coasts because sand makes it more difficult for plants and animals to gain a permanent foothold at a single spot.

Coral Reefs

Coral reefs are an amazing interplay between the organic and inorganic realms. Coral reefs result from the excrement of trillions of various species of small animals known as coral polyps (generally less than 5 mm [2 in] long). Living in colonies, the polyps excrete calcium carbonate, which becomes rocklike. These rigid structures protect the polyps against waves, tides, and predators. The reef grows from the seafloor upward and can make for a complex seafloor bottom with quiet water and sanctuary for small animals and plants. Coral reefs are among the most biodiverse places on Earth. Geographically, there are two requirements for coral reefs: (1) shallow seawater less than 50 m (160 ft) deep; and (2) water temperatures above 20°C (68°F). Coral reefs are found in all the shallow tropical waters of the world. The largest reef is the Great Barrier Reef (actually a complex of many thousands of individual reefs off of the northeastern coast of Australia); it has an extent of 2,600 km (1,600 mi). Fossilized reefs found underwater near cold places such as Greenland, indicating a history of continental drift/change in climate. Additionally, mountains such as the Guadalupe Mountains of Texas (Permian) and the Hochkonig Massif of Austria (Triassic) were once reefs, hinting at the enormous changes wrought by **plate tectonics**.

The second marine biome is composed of coral reefs. Near shorelines, coral reefs provide specialized situations where life thrives. Coral reefs are actually built by the lime-rich excrement of tiny colonial animals called coral polyps. These reefs are found worldwide in shallow, warm waters. The presence of reef structures allows a multitude of symbiotic relationships between plants and animals fostering significant diversity of life forms.

The third marine biome is found in Earth's estuaries. These are constricted ocean arms into which freshwater streams debouch and are characterized by the mixing of freshwater with saltwater. A large number of plant and animal species can be found and, typically, biomass and biodiversity are much higher than in the open ocean. Nutrients emanating from rivers and the various salt contents up and down the estuary make for an energetic chain of life with reeds and other plants anchored in the shallows and ocean fish coming into protected waters to spawn.

Another class of aquatic biomes is associated with fresh water. The freshwater class includes lakes, ponds, and streams. Combinations of size and flow conditions make for rich biodiversity. Particularly fecund in the production of biomass are wetlands (swamps and marshes) that are covered by shallow water for all or part of the year. They are delineated with trees dominating swamps and grasses dominating marshes.

Land biomes roughly correspond to world climate zones. For clarity, world biome maps are usually limited to a dozen or less biomes. Most plant and animal species are not exclusive to single biomes although they tend to be more or less prevalent in certain biomes. The following biome description is based mainly on vegetation, the most noticeable landscape feature of each biome. The natural areas of these biomes are given in the map, although there has been major anthropogenic disturbance in boundaries and composition of most of them.

First, there are some land areas virtually uninhabited by life forms. These are the permanent snow and ice areas found at high altitudes and, most expansively, in Antarctica and Greenland.

Tundra is a low-energy biome related to high latitudes and altitudes. Corresponding to tundra climate, the world's largest extent of the tundra biome is found fringing the Arctic Ocean. The growth of plants is slow because of low temperatures and seasonally short/nonexistent sunlight periods. The landscape is dominated by lichens, sedges, short grasses, and herbs that can withstand an environment that is frozen much of the year. These plants are relatively simple and the biome is not very diverse in plant and animal species.

Boreal forest (taiga) is found equatorward or at lower elevations than the tundra biome. The forest is coincident with subarctic climates and is only extensive in the Northern Hemisphere because of the lack of continental surfaces at accordant latitudes in the Southern Hemisphere. The boreal forest has a considerable cold season making photosynthesis all but impossible except for the short, cool summer. The typical forest species are evergreens like spruces and firs and deciduous like birches and maples. Underlying soils tend to be acidic and infertile.

Temperate forest extensively occurred in the more humid portions of the middle latitudes, but these areas have been heavily altered through deforestation and non-native species introduction. The wetter portions of humid continental, humid subtropical, and marine west coast climates were once dominated by the temperate forest. The temperate forest has a summer season with little moisture or **temperature** stress and a cold winter season (in some places snowy) precluding intensive photosynthesis. The forest is dominated by large trees that vary in species mix by location. The mix tends to be very biodiverse and encompass a large amount of biomass. Usually, there is a forest canopy layer with understories of smaller trees, bushes, and shade-tolerant plants. Most tree species are deciduous, but there are notable areas of needleleaf evergreen dominance in the southeastern and northwestern United States.

Temperate grasslands are areas dominated by grasses with a relatively few trees relegated to watercourses. Temperate **grasslands** are the vegetative response to cold winters and warm/hot summers with evapotranspiration stress. Grasses store energy in their root systems in preparation for winter and regenerate their above-ground portions in the early warm season. They are quite conservative of moisture and nutrients so that they present fertile soil when first tilled. Geographically, temperate grasslands are widespread between humid areas of forest and arid areas of temperate deserts. Humid grasslands have taller and more diverse grass species than their drier counterparts. Major temperate world grasslands include prairie, veldt, pampas, and steppe.

The Mediterranean biome is sometimes classified as a woodland and sometimes as a grassland. In reality, it is area dominated by shrubs and grasses with some trees interspersed. Coincident with the Mediterranean climate, this biome is found in selected coastal areas of the middle latitudes. The largest extent of this biome is around the Mediterranean Sea, but it is also found along the California coast, southwestern Africa, southwestern-most Australia, and in southwestern South America. Unlike the other biomes, the Mediterranean biome receives most of its precipitation (rainfall) during the winter and has persistent drought in the summer. Indeed, the biome is brownest in the summer and this leads to a pervasive risk of fire. Although the native plants are drought- and fire-tolerant, the fire regime does not meld nicely with human activities. Animals tend to be nocturnal and small so as to avoid the high heat of the summer. This thermally mild climate has been heavily inhabited for thousands of years and has undergone substantial alterations by humans. In the case of the Mediterranean basin there are few areas that can be called natural.

Desert biomes are regions with sparse vegetation and animal life but life does exist. There are subtropical and temperate deserts and, so, great aridity can occur over a wide range of latitudes. There is no yearly precipitation amount defining the borders of deserts. Rather, deserts are the results of great moisture stress occurring in climates ranging from hot to cool. The key to moisture stress is the balance between **precipitation** and potential evapotranspiration. Generally, deserts are places where the annual precipitation is less than half of the potential evapotranspiration. Plants are xerophytic, meaning that they are adapted to perennial drought through mechanisms such as deep root systems and thick, leathery leaves. Desert plants are usually spaced far apart to lessen the competition for the scarce water resource. Animals are few and nocturnal. Thus, the nature biomass is low compared to most other biomes. Ironically, the nutrient desert soils can be very productive for agriculture if properly irrigated.

The tropical savanna is a biome consisting of open grassland with scattered trees. This biome occupies intermediate regions between tropical rainforest and subtropical deserts. Temperatures are always tropically warm. The tropical savanna has a wet season near the time of the high sun and a long dry season the rest of the year. The landscape browns out and is subject to fires during the dry season. Biodiversity and biomass are both greater than in the desert. This is the region of herding big game animals. Indeed, over half of Africa is covered by savanna.

Tropical rainforest is a warm, wet biome usually found close to the Equator. Not as hot as the desert, the year-round warm temperatures encourage plentiful plant growth because of unlimited water availability. Precipitation comes from afternoon thunderstorms caused by local instability and the precipitation is copious. In some places, precipitation can average upwards of 800 cm (314 in) per year. Plant growth is prolific and represents the greatest diversity and yearly biomass additions on the planet. The forest is quite complex with hundreds of tree species per hectare. The tallest trees sometimes exceed 50 m with several other canopy layers resulting in a perpetual dimness on the rainforest floor.

Boundaries

Boundaries are lines that separate space into defined, discrete units. They may be tangible or imaginary, may be visible upon the landscape or exist quite literally in thin air, and may appear at any scale, from a few inches or centimeters in length to many thousands of miles. In many instances, boundaries are not actually lines, but function in two dimensions as a plane, because the division marked by the boundary extends below the surface of the earth and into the **atmosphere** above the boundary marker. Boundaries are set by humans, and many other species, and can be associated with plant life as well. They can be naturally occurring features, or may be constructed or imposed by individuals or groups. Boundaries are frequently established due to **territoriality**, the human or animal desire to control a given space. Often boundaries are absolute and rigid, as in the case of an international border, but they may also be gradual and porous, as are the boundaries dividing the various dialects of a language, for example. A boundary may be created to enclose a location to identify it, or its purpose may be to prevent some exterior force or element from entering a specific space.

The concept of a boundary for human society has probably been in place since well before the dawn of civilization. Most people have a subconscious sense of a personal boundary, enclosing a personal space. Unauthorized or unexpected intrusions across this division into one's personal zone typically lead to feelings of discomfort, anxiety, fear, or aggression, depending on the type of incursion and who is violating the boundary. The precise location of the personal line differs from society to society, and often between genders, but represents perhaps the most fundamental boundary that guides human relations, even if many are not fully aware that such a line of defense surrounds them. Other boundaries are more overt and obvious. Some boundaries may function as **buffer zones**, which serve to separate two actual or potential adversaries. A demilitarized zone, such as that which exists between North and South Korea, is a good example. A type of boundary that most people have encountered in some form is a border. These lines of division operate on many scales. Borders are established to legally demarcate jurisdiction over a defined, specific space. Property lines separating lots in a subdivision, and the fences that may correspond to such a line, represent short, legally defined borders drawn by certified surveyors, whose methods, standards, and techniques are accepted by courts of law in many countries. Borders at this scale, which usually are only a few hundred feet in length, can be defined with considerable precision. Borders at a larger scale, such as international boundaries, frequently exhibit much less precision, and are many times disputed by one or both countries separated by the border. On the other hand, in a few locations borders between sovereign states remain undefined and unmarked, even in the 21st century. Borders not only define zones of control, they also affect the rate of many geographic processes, such as **migration** and **cultural diffusion**, because they tend to delay or completely prevent movement across space.

The practice of clearly identifying the location of international borders has become commonplace only in the last two to three centuries in many parts of the world. Previously, borders were frequently defined by less distinct boundaries known as marchlands or frontiers, which provided only a general idea of where the rule of a sovereign or state came into effect. Of course, some borders were marked in ancient time by defensive walls that were virtually impenetrable and reached lengths of many miles. The Great Wall of China and Hadrian's Wall in Great Britain are two of the most famous examples. These are also examples of what political geographers term *relict* boundaries, meaning that while they still appear on the **landscape**, they no longer function as a boundary. There are more modern relict boundaries as well, including the Berlin Wall and the famous "Iron Curtain" lying between Eastern and Western Europe during the Cold War. The latter was an imaginary boundary, in that there was no real curtain, although the border certainly was sealed and operated as a real barrier, keeping millions of people inside. Relict boundaries serve to remind us that the location of boundaries is never permanent, and political borders in particular tend to change or lose their relevance over time.

In addition to relict boundaries, geographers identify political boundaries based on various criteria. A common type of international border is the *natural* boundary, which is represented by a naturally occurring physical feature. The Danube River in Eastern Europe, for example, serves as the international border between several countries as it flows from Germany toward the Black Sea. The Himalaya range in southern Asia is another natural feature that serves as a political boundary. The term *physiographic* boundary is also applied to this type of boundary. On the other hand, *geometric* political boundaries are drawn artificially without consideration of the natural terrain. A geometric border must be surveyed, and in many cases is indicated by a fence, wall, or some other constructed marker. These types of divisions often correspond to line of latitude or longitude, and are usually created

Hydropolitics

Water has become a scarce and precious resource in many **regions**. Where bodies of water are shared by more than one state, conflict may arise over the allocation of water. This is increasingly the case where a river crosses an international **boundary**. Some of the world's major river systems have now become the focus of geopolitical tensions because of increasingly pressures on the use of water. The Nile, Jordan, and Tigris-Euphrates river systems in the Middle East, the Indus River in South Asia, and the Amu Darya in Central Asia all are at the center of potential regional water conflict. It is not just the *quantity* of water that is at issue, but often the *quality* of water is a concern as well, especially if the water will be used by humans or livestock for drinking. Growing **populations** in some of the river basins where friction is highest compound the problem, because the additional people of course require additional water. Countries that pursue development on their stretch of the river, such as building a hydroelectric dam for energy production, may instead generate tensions with a downstream neighbor, who is concerned about reduction in flow. The politics of water will likely be a major problem in the forthcoming century.

by the mutual consent of the states sharing the boundary. Such borders are quite common in the Western Hemisphere, and some parts of Africa. Borders are also drawn along *ethnographic* lines of division. These borders separate groups based on distinctions in language or religious affiliation, and are the dominant kind of political boundary in much of the world. New countries that have appeared on the map in recent years have formed along ethnographic boundaries: Kosovo, Timor-Leste, and Eritrea are all countries that have become independent since 1993, and are defined predominantly by ethnographic boundaries. It is often the case that the borders between states are composed of all three types of these boundaries. The border between the United States and Mexico, for example, contains segments that are both natural and geometric, and the entire length may be said to represent an ethnographic boundary as well.

Political boundaries may also be classified in a system based on when the boundary was established. The eminent political geographer Richard Hartshorne suggested such a system in the early part of the 20th century. In this system, a boundary that appears in a region that is devoid of settlement is termed a *pioneer* boundary. Such a boundary may become a relict boundary relatively quickly once the region it lies within becomes inhabited or, on the other hand, may prove to be resilient and continue to serve as a functional boundary for a considerable time. An *antecedent* boundary is one that is set prior to the establishment of the contemporary settlement pattern. Many of the boundaries in modern Africa are of this type, because the colonial powers that pursued a policy of **imperialism** there in the 19th century drew boundaries between their territories well in advance of any significant occupation. Boundaries that are formed simultaneously with the process of the development of the human landscape are classified as *subsequent* boundaries. A subsequent boundary may follow the ethnographic landscape of a region, holding to spatial divisions in the linguistic, religious, or ethnic geography. However, boundaries may also be created in regions that already feature a sophisticated pattern of human occupation. Such a boundary is labeled a *superimposed* boundary, and frequently results from conflict. A contemporary example is the border between North and South Korea, a boundary that divides a people who speak a common language and share a common heritage, and who for centuries were part of the same political space. Many current political boundaries in Africa fall into this category as well.

Boundaries are drawn not only on solid land, but also in the air and on the high seas. All modern sovereign states claim airspace over their territory, and unauthorized violation of this space by another country is considered the equivalent of a border incursion, and technically an act of war. By international agreement, countries may claim jurisdiction over the airspace contained by their borders, or out to a distance of twelve miles along a coastline. There is no internationally recognized standard for altitudinal limits to airspace, however, and some countries theoretically claim control over the airspace above their territory to outer space, although such a claim is effectively impossible to enforce. According to the **Law of the Sea**, sovereign states that possess coastlines on an ocean or sea may claim a *territorial sea* to a distance of 12 miles, over which the state may claim legal control. Some countries attempt to claim a much larger area, in some cases out to a distance of 200 miles, but such boundaries are not generally observed by other seafaring powers. Boundaries that govern the right to control and recover natural resources extend further than the territorial sea, allowing coastal states to claim an Exclusive Economic Zone (EEZ) to a distance of 200 miles from the outer margin of the territorial sea boundary.

Break-of-Bulk Point

"Break-of-bulk point" is a term from the **geography of economic development** that refers to a location where cargo transported in bulk must be broken into smaller units so the cargo may be moved further and generally distributed over a larger space. Usually the mode of transportation changes as well; for example, from ship to railroad, or from rail to trucking. This results in a concentration of economic development at the location, as storage, processing and transportation facilities all cluster near the point of delivery. In essence, every port city is a break-of-bulk point, because many commodities delivered to such locations are transported in bulk and often must be processed or refined. An example is crude petroleum, which is delivered to ports utilizing enormous vessels called supertankers. Some of the largest supertankers may carry up to 2 million barrels of oil, nearly the entire daily demand of some smaller industrialized countries. The vast volume of oil carried by these ships must be off-loaded and stored in large tanks, before it is moved by pipeline to refineries where it is made into heating oil, gasoline, and other petroleum products.

Physical barriers also can serve as break of bulk points. This was the case in the early urban development of the eastern United States, when streams were used to penetrate the interior of the region. As larger ships moved upstream, they typically encountered rapids or waterfalls at a point where the streambed crossed an abrupt change in elevation. This point where the elevation changed is known as the fall line, indicating the imaginary line in the eastern United States separating the Gulf Atlantic Coastal plain from the Piedmont, a region of rolling hills and rugged terrain. At the fall line, a change in the mode of transportation was required—in this case, moving goods from a larger ship to smaller barges or canoes to send the cargo further upstream. Because a change in the means of transport was needed that required shifting products from vessels of larger bulk to smaller carriers, storage and processing facilities were also typically required at the point of the fall line along streams. This led to the **agglomeration** of economic activities at the break of bulk location, and the subsequent development of urban centers, as labor and capital were drawn to these points. Break-of-bulk points, especially port cities, tend to develop into industrial centers since there is always an additional cost involved when transferring to a different means of transport. This is especially true in the case of bulk commodities like petroleum, iron ore, coal, etc. Thus, break-of-bulk locations frequently develop into processing centers for raw materials, as it is more cost effective to refine the materials into higher-value products at that location rather than transport them further in unprocessed form. Finally, break-of-bulk points may be established by changes in political space, such as at a border crossing, where legal limits on the weight of bulky cargos force division into smaller parcels.

Buffer Zone

A buffer zone is a space placed between two or more locations increasing the degree of spatial distance between them. In effect, a buffer zone functions as a geographic insulator, separating zones of differing use, various regions, or countries that may potentially be in conflict from one another. They may be created for reasons of security, convenience, environmental preservation, or for aesthetic purposes. Buffer zones are increasingly used in urban planning, especially in large cities, and also have played an important role in **geopolitics**, in the form of buffer states.

Buffer states are weaker countries sandwiched between two or more adversarial states. Many buffer states were formed during the era of imperialism in the 18th and 19th centuries. In principle, they are created to spatially insulate potential combatants by placing distance between them. A classic example of a buffer state is the country of Afghanistan. The **boundaries** of the country were established primarily by outside imperial powers (Great Britain and Russia) to avoid sharing a common border and thereby limiting the potential for conflict. Afghanistan's role as a buffer can be readily seen in its unusual prorupted shape in the northeast, along what is called the Wakhan Corridor. This portion of the country, consisting of a narrow strip of territory extending about 200 miles to the east, was created by British and Russian diplomats in 1895, who insisted that it be added to the boundary of Afghanistan to separate their respective empires. A second example is Thailand, a country that served as a buffer between French Indochina and British colonial holdings in Burma and India. Functioning as a buffer state, Thailand managed to maintain its independence and avoid colonization by either European power. Over time, buffer states may lose their function, as the political geography or political relationships that brought them into being change. Thailand is no longer considered a buffer, since the colonial powers it separated have withdrawn from the **region**, and it does not stand between countries that have a history of conflict.

Buffer zones designed to avoid conflict may be much smaller than buffer states. Examples are demilitarized zones and no-fly zones. At the conclusion of organized hostilities in the Korean War, a DMZ (demilitarized zone) was established between North and South Korea along the 38th parallel. This zone is approximately 2.5 miles wide and crosses the entire Korean Peninsula from west to east, a distance of over 150 miles. The Korean DMZ has been in place since 1953, and is the longest-serving demilitarized zone in the world. Other DMZs created after recent hostilities stand between Israel and Syria in the Golan Heights, between Iraq and Kuwait, and between the newly established country of Kosovo and Serbia. No-fly zones are air spaces in which only certain aircraft may operate and again are frequently established between hostile states.

Buffer zones also appear on a smaller scale. For example, in local zoning, a metropolitan administration may create a buffer between an industrial area and a residential neighborhood, restricting both types of development within the zone. This is typically done to separate the two areas to protect property values in the residential area by reducing the impact of noise, pollution, and congestion generated by industrial activity. Another type of buffer zone used in urban planning is a green belt, which is typically a circular zone around or within the city in which development of any kind is prohibited or curtailed. Such belts are frequently placed around residential areas to provide recreational space in the form of parks and wilderness areas and to shield the residential zone from transportation or commercial activity on the far side of the green belt.

C

Capital Leakage

Capital leakage, sometimes referred to as *capital flight,* occurs when wealth, generally in the form of investment capital, is moved from the source of its creation to another location, rather than being reinvested in the original location. This may happen at a number of scales—from the individual company to an entire country or region. In the developing world, capital leakage is an issue for many countries, which may successfully attract Foreign Direct Investment (FDI) in sizable amounts, only to see most of the profits from that investment siphoned away into external markets such as foreign stock exchanges, government bonds, insurance annuities, or even investment in hard assets abroad, such as real estate holdings or new production facilities like additional factories. For economies that are attempting to build wealth, this process is problematical, because although investment in the country creates employment, it does not result in a proportional expansion of the pool of capital and a subsequent rise in investment, meaning that economic growth falls below levels that would otherwise result from the initial investment. In other words, **sustainable development** is quite challenging under conditions where significant capital leakage is occurring.

The domestic capital leaks back toward more developed economies for a variety of reasons. First, capital flows tend to be directed toward more stable economies and political systems where the risk of economic loss is lower. Second, investment opportunities are greater in the developed world, because the stock markets there offer investment in more established companies, the potential for profits is greater, stock prices are more stable in many cases, and for a number of other reasons. Many developing countries do not even have fully functioning domestic equities and commodities markets, so capital tends to move toward the larger, external investment markets. Political instability, high tax rates, and unfavorable currency exchange rates all drive capital toward the more established markets as well. In some cases, the amount of capital leaking from the developing world is enormous, on occasion nearly equal to the entire international debt of some lesser-developed countries. Capital leakage may take place legally or illegally. Examples of illegal capital leakage include bribery, evasion of tax payments, falsifying of import/export documents or invoices, etc.

The effects of capital leakage are magnified in developing economies that lack economic diversity (as many do), and that are dependent on only a few sectors to support the overall economy. In the case of countries whose economies are highly dependent on tourism, for example, the construction of tourist facilities and infrastructure results in economic growth, as jobs are added to the economy and capital in-flows increase to finance the costs of construction. But in many cases the profits generated to both foreign and local investors in such projects are seldom invested in the domestic economy, but are sent back to markets in the economically developed world. Emerging countries must develop incentive strategies to induce capital to remain in the local economy, but such strategies often lack effectiveness due to the great allure of investment options in the industrialized countries.

Carrying Capacity

Carrying capacity is a quantitative measure of the human, animal, or vegetative population that a specific environmental space can support. It may be calculated for various scales, from the local to the global. It is an important concept in **cultural ecology** and environmental studies, is a centerpiece of the notion of **sustainable development**, and is derived from **Malthusian Theory**. Carrying capacity represents equilibrium between a population and its environment: A population that is below the carrying capacity of its environment will in theory increase to the number represented by the carrying capacity, while a population that is above the threshold of the carrying capacity will die back to that level.

Carrying capacity is a dynamic concept and may be affected by many different factors. In the case of a human population, the carrying capacity of some regions may change dramatically in a short period due to changes in technology or variation in local climate. The carrying capacity of a certain hectare in Africa's Sahel region, for example, may be ten adults under current conditions. This means that enough food can be produced using current technology and under current climatic conditions to support ten grown people. But, if the following year is particularly dry and a drought occurs, the carrying capacity for that year may drop to seven, meaning that famine will likely be prevalent in the region. On the other hand, if climatic conditions are good and the farmers working the hectare are presented with a tractor, an improved hybrid that is more productive or more resistant to insect damage, or some other technological innovation, this change will suddenly increase the carrying capacity. For example, the carrying capacity of some farmland in India, the Philippines, and other developing areas dramatically rose between the mid-1960s and the early 1990s from the technological success of the Green Revolution, as yields per acre in some areas increased by a factor of three. In some areas of south Asia yields for wheat doubled in only five years. On the other hand, the carrying capacity of much of the Sahel has declined during the same period, due to overgrazing, drought, and soil erosion. Carrying capacity for nonhuman species, an important concept in wildlife management and ecology, tends to be more stable because such species do not introduce technological innovations that alter the food-producing capacity, although changes in the environment can affect the carrying capacity.

Some scholars have criticized the notion of carrying capacity as a measure of gauging sustainability, because of the highly dynamic character of the human-environment relationship. They argue that the carrying capacity of any given region is so fluid and so subject to alteration in the short term that to set policy based on the concept is misleading and inaccurate. In other words, projections and estimates of what is sustainable population growth for a region, or for the planet as a whole, that are based on carrying capacity calculations are inherently fallacious, because they have historically failed to account for human innovation and technological change.

In tourism geography, carrying capacity is used in a somewhat different manner than in ecological or demographic studies. For geographers studying the spatial aspects of tourism, the term is used to estimate the maximum number of visitors a given location may support, without diminishing the location's attraction for tourists. More formally, the World Tourism Organization defines carrying capacity as "the maximum use of any site without causing negative effects on the resources, reducing visitor satisfaction, or exerting adverse impact on the society, economy and culture of the area." Mathematical formulas are used by scholars and managers to determine the actual carrying capacity of a tourist site.

Cartography

Cartography is the art, the skill, and the science of making **maps**. Human beings have been creating pictorial representations of spatial information for thousands of years. The earliest maps were probably crude temporary drawings made on the ground in dust or mud, designed to show simple landmarks guiding a journey, the position of a herd of animals and the points of attack a group of hunters might pursue, or other simple data and features. Over time, the necessity of illustrating more sophisticated and detailed spatial information became common, especially with the rise of cities, which eventually became so large that even many of the permanent residents were not familiar with some parts of the urban setting. The

Babylonians, for example, made maps of their cities and, like a number of early civilizations, also studied the stars. Astronomical drawings and charts were also an important ancient application of cartography, often used not only for scientific applications but for religious purposes as well. The Babylonians may have produced the first map of the "world," at least the world as they knew it, in the Imago Mundi, a geographical depiction of the Tigris-Euphrates river valley and its environs.

The ancient Greeks were the first culture to develop scientific techniques for making accurate maps. Two Greek scholars have a particular importance in the history of cartography: Eratosthenes and Ptolemy. Although they lived approximately three centuries apart, both men played a vital role in advancing cartographic techniques that would pave the way for much more accurate and detailed maps. Eratosthenes, using the principles of geometry, derived a highly accurate measurement of the **Earth**'s size almost two centuries before the birth of Christ and over a millennium before the planet was circumnavigated. He is also credited with constructing the first grid system for cartographic representation, a major achievement in that it established a means for accurately determining both distance and direction from a map. Ptolemy modified the grid system of Eratosthenes and offered his own calculation of the earth's circumference, although ironically, his estimation contained a much greater error than that of his predecessor. His coordinate system, however, provided the basis for the modern application of

Eratosthenes

Eratosthenes was a Greek geographer who may have been the first scholar to accurately calculate the circumference of the earth. Remarkably, he was able to estimate the size of the planet to within one percent of its actual circumference at the equator, depending on the exact units of distance he may have used. His method was ingenious—by measuring the difference in the angle of elevation of the sun at solar noon between two locations on the summer solstice, and assuming that both lay along the same longitude, he was able to deduce that the distance between the two locations must be equal to one fiftieth of the earth's circumference. He also derived a system of latitude and longitude, which was later perfected by Ptolemy and used in the West by mariners and geographers for centuries, and he may have calculated the distance between the earth and the sun with reasonable accuracy. Eratosthenes also was skilled in the science of **cartography** and composed one of the first maps of the world in antiquity, based on the limited geographical knowledge at his disposal. Eratosthenes served as the head librarian for some time at the Library of Alexandria, one of the ancient world's most important scholarly collections.

latitude and longitude, the Universal Transverse Mercator (UTM) system, and other grid patterns used today.

Religion often motivated early European mapmakers and others, who sought to spatially locate not only known points but also identify the location of places from scripture. Furthermore, locations having great religious significance figured prominently on maps and were often represented at the center of the illustration, with all other points clustered around them. The classic example of such a map is the "T and O" map drawn by European cartographers of the Middle Ages. These maps represented the world as it was known to the Europeans, with the Mediterranean Sea forming the "T" portion of the map, which divided the depicted land masses of Europe, Africa, and Asia. Jerusalem or Rome was generally located at the center of these maps, which, unlike modern maps, were not oriented with north at the top of the map, but rather east. The Garden of Eden was also frequently included on T and O maps, usually placed somewhere on the Asian continent. This cartography was not very useful for navigation or other practical purposes, but nevertheless illustrates the continued desire of humanity to visually represent the world, even in the face of widespread geographic ignorance.

By the late Middle Ages, exploration, an activity humans pursued instinctively, demanded a visual guide whenever possible. The advent of oceangoing travel and trade required an accurate record of landmarks and hazards for mariners, who needed to retrace the journeys they had survived either for conquest or maintaining trade. Such early ships' logs were simply written commentaries, but these were eventually replaced by charts that contained both written and visual details. Called portolan maps, from the Italian word *portolano* (port chart), these maps represented a major step forward for cartography. Such maps frequently featured a crude grid, based on "rhumb lines," that provided some basis for calculating distance. Although not particularly useful for navigating large distances across open water, portolan maps could be utilized in smaller bodies of water, especially the Mediterranean Sea, where they were frequently employed by merchants and traders. Cartographers of a few centuries later, like Gerardus Mercator, would use the portolan charts and much of Ptolemy's work as starting points for much more detailed and useful nautical maps, opening the way for modern cartographic techniques. The projection of the surface of the earth that Mercator developed in the 1500s remains widely used today in nautical and aeronautical navigation.

The advent of the printing press, the development of new techniques for gathering spatial data, and the discovery of the New World at the end of the 15th century heralded a flowering of cartographic expansion. As new lands were discovered, conquered, and settled, the need for maps increased dramatically. Cartographers became not only recorders of spatial data but also agents of political and social change. Maps were vital instruments in determining **boundaries** between new colonies and countries, in displaying and tracking demographic and economic expansion, and even in promoting national and **cultural identity**. Maps on occasion became tools of political propaganda, sometimes employed to illustrate an alleged threat, or promote a territorial claim. At the same time, advances in surveying meant that every piece of property was legally defined via a survey map, so that accurate cartography became a crucial element of property ownership. At the beginning of the 20th century the invention of the airplane led to the new discipline of **remote sensing**, adding yet another tool to assist the cartographer in making increasingly detailed maps. Maps could be both verified and composed using aerial photography, and the launching of data-gathering satellites after 1960 only enlarged the universe of spatial information that cartography could express.

The modern cartographer has at his or her disposal many tools and techniques that were unavailable to the pioneers of the field. But today's cartographer also has new challenges—the vast array of cartographic data produced on a daily basis means that cartographic design is a major consideration in making maps, and what information to leave off the map is as vital to the map's success as what details to include. Many different considerations must go into designing a modern map. It is essential to base design on the potential uses of the map. This will often determine the **map projection** that the cartographer employs. This is important because no map is a perfect representation of reality; all maps contain some elements of distortion. This is the case because the map, a two-dimensional surface, cannot fully represent the three-dimensional space of the world without presenting some degree of error. Not only must the cartographer select an appropriate projection for the spatial information to be conveyed, the proper **scale** must also be chosen. Some of these decisions will be obvious, depending on the purpose of the proposed map. For example, a Mercator projection shows true direction; thus, it is ideal for use when making navigational charts and maps. But the Mercator projection distorts areas as distance increases from the equator, meaning that if the intent of the map is to show areal relationships in true proportion, the Mercator projection is a poor choice and will convey information that is inaccurate.

Cartographers typically must generalize the spatial data they intend to present on a map. Exactly how this process is achieved depends on the type of information the map should illustrate, the scale at which the information will be displayed, and which data are the most relevant to the purpose of the map. Those designing maps follow four steps when preparing data for presentation on a map. The first step is *simplification* of the data. In previous centuries, when accurate cartographic data were scarce, this process was unnecessary, but with modern information-gathering techniques, simplification is usually required to avoid cluttering the

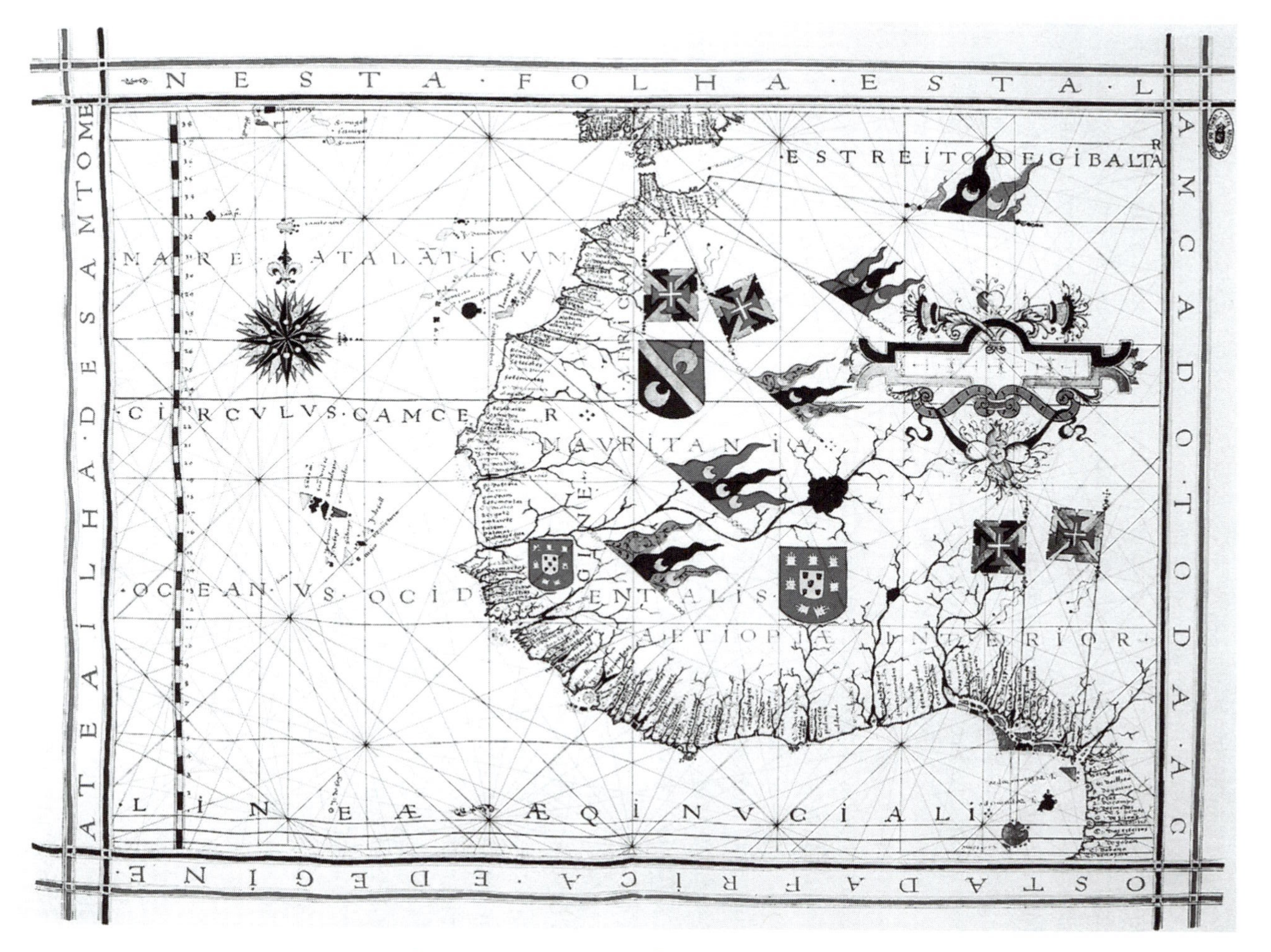

An early map showing the west coast of Africa by the Portuguese cartographer Fernao Vaz Dourado. The map was produced in 1571, and is remarkably accurate and detailed. (National Archives of Torre do Tombo, Lisbon)

map with irrelevant detail. Second, the cartographer must *classify* the data, especially if a thematic map is the goal (see the **Map** entry for a discussion of map types). The type of classification system employed must be chosen with care, as the type and number of classes may affect the appearance, and thus the interpretation, of the information on the map. The classification scheme used is usually shown using a map legend, a separate component of the map that also illustrates the next phase of design, *symbolization*. The various classes of data shown on the map must be represented in some way. This may be accomplished using different colors, patterns, or depending on the nature of the classification system and the data, lines or areas. Finally, the process of *induction* may be used to fill in gaps on the map where data are lacking. This must be done carefully, as the cartographer is using known data points to interpolate new data points, a process that may lead to inaccuracies if not performed with deliberation and caution.

Technology has become enormously important to cartography. Cartographers in the 21st century produce maps using computers, which greatly enhance both the appearance and functionality of modern maps. Data-gathering techniques have also become increasingly sophisticated, complex, and accurate. For example, a Global Positioning System (GPS) can be employed to gather incredibly detailed and positionally accurate information—some GPS receivers can record locations of features to within centimeters of their actual location on the earth's surface (or above the surface, in some instances). Such detailed data, gathered so readily, have never before been available to cartographers, and make the process of map design and presentation even more crucial. Computer cartography also allows much greater flexibility in the production of maps, in that a map's classification scheme, symbols or even scale and projection may be changed with a simple click of a button. Previously, once a map was produced, it represented a static presentation of reality because the characteristics of the map were fixed, and could be altered only by making and printing a completely new map. The advent of computer cartography software packages, which store cartographic information digitally for easy and rapid retrieval, permits the production of a host of maps from the same data set, and makes possible the manipulation of any or all of the map's basic characteristics.

In fact, many maps available today are not even produced and disseminated using paper, but rather appear as cybermaps, which are stored in a cyberatlas, available only on the Internet. Moreover, the appearance of **Geographic Information Systems (GIS)** has revolutionized cartography. A GIS is a database, often containing many different types of spatial information, which may be displayed, analyzed, and published in any combination. A large GIS may be capable of producing thousands of different maps based on various data, or of displaying the same data set in many different ways by changing the scale, projection, symbology, or other cartographic characteristics. The increasing application of GIS technology has led to a parallel expansion and use of cartographic analysis. Maps have become essential tools in such diverse fields as natural resource management, retail marketing, and medicine. Indeed, one may readily find maps of ground water supplies, locations of retail outlets, and maps of the human body all with a quick Internet search. But simple mapmaking is also becoming easier, and does not require years of training or thousands of dollars worth of sophisticated equipment or complicated data sets. Simple GPS receivers today allow the user to make maps of a favorite park, fishing spot, or the route to a favored restaurant. Thus, everyone can make use of cartography, from an unsophisticated personal level to one involving intense scientific analysis. The journey from drawing simple, crude diagrams on the ground to mapping using satellites feeding data to a GPS receiver may appear to be a long one, but the principle behind the two processes is the same, as is the motivation. Cartography is the reflection of the world's spatial complexity and how humans represent it.

Central Place Theory

A theory aimed at describing and explaining urban settlement patterns. Central Place Theory provides one of the most influential philosophical frameworks to emerge from the study of urban geography in the 20th century, and continues to be debated and refined. The foundations of the theory were first laid out in the doctoral dissertation of Walther Christaller in the 1930s. Christaller had studied settlement patterns in his native Germany and concluded that the arrangement and size of urban places was directly related to the economic services and functions the various locations offered. Each urban center was associated with an economic **hinterland**, sometimes referred to as the market area, resulting in a regular pattern of the growth and spacing of settlements. Christaller's great contribution lay in explaining the economic fundamentals behind the formation of the hinterlands, their basic shape, and how certain principles might affect the urban spatial pattern.

Central Place Theory assumes that the physical geography considered is uniform—a completely flat plain exists, with no hills or mountains present, and no rivers, streams, or other features that would impede motion. Distance, therefore, is the only factor when considering transportation cost and accessibility to markets. Moreover, the plain holds an evenly spaced population, and no area has an advantage in terms of resource endowment—labor, capital, and raw materials are all equally available and of identical quality. Under these conditions, demand for goods is also identical at every **location** on the plain, and the only difference in cost between similar types of goods is the additional transport cost associated with traveling a greater distance to obtain the good, a concept linked to **distance decay**. That is, each good or service offered at a location on the plain would have a specific *range*, or distance that consumers would be willing to travel to procure the good or service. More expensive goods or services that would be purchased less frequently, which Christaller called "higher-order" goods and services, would have a larger range than so-called "lower-order" goods and services that would be cheaper and purchased on a regular and frequent basis. For example, a higher-order good might be a diamond ring; a lower-order good would be a bottle of milk. This would result in the formation of a specific hinterland in the shape of a circle for each good or service, extending outward from the place it was offered. The radius of the circle would equal the range of the particular good or service. Any consumers located beyond the **boundaries** of this hinterland would not travel to the place to obtain the good or service, as the transport cost would exceed the value of the good or service.

Each type of good or service offered at locations on the plain would have a specific *threshold*. The threshold is the minimum number of consumers necessary to offer the good or service at that location. This can be represented by an area, because the theory assumes that the population is uniformly distributed across the space of the plain. Higher-order goods will have much larger thresholds, because they are goods or services that are only purchased infrequently, but lower-order goods or services will have much lower thresholds, due to the higher demand for them. This can be easily visualized by using the examples provided above—one may purchase milk on a weekly basis, but a diamond ring only once in a lifetime. A jewelry store requires a much higher threshold (number of consumers) than a supermarket. The result of this system is that the plain now is covered by imaginary circles radiating outward from central locations, each symbolizing the hinterland of a type of good at that location. This means, however, that some spaces will remain between the hinterlands, since circles do not fit tightly together. Central Place Theory suggests that the hinterlands are not circular, but rather take the shape of hexagons that then follow a beehive pattern covering all of the space of the plain. Furthermore, there exists a hierarchical ordering of the hexagonal pattern, based on the varying sizes of hinterlands. Central places offering the highest order goods and services will have the largest hexagonal hinterlands. For lower-order goods and services, these large hinterlands will be subdivided into a series of smaller hexagons, representing the next order of goods and services, offered at smaller central places. The number of hexagon sequences in the hierarchy is dependent on the total number of different orders of goods available on the plain.

The exact spatial configuration of the hierarchical structure of central places and their hinterlands may be modified by applying certain characteristics to the economic environment, factors that Christaller called "principles." If one assumes, he argued, that the system is dominated by a *marketing principle*, which provides the greatest profit margin to businesses operating at central places, then it is ideal to keep the number of central places to a minimum—this results in the lowest level of competition. The pattern created by the application of this principle would result in a k = 3 solution, meaning that each hexagonal hinterland would contain the equivalent market area of three hinterlands of the order below it. If, on the other hand, the central places of lower order were placed at the linear midpoint between places of the next highest order, this would result in not only more central places, but would maximize the number of central places along potential transportation routes, using the *transportation principle* (sometimes referred to as the traffic principle), or k = 4 pattern, in which each hinterland of a higher order contained the area of four hinterlands of the next lowest order. Finally, Christaller stated that an *administrative principle* might be applied, which would be designed to ensure the maximum level of governmental efficiency in managing economic activity. In this case, each higher order central place would hold six central places of the next lowest order, and its hinterland encompasses the equivalent of seven hinterlands (the six lower-order hinterlands and its own) of the level directly below it. Christaller proposed that at least one of these principles was at work in the spatial ordering of urban centers.

Centrifugal and Centripetal Forces

Collectively, the social, political, cultural, or economic forces that weaken (centrifugal) or reinforce (centripetal) the geographic integrity of a **nation-state**. Centrifugal forces may result in political fragmentation or **balkanization** of a state. All states contain elements that contribute to centrifugal tendency, although these may be quite weak in some cases, especially in countries that are culturally homogeneous, or nearly so. Japan, for example, a country that holds few religious or ethnic minorities, evinces little centrifugal tendency. Countries containing numerous minority groups (India, Nigeria, Malaysia), or a single large minority that is regionally concentrated (Canada, Belgium, Kazakhstan) often experience significant centrifugal forces. In other states, control over resources or differing levels of economic development may trigger the expression of such forces. Such was the case in the United Kingdom in the 1980s and 1990s, when Scotland gained significant political autonomy as a result of centrifugal tendency generated by desire to gain a greater share of the oil and gas deposits in the North Sea, located off Scotland's east coast. Even a single political issue may serve to generate strong divisions in regional identity and loyalty, resulting in violent expression of centrifugal tendency that may threaten the survival of the state. Such was the case with the issue of slavery in the United States in the first half of the 19th century. In some instances centrifugal force is generated because of friction generated by an exaggerated **core and periphery** relationship.

If **cultural identity** in a country is not equivalent to a unitary national identity, centripetal forces can then become so strong as to result in complete disintegration of the state. Centrifugal forces have the ability to radically change the geography of regions in a very short period of time. In the early 1990s, two large countries, including the largest nation state in the world, collapsed due to a rising tide of centrifugal forces within their **boundaries**, indicating a trend toward **devolution** with the region of east-central Europe and the Soviet Union. Yugoslavia, the largest country in Eastern Europe, and the Soviet Union, the largest country in the world, fragmented along the lines of ethnolinguistic and religious distinctions, resulting in a total of 20 new countries on the Eurasian landmass, almost all of which appeared in the two years of 1991 and 1992. Strong centrifugal forces can exist between groups which, from an outside perspective, should share a strong cultural and political bond. A case in point is the animosity between the Croats and Serbs in former Yugoslavia. Both groups are Christians, although they follow differing branches of that faith, and ironically, both peoples share a common spoken language. The language once universally called Serbo-Croatian is written using two alphabets—the Croats write the tongue using the Latin alphabet, while the Serbs utilize a modified Cyrillic script to write the language. The cultural differences between the Croats and the Serbs seem minor to many outsiders, but each group considers the distinctions to be significant. Historical conflicts between the two peoples also contribute to a common hostility in many instances. And the Croats and Serbs represent two groups with much in common, compared to the remaining ethnic groups of former Yugoslavia! Yet, it should be pointed out that for half a century, the centrifugal forces in Yugoslavia were held in check, and, at least for some, a strong sense of Yugoslav identity was achieved.

The influence of centrifugal forces may be released upon the collapse or removal of a central authority, the presence of which previously held such forces in check. In the 20th century this led in many cases to catastrophic loss of life. Numerous examples of this process may be found after decolonization of much of the developing work in the wake of World War II. India, a colony under British government rule for nearly a century, was rent into two states following independence in 1947, due to the religious animosity and distrust between the Muslim community and the Hindu majority. The subsequent split of colonial India into two sovereign states, and the accompanying forced migrations across a newly established border that had never previously existed, resulted in the death of perhaps as many as 3 million people. The remnant centrifugal forces from the partition continue to be expressed in the antagonistic relationship between India and Pakistan. Almost a decade and a half later Nigeria experienced severe centrifugal forces in the wake of decolonization. Granted independence in 1960, the Nigerian government faced the daunting task of creating a unitary national identity in a state that held perhaps as many as 250 distinct ethnic groups. By 1967, centrifugal forces in the southeastern portion of the country led to the attempted secession of Biafra, a region dominated by the Igbo people. In this instance, centrifugal force did not result in a division of the country, but the civil war that erupted as a consequence of the Biafran secession led to widespread devastation and possibly 3 million dead, mostly non-combatants.

Centripetal forces may counter the effect of centrifugal forces. These are the glue that holds a nation-state together, and confers a shared sense of identity, expressed as national identity. Centripetal forces are represented by a strong sense of shared historical and cultural origins, a common language or religion, or an economic motivation toward unity. In the United States, as least two of these factors historically helped generate centripetal force. The notion of *manifest destiny*, a common purpose to settle and "civilize" the North American continent, played a strong role in creating an American national identity in the 19th century, and the use of English as a national tongue, although not a legally defined national language, also contributed greatly to centripetal force in the country. Japan represents a nation-state that has a long pattern of strong centripetal linkages, including a common language, a national identity closely connected (at least until recent times) to the indigenous faith of Shinto, and an enduring and well-indoctrinated national myth. Moreover, the country has been successfully invaded and occupied by a foreign power only once in history, and has not experienced any significant waves of immigration of minority groups. As a result, Japan has faced few episodes of centrifugal force that threatened its spatial integrity.

Centripetal forces are evident in efforts at **supranationalism**. In such cases, the centrifugal forces that must be countered are the respective national identities of the member states. The major centripetal force behind the formation of the European Union, often cited as a prime example of supranationalism, is economic motivation—the goal of constructing a larger economic entity that will benefit all the member states. The projected benefits of belonging to a larger economic organization have obviously provided grounds for many European nation-states to abandon a portion of their sovereignty (a centrifugal force in such an organization) as the number of member states now exceeds 25, and appears set to grow larger. Interestingly, centrifugal *and* centripetal forces appear to be vital influences shaping the political geography of many world regions, and much work remains to be done in coming to a full understanding of the dynamics behind both.

Choke Point

"Choke point" is a term used in **geopolitics** and military geography to denote a narrow passage, either on land or water, through which a military force or economic resource is forced to pass, and that may be easily controlled by one or a few countries. Choke points can carry great military strategic importance, because the geography of such locations may be utilized by a smaller force to offset a disadvantage in size when facing a larger army. The classic example of a smaller force using a choke point to great advantage is at the Battle of Thermopylae, when a small army of Greeks, led by the Spartan king Leonidas, encountered a much larger Persian army led by Xerxes I. The pass at Thermopylae at the time of the battle in 480 BCE was extremely confined, wedged between the sea on one side and steep hills on the other. The Greek historian Herodotus described the passage as so restricted that only a single chariot could move through it at a time. Leonidas, recognizing the advantage the topography would give his outnumbered army, used this choke point to supreme advantage, stopping the advance of the Persian soldiers for several days, and allowing the Greek forces behind him to organize a defense against the invading Persians. Indeed, had it not been for the betrayal of a local herdsman who showed the Persians a path through the hills that enabled them to bypass the Greeks and attack them from the rear, Leonidas and his followers might have held the position for many more days. Considering that modern scholars estimate that the Greek fighters were facing an army that was likely *50 times* larger, the strategic advantage offered by a choke point is obvious.

In geopolitics, choke points on the world's sea lanes are of great strategic importance as well. This is because such constricted waterways could be closed or restricted by a hostile power, essentially pinching off the supply of a vital resource, such as petroleum. Perhaps the most crucial and patrolled choke point on the globe is the Strait of Hormuz, lying between the Arabian Peninsula and Iran and linking the Persian Gulf to the Indian Ocean. The strait carries an enormous amount of oceangoing traffic, with much of it in the form of large supertankers hauling petroleum. Some experts estimate that more than a third of the world's petroleum supply passes through the strait in a given year. The U.S. Navy and other navies maintain a strong presence in the strait due to its vital role in supplying the global economy with energy, and recent years have seen a number of hostile incidents involving western naval forces and Iranian gunboats. Many similar choke points may be found around the world, including the Strait of Gibraltar, the Turkish Straits, and the Strait of Malacca. Waterways built by humans can be considered choke points as well—the Panama Canal and Suez Canal are both prime examples of such stretches of water that serve vital strategic and economic functions.

Climate

Climate is the characteristic state of the **atmosphere** at a **location** or in a **region**. It is the entirety of weather conditions over the long term. It is composed of averages, extremes, and variability. The World Meteorological Organization specifies climate-length weather records to be 30 years and more. Climate is derived from the Greek word *klima*, signifying the angle of the sun above the horizon. The input of solar energy is the prime driver of climate. Solar angles vary by latitude with lower latitudes having the highest solar angles and the most solar radiation at the top of the atmosphere.

From the times of the classical Greeks it was known that **Earth** was spherical and that varying solar angles caused varying climates. So, two-and-a-half millennia ago climate was classified into torrid, temperate, and frigid zones. Although the Greek classification was largely correct, they had never traveled to the Equator to discover temperatures are not as hot as in other parts of the tropics where the sun is not as high in the sky. The erroneous Greek conclusion points to the existence of non-solar influences on climate. In this case, the presence of the clouds in the Intertropical Convergence Zone (see **Winds and pressure systems**) moderates equatorial temperatures.

Latitude is clearly a key factor, but there are a number of factors that, when integrated together, determine the climatic nature of Earth's regions. Six other factors are generally given as climatic controls. (1) The distribution of land and water is quite uneven over the planet and land heats and cools so much more rapidly than oceans that annual temperature ranges are significantly impacted. (2) The circulation of the atmosphere circulates large amounts of air. For example, in the middle latitudes, the surface and upper westerlies bring oceanic air to the western parts of continents thus moderating their temperatures. (3) There are pronounced **ocean currents**, and these are capable of modifying climate. For instance, the cold Benguela Current off of southwestern Africa stabilizes air passing over it to help create the exceptional dryness of the Namib Desert. (4) Storminess and lack of storminess play a considerable climatic role. Places prone to storminess are subject to much more cloudiness and precipitation than places lacking storminess. In that the various types of storms are mechanisms by which the atmosphere's

Ice Shelves

In the polar lands are ice shelves that are permanent ice floating on the water while attached to the land. The thickness of the shelf can be as much as 1,000 m (over 3,000 ft). Most ice shelves are the result of the flow of glacial ice reaching the sea and continuing on because of the gravity-cause movement of ice behind it and because ice floats on liquid water. The size of an ice shelf depends on the amount and speed of flow of the glacial ice and the **climate** and water motion on the ocean. The age of ice shelf ice can reach to thousands of years in places protected by coastal configurations. Ice shelves occur off of Canada's Ellesmere Island and off of mountain glaciers in North America and Eurasia. These areas of shelf ice pale in comparison to Antarctic ice shelves, the largest of which—the Ross Ice Shelf—approaches a half million square kilometers (193,000 sq mi). Ice shelves do not affect sea level until their water melts. In recent years, Antarctica has had notable collapses of ice shelves each occurring within a few months. The largest was the disappearance of 8,000 sq km (3,100 sq mi) in two months. Ice shelf disappearance may well be a signal of global warming.

energy budget is maintained, storminess helps to moderate extreme surface temperatures by forcing horizontal and vertical mixing. (5) Elevation is well correlated with various weather elements like pressure, moisture and temperature which all decrease with altitude. Significant altitudinal differences make for different climates in adjacent, whether or not there are mountains present. (6) Topographic blockage is the effect of surface weather unable to bypass ranges of mountains and large hills. The climates can be quite different on the different sides of the topographic impediments. For instance, the frigid wintertime Siberian air is unable to pass over the Tibetan Plateau with an average elevation of 4,500 m (14,000 ft) and the Indian Subcontinent to its south enjoys a mild winter.

Vladimir Köppen (1846–1940) was first to numerically regionalize climate thus revolutionizing climate science. His work was inspired by the world vegetation maps then available; Köppen attempted to fit numerical boundaries. More subtly, his classification scheme of the early 20th century was made possible by the increasing availability of climate-length weather records. After the middle of the 19th century, weather instruments had become standardized. Numerous world cities boasted published climatological summaries. The summaries were mainly temperature and precipitation averages. Köppen used these two dimensions in constructing his classification system. In modified form, this classification has been well used by geographers and other scientists for the last century. The Köppen system is a relatively simple system with which to understand the large regionalities of Earth's climate. It has continued appeal because of its correspondence with **biomes** and **soil** regions.

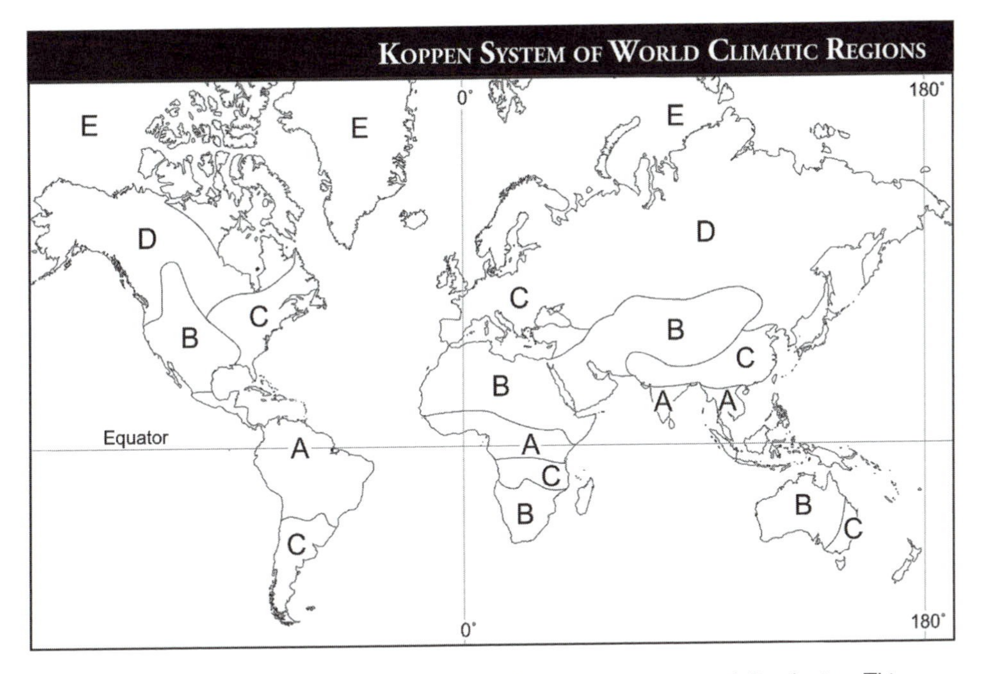

The Koppen system of climate classification is based on vegetation type and distribution. This map shows the major categories of climate according to this widely used system. (ABC-CLIO)

Below is a listing of modified Köppen climate zones that are also displayed on the accompanying **map**. Köppen's original work was modified several times as the boundaries of the world's major vegetation regions became better known and small weaknesses in the original scheme became evident. Nonetheless, the popularity of Köppen's classification has never been supplanted. Several other classifications exist for specialized uses.

Climate maps are intellectually appealing to geographers because they invite place-to-place comparisons. This is vital in the geographer's understanding of similarities and differences on the physical Earth. In the broadest sense, climate is an explainer of other world patterns. Earth's climate is the delivery system for energy and water. Thus, climatic knowledge is a potent tool in understanding the world distribution of biomes and soils. One can surmise that the soils are similar enough that a crop adapted to one of these climates would do well in broad zones of the analogous climate of another climate. However, the map user must exercise caution in using these maps in settings beyond the classroom. World climate regions are broad zones with climatic nuances that might be quite important for crops. For instance, the humid subtropical climate (Cfa) occurs from western Oklahoma to northern Florida. Oklahoma has significantly hotter, drier summers than Florida so that it does not follow that a successful crop in Oklahoma *must* also be so in Florida. We must remember that the attempts to classify the invisible entity of "climate" are based on similarities so that differences are downplayed.

The Köppen classification is based on letter codes (Table 1). The code is composed of six major climate types: A (tropical), B (dry), C (mild middle latitudes), D (severe middle latitudes), E (polar), and H (highland). Individual climate types are designated with two and three letters. The secondary and tertiary letters are w (winter drought), s (summer drought), f (year-round precipitation), W (desert), S (steppe); a through d (increasingly cool summer temperatures), h (subtropical variety of desert), k (middle latitude variety of desert), T (tundra), and F (icecap).

Table 1. Climate Typology

Köppen code	Climate name	Climate characteristics	Largest extent
Af	Tropical rainforest	Always warm; copious rainfall throughout the year	Brazil
Am	Tropical monsoon	Always warm; short dry season during low sun; very wet high-sun period	Southeastern Asia
Aw	Tropical wet-and-dry	Always warm to hot; low-sun season dry; overall drier than Am climate	Africa south and east of the Af climate
BWh	Subtropical desert	Hottest temperatures on Earth; precipitation scarce and irregular	Sahara of Africa
BWk	Middle latitude desert	Very hot in summer with cool winters; precipitation somewhat more than in BWh and includes winter snow	Central Asia
BSh	Subtropical steppe	Marginal to BWh climates with more moderate summer temperatures and greater precipitation	African Sahel
BSk	Middle latitude steppe	Marginal to BWk climates with more moderate summer temperatures and greater precipitation	Steppes of southern Russia
Csa	Mediterranean	Modest winter precipitation and dry summers; milder summer temperatures than inland climates with winters above freezing	Coastal lands fringing the Mediterranean Sea
Csb	Mediterranean	Summers not quite as hot as Csa	Coastal lands fringing the Mediterranean Sea
Cfa	Humid subtropical	Summers warm to hot with precipitation all year; winter temperatures sometimes below freezing	Southeastern United States

Table 1. Climate Typology (Continued)

Köppen code	Climate name	Climate characteristics	Largest extent
Cwa	Humid subtropical	Summers warm to hot with plentiful precipitation; winters dry with temperatures sometimes below freezing	Southeastern China-Northeastern India
Cwb	Humid subtropical	Same as Cwa except for a cooler summer	Southeastern China-Northeastern India
Cfb	Marine west coast	Many cloudy days with light precipitation year round; moderate summer and winter temperatures	Western Europe
Cfc	Marine west coast	Same as Cfb except for cooler summers	Northern Europe
Dfa	Humid Continental	Hot summer, cold winter, year-round precipitation	U.S. Midwest
Dfb	Humid continental	Summer briefer and winter colder than Dfa; significant length of snow cover	Western Russia
Dwa	Humid continental	Hot summer, cold winter, small amount of winter precipitation	Southwestern Russia
Dwb	Humid Continental	Summer not as hot as Dwa, small amount of winter precipitation; significant length of snow cover	Southwestern Russia
Dfc	Subarctic	Moderate precipitation; summer brief and cool	Northern Russia
Dfd	Subarctic	Moderate precipitation; summer brief and cold	Northern Russia
Dwc	Subarctic	Winter very dry; summer brief and cool	Eastern Siberia
Dwd	Subarctic	Winter very dry; summer brief and cold	Eastern Siberia
ET	Tundra	Up to four months average above freezing; modest precipitation	Northernmost Eurasia
EF	Polar ice cap	No month averages above freezing; very little precipitation	Antarctica
H	Highland	An accumulation of heterogeneous mountain climates present with all the other climate types; H climates are found at all latitudes and are cooler and moisture than their lowland surroundings	Tibetan Plateau

Clouds

At any one time, about half the planet is shrouded by clouds. In the estimation of many, they are beautiful and, at times, awesome. Everyone appreciates the importance of clouds as bringers of precipitation and providing shade from the sun. Yet, clouds also play vital roles in the transport of energy in latent form and in maintenance of the planet's greenhouse effect. Clouds are visible indicators of ongoing physical processes in the atmosphere and harbingers of weather to come. Through observation of clouds, it is possible to predict weather a day or more in advance at an individual location.

Clouds are composed of suspended water droplets and ice crystals that are individually almost too small to see. Although each piece literally floats on air, collectively they can represent considerable mass. A cumulus cloud on a summer day might hold as much as 50 tons of water. As entire clouds, these liquid and crystal pieces are quite efficient in reflection and refraction of **solar energy**. Moreover, clouds represent phase changes in water with accompanying gains and losses of energy. Crucially, clouds absorb longwave energy from **Earth's** surface and then reradiate it—the

Clouds are visible evidence of processes in the atmosphere. Here, stratus and cumulus clouds hug the mountains along College Fjord, Alaska. (Photo courtesy of Steve Stadler)

basis of the greenhouse effect. Moreover, clouds are an integral part of the atmosphere so wind moves them over long distances. Clouds are measured by eye and by human observation. Usually measured are cloud type, the height of the cloud bases, and portion of the sky covered. The international standard is to report cloud coverage in eighths known as octas. Coverage, therefore, is reported on a 0 to 8 scale.

Clouds are made because there are impurities in the air. Even before the advent of human-caused air pollution the air was never pure. There are a number of solid and liquid substances that serve as condensation and freezing nuclei around which pieces of water can gather. These nuclei are so light that they float on air and are composed of many substances, including dust from weathered soils and rocks, pollen, sulfate particles from oceanic phytoplankton, and sea salt. The nuclei best suited to gather water have radii of 0.1 millionths of a meter and larger. Most places in the lower troposphere have on the order of 100 to 1,000 of these nuclei in a cubic centimeter, although their density decreases away from the surface sources. Some of the nuclei are hygroscopic and present surfaces on which water can gather and start to make pieces of clouds at relative humidities much lower than the 100 percent defining saturation. Indeed, on summer days with plentiful water vapor it is common to experience haze, which is the beginning of water agglomerating onto condensation nuclei.

Clouds are made as substantial amounts of water are gathered around the nuclei. The atmosphere needs to be saturated for clouds to form. Increases in relative humidity to saturation can occur in two ways. The first is by the addition of water vapor into the air. This goes on via evaporation or sublimation and is most important where there is plentiful water vapor. This takes place on the order of hours and can produce some substantial low clouds but rarely significant precipitation. The second method is associated with the rise of air. Rising air dramatically cools via the decompression of air molecules. Here, air cools until it becomes saturated and this takes place on time scales of only a few minutes and the air does not initially need to be close to saturation to have this take place. Four basic mechanisms make air rise. The first is the orographic effect, which makes air rise because of topography in the path of air flow; note that mountain peaks are cloudier than their surroundings. The second is the lifting of air along **fronts**. The third is the rise of air because of heating of the surface—convection. The fourth is the rise of air in the centers of low pressure as air streams meet and are forced upward. At individual locations they can occur singly, in combination, or not at all. Any one of them can cause cloudiness.

In that the presence of significant cloudiness is dependent on the rise of air, topography, and presence of water vapor, there is a definite geography to cloudiness on Earth. Cloud geography is well related to global winds and pressure belts. Latitudes with average low pressure—the Intertropical Convergence and the polar front—engender the rising air necessary for cloudiness. Topographic barriers are places for greater cloudiness on the upwind side because of rising air and decreased cloudiness on the downwind side because of the sinking of air. Continental interiors generally, but not universally, are less cloudy then their oceanic and coastal counterparts because of the relative lack of water vapor from oceanic sources.

Vertically, clouds are common only in the troposphere. The water vapor that makes clouds is most plentiful in the lowest portions of the troposphere because the water source is evaporation from oceans. Few clouds exist above the tropopause and these involve so little mass that they do not provide precipitation to the surface. It has been suggested, however, that the presence of these thin, ice crystal clouds can impact the amount of solar energy reaching the surface and, so, changes in their distribution and mass might be related to subtle changes in climate.

Clouds are classified like other features of physical geography and the classification of clouds used by the World Meteorological Organization today is directly descended from the work Luke Howard published in 1802. Although there are hundreds of variant cloud types in the modern classification, the most common cloud types can be readily given. Clouds are arranged into four families: (1) low clouds, (2) middle clouds, (3) high clouds, and (4) clouds of vertical development. While middle latitude altitudes will be given here, these families occur at higher altitudes in the tropics and lower altitudes in the colder air over the poles.

Low clouds occur from the surface to 2,000 m. They tend to have a lot of water in them and are ice-free in many places even during the winter season (not all cloud droplets freeze until the air is at –40°C). The primary types are stratus (from the Latin for "layer") and stratocumulus, which is an intermediate form between stratus and cumulus clouds. Low clouds form dense overcasts. They differ because the rate of rising air is greater in stratocumulus clouds. An important subtype is nimbostratus, which is a stratus cloud that is precipitating. Fog is a stratus cloud with its base at Earth's surface.

Middle clouds range from 2,000 through 7,000 m in the middle latitudes. They are a mixture of ice particles and liquid water droplets. They tend to be less massive than low clouds. The two principal types are altostratus and altocumulus. Altostratus clouds appear layered while the faster rise of air in altocumulus clouds makes the individual clouds look puffy and organize them into bands and waves. Middle cloud decks sometimes extend over hundreds of thousands of square kilometers. Even though these clouds can precipitate small amounts, this moisture rarely reaches the surface.

High clouds range from 5,000 through 13,000 m (note the overlap of altitudes with middle clouds). These clouds are composed entirely of ice crystals and have little mass, rendering them incapable of precipitating to the surface except minimally in some high mountain locations. The three principal types are cirrus, cirrostratus, and cirrocumulus. Cirrus clouds are formed from individual, detached elements that appear white and fibrous. The word "delicate" is frequently applied to this type. Cirrostratus clouds frequently cover the entire sky and seem to form a "veil" high in the sky. These first two types are formed at elevations where there is slowly rising air or into air that has acquired moisture. They are sometimes difficult to identify from each other because both cirrocumulus and cirrostratus clouds frequently change into each other. The cloud decks made by stratus and cirrostratus clouds can be geographically extensive but are thin and allow considerable sunlight to penetrate them. They are often omens of surface weather disturbances following a day or more behind. Cirrocumulus clouds appear as small units organized as patches in the sky. They denote a faster rise of air than in cirrus or cirrostratus and their individual cloud elements are smaller than those in altocumulus.

The final cloud family is composed of clouds with vertical development. The two principal types are cumulus and cumulonimbus clouds. These clouds develop in air that is unstable and rising quickly. Cumulus clouds are the smaller of the two. Their bases are usually at modest heights from the ground and define the lowest level at which the rising air has become saturated. They consist of large, puffy elements that have sharply defined edges sometimes likened to cauliflowers. These clouds are the result of convection caused by surface heating and are common sights in summer. Cumulonimbus form from the towering and merging of individual cumulus clouds.They are also known as **thunderstorms** and are clouds made by the rapid rise of air to very great heights up to and sometimes punching through the tropopause to enter the lower stratosphere. These clouds are the focus for tremendous amounts of latent heat exchange and a variety of severe weather effects including immense amounts of rainfall.

Coastal Erosion and Deposition

Coastlines are fascinating in mass and energy terms because it is here that **atmosphere**, lithosphere, biosphere, and hydrosphere vigorously interact. They are also of great importance to humans because of our historic attraction to moderate climates, ocean trade, and fishing.

Water can erode shoreline material in several ways. The most significant method of erosion is by the tremendous hydraulic force provided by waves. Even modest waves can focus enough energy to break apart rock material and transport the loose remains. At the extreme, some coastlines are subject to large tsunamis generated by earthquakes. These are rare but can accomplish considerable erosion. The chemical action of water is considerable and able to break down rock by several means as part of **weathering and mass wasting**. Ice along shorelines is capable of considerable leverage in prying apart rocks.

Movement of materials along the shoreline is quite obvious by close observation. Sand moves in rhythm shoreward and seaward with the break and ebb of the waves. These distances are, at best, a few meters. There is considerable work that is performed by the abrasion of small rock and sand pieces one against the other; they are lessened in mass and smoothed. On closer examination, sand and other small materials exhibit "odd" behavior as waves break along the shoreline. A small portion of the breaking wave water is absorbed into the wet beach while most of the broken wave flows down the beach slope in the direction of the ocean. Although this might seem obvious, the net motion of materials suspended in the waves is not. Usually, waves approach the shoreline from a direction other than straight on to the beach slope. While the local beach has a single slope direction, the wind can push waves from various directions, which breaks waves onto a beach from directions other than that of the slope. The suspended materials travel with the waves onto the beach from the direction from which the waves originate. The water that has broken onto the beach travels directly down the beach slope via gravity. The motion is a zigzag transport of materials causing net transport in parallel to the shoreline, which is known as beach drifting. Some of these materials are picked up in longshore currents and can be transported great distances.

The beach is a depositional form. Erosion is prominent along coasts but beaches are, perhaps, the most obvious coastal landform. They represent depositional materials in "temporary storage" until such time as the materials can be re-suspended. As such, they are continuously being built and/or eroded by water. Tides have surprisingly little net landscaping influence on beaches and are greatly overshadowed by events such as storms or tsunamis or long-term rise and fall of sea level.

Beach sand is not always white and varies considerably in color, including reds and blacks, as the result of the erosion of volcanic rocks. Though very common, sand is not the only size material found along shorelines. Cobbles, rocks, pebbles, silts, and clays comprise beach along various coastlines and their occurrences are dictated by the particular combination of eroded materials and energy of the water action at each location. In areas with plentiful shoreline sand and enough wind, dry sand can be suspended in the air and moved inland. Progressive wind erosion of the beach sand causes dunes inland. Some coastal dunes rise many tens of meters and their presence is a protection against higher water levels in major storms.

There are numerous other depositional forms. When sand piles on the sea bottom offshore it is known as an offshore bar. Sometimes, offshore bars are deposited above sea level and become barrier islands. The Gulf and southern Atlantic coasts of the United States are flanked by thousands of kilometers of barrier islands with shallow, protected lagoons lying between them and the mainland. As longshore currents encounter embayments there is the tendency for them to slow and drop some of their sand loads into deeper water. This causes elongated strips of sand known as spits to build out from the shore.

There are three types of coastlines delineated with respect to sea level. They are emergent, submergent, and neutral. Tectonic forces make land rise or fall and sea level can rise and fall because of changes in ocean water volume. At this point in **Earth** history we observe many submergent shorelines. This is because of the rise of sea level resulting from the melting of ice of the Pleistocene glaciers. The east coast of the United States is dominated by submergent coastlines with Chesapeake Bay a prime example of a river system that was drowned by the ingress of ocean water. Conversely, the Island of Arran in western Scotland is home to King's Caves, which were formed by erosion of water at sea level but are now several meters above sea level as the result of the isostatic uplift of land resulting from the unloading of the mass of Pleistocene ice.

Coastlines are not all inorganic. Coral coasts are common in shallow tropical oceans. Coral polyps are simple, soft creatures a few millimeters tall and a few millimeters in circumference. They live as huge groups of genetically identical animals and excrete calcium carbonate, which makes up the hard substance commonly called coral. Besides providing support for the coral, the matrix of calcium carbonate frequently becomes a reef-provided habitat for a rich diversity of life.

Humankind has always been drawn to coasts. We have built prolifically along shorelines to sometimes disastrous effects. For instance, the population of U.S. coastal counties has more than doubled since World War II. The building of housing, towns, marinas, and golf courses has exponentially increased the chances for property damage during storms and played havoc with some parts of the natural system. By building on coastal sand, the local sand supply available to nature is effectively lessened and beaches disappear because beach drifting far exceeds the local storage. In the modern era, with good knowledge of the physical geography of coastlines, it is unfortunate that we do not fully respect the power of nature along the coastline.

Comparative Advantage

First described in detail by the British economist David Ricardo in the early 18th century, the theory of comparative advantage holds that it may be advantageous for two countries to trade with one another, even when one holds an absolute advantage in the production of the traded goods. Comparative advantage is linked to the notion of **areal differentiation**, in that it is assumed that countries that are potential trading partners have different allocations of natural resources, differing levels of labor productivity, capital investment, etc.

In his classic explanation of the concept of comparative advantage, Ricardo notes that both England and Portugal produce wine and cloth. Portugal, according to Ricardo, has cheaper labor costs in the production of both commodities, and therefore has an absolute advantage in the production of both, i.e., Portugal is able to produce both wine and cloth in greater quantities than England. However, Ricardo points out that in England, wine is relatively more expensive to produce, while making cloth is relatively inexpensive. This situation then gives England a comparative advantage in the production of cloth, and Portugal a comparative advantage in the production of wine. It will be more economically advantageous for each country to specialize in the commodity for which it has the comparative advantage, and trade that commodity to the other country in exchange for the good it does not possess an advantage in producing. Portugal should devote its economic resources to wine production; England should commit its resources to spinning cloth.

To illustrate this using numbers, let us consider the following situation. It is known that if Portugal devoted all its resources to producing wine it could produce 8 tons; or it could direct all resources to making cloth, which would result in 4 tons of output. England's economy could produce 2 tons of cloth by allocating all resources to that commodity, or 2 tons of wine. The cost of making 1 ton of cloth in Portugal is 2 tons of wine, but in England the cost of 1 ton of cloth is only 1 ton of wine. England has a comparative advantage in the production of cloth, while Portugal has the comparative advantage in the making of wine. Without specialization, England produces 1 ton of cloth and 1 ton of wine. Portugal makes 4 tons of wine and 2 tons of cloth. Total output of the two countries is 3 tons of cloth and 5 tons of wine. If each specializes in producing the commodity it has a comparative advantage in, however, then Portugal can make 6 tons of wine and 1 ton of cloth, while England will spin 2 tons of cloth. Total output will now be 6 tons of wine and 3 tons of cloth, so production of wine has increased. Moreover, England can trade 1 ton of cloth to Portugal for 2 tons of wine, meaning that England now has 2 tons of wine to consume, instead of the single ton it had before specialization. Portugal receives a ton of English cloth for the wine it trades, meaning that in Portugal there will be 5 tons of wine and still 2 tons of cloth, instead of the 4 tons of wine before specialization. Both production and consumption have increased, because each country has pursued its comparative advantage. Although in theory this simple example illustrates why countries trade under such conditions, the concept of comparative advantage does not take into account barriers to trade, such as differences in the quality of goods or consumer preferences, variations in exchange rates between the currencies of the trading partners, or consideration of the fact that countries may not wish to be wholly dependent on another for some goods (as England is for wine, in the example above).

Complementarity

This is a concept utilized in economic and transportation geography that emerges from the notions of **areal differentiation** and **spatial inequality**. Complementarity arises between countries or regions due to differences in resource allocation, variations in levels of economic development, or advantageous relationships in economic production. Because resources, labor, capital, and other economic factors are not evenly distributed across the surface of the **Earth**, some regions will possess or produce goods that other regions need, and vice versa. That is, a country or **region** may complement others in terms of its resource endowment and its economic production. This complementarity forms part of the basis for spatial interaction, in that it provides the motivation for economic exchange—countries or regions must form relationships with those that complement them to obtain the goods and resources they need or desire. It should be emphasized that simply the spatial occurrence of surpluses and deficits is not enough to result in complementarity between regions—a mutual desire and ability to satisfy the requirements of supply and demand must be present. Furthermore, complementarity by itself is insufficient to stimulate interaction between two or more places, as transferability and the lack of an intervening opportunity must also be considered; but without complementarity the other factors will not be relevant. It may be that complementarity exists between two regions due to the **comparative advantage** one or both have in regard to economic production and trade. In this instance, although both regions in question produce the goods that are traded (at least initially), complementarity is achieved by both specializing in the production of commodities in which they hold a relative advantage in comparison to the trading partner.

Complementarity is of course a dynamic concept, and the complementary relationships between regions and countries can fade away, or may even be completely reversed over the course of time. Nor is complementarity necessarily advantageous to all countries or regions in the relationship—during the era of mercantilism, regions that were held as colonial possessions by more economically advanced countries, and which therefore had a high level of complementarity with them, were exploited for their resources and denied the opportunity to develop industries that might compete with those of the colonizing state. The decline of the colonial era has not meant a diminution in complementarity, however. The modern global economy is replete with examples of complementarity. Many examples stem from the unequal distribution of crucial energy resources, vital strategic metals, and the uneven nature of the **geography of economic development** that has divided the world into "developed" and "underdeveloped" worlds for the past three centuries. A high degree of complementarity exists between the oil-producing countries of the Middle East, for example, and the industrialized world, which relies on the hydrocarbon resources of many Middle Eastern countries for energy. The complementarity between the regions of supply and those of major demand in the economic geography of the global petroleum trade are mirrored in the complementarity between the oil producers and the industrialized world for manufactured goods, except that the supply and demand relationship is reversed.

Core and Periphery

A generalized, spatial relationship model that is a central component of many subfields of human geography. The peripheral region, sometimes referred to as a **hinterland**, is dependent on the centralized core region for the spatial dissemination of all, several or one of a variety of influences: economic, cultural, or political. The core represents a base of power or a point of origin, the periphery represents a surrounding region that is reliant on the core region. The core and periphery model has applications in the **geography of economic development**, the concept of **cultural diffusion**, and many other geographical theories. In addition, the core-periphery approach may be applied at a variety of scales. For individual states and small regions, the core may be represented by a well-developed coastal region that holds most of the country's economic and political influence. The periphery, on the other hand, is an underdeveloped hinterland lying further inland. In economic geography, classic growth pole theory is based on the theoretical assumptions inherent in the core-periphery model. At a much broader scale, the **World Systems Theory** articulated by Immanuel Wallerstein also relies heavily on the core-periphery model. In this context, the economically developed countries of the world are the "core," and the underdeveloped regions of the planet represent the "periphery." Several examples of how the core-periphery concept is applied to some of the subfields of geography follow.

In economic geography, the core-periphery model lies at the heart of much of the theoretical discussion regarding spatial interaction and economic development. It is obvious to even the casual observer that at almost any scale, the attributes of economic space are unevenly distributed, which is a situation that frequently results in **complementarity** and subsequent economic exchange. Some places exhibit higher rates of productivity, higher standards of living, greater per capita income, etc., than adjacent locations. This uneven pattern may be observed from the global scale to the local, as even within metropolitan areas such differential development is frequently apparent. Some view the economic relationship between core and periphery as essentially antagonistic, and as an expression of an imbalance of not only economic but also social power. Thus, core and periphery relationships are conceived of as undesirable and exploitative, with spatial integration and economic interdependence seen as "solutions." This view is evident in the writings of many neo-Marxist theoreticians, who generally see core and periphery development as symptomatic of what, in their perspective, are the broader distortions of capitalistic economic systems. Other theoretical answers have been proposed to redress the imbalances wrought by core and periphery development, including growth pole theory. Proponents of the growth pole model hold that economic development can be stimulated by the establishment of urban settlements in the periphery, which, through the process of **agglomeration**, stimulate further development across the entire region. The success of growth pole strategy is open to debate, but additional models and strategies designed to alleviate the worst characteristics of core-periphery development are the focus of much of the work of economic geographers.

Political geographers often make use of the core and periphery model in forming theories about political integration, state cohesion, national identity, and other aspects of the subdiscipline. Indeed, the nature of the core and periphery relationship in regard to control of political space has been viewed as playing an integral part in the generation of **centrifugal and centripetal forces**. For political geographers, the core region is characterized by the spatial concentration of political power and influence, often represented by the capital city of a state. Political control, at least in theory, diminishes as distance from the core area increases, so that in the peripheral regions of the state, the manifestation of state authority is minimized. As was the case with economic geography, this nonuniform distribution and reduction in the political influence of the state apparatus in the hinterland of the country is viewed as negative, and states sometimes seek spatial remedies. In recent decades, for example, a number of countries relocated the capital city from a marginal core region to a more centralized location. In theory, by shifting the seat of political authority to a more central location, those regions in the hinterland that were less integrated into the country's political structures would become more tightly united to the state, by dint of simple proximity to the capital. An excellent example of this strategy is the case of Kazakhstan, which in 1997 relocated the capital city from Almaty, a marginal location in the southeastern corner of the country, to Astana, a city positioned near the geographical center of Kazakhstan. Although several reasons were provided by Kazakh authorities as to why the move was necessary, many observers hold that the motivation was to secure the loyalty of the northern regions of Kazakhstan, which hold a large Russian minority. Core and periphery theory may be found in much of the theoretical structures of political geography, including both **Heartland theory** and **Rimland theory**.

Much of the theoretical basis for cultural geography also rests on the core and periphery spatial relationship. Perhaps obviously, the notions of **cultural hearths** and cultural diffusion carry an implied core and periphery dynamic. The cultural core areas are those where cultural innovation emerges, which then diffuses to the cultural hinterland surrounding the core. In addition, cultural core areas may play a vital role in solidifying **cultural identity**. The late Terry Jordan-Bychkov, a leading cultural geographer in the United States, noted that the lack of a cultural core area often left a void in the formation of cultural or national identity, and such states frequently were politically unstable, because regions lying in the periphery were not culturally bound to the state, and often formed allegiances to external cultural influences. In states where a strong cultural core area exists, cultural identity is likely to be stronger because the elements underpinning that identity diffuse to the periphery and are incorporated by residents there. The core and periphery structure is also encountered in the study of **linguistic geography**, another subfield of cultural geography. Language regions rarely are characterized by distinct, clearly defined **boundaries**. Rather, the geographical expression of a language is strongest in a core area and then diminishes into a peripheral zone—the greater the distance from the core area, the lower the instance of usage of the language in question. A language may be spoken on a daily basis by 90 percent of the population in the core areas, for example, but by only 10 percent in the periphery, where it is displaced by another tongue, or is not spoken in the home, but only as a second language.

Cryosphere

The term cryosphere comes from the Greek root *kryos* and means "cold" or "frost." In physical geography, cryosphere is a general term referring to the worldwide occurrence of frozen water in its snow and ice forms. These cold **landscapes** and seascapes are important on their own but also help to regulate **temperatures** for the planet as a whole. The high reflectivities of snow and ice are considerable greater than liquid water and land surfaces, so they redirect **solar energy** back toward space preventing part of that energy from warming the surface via absorption. Snow and ice are also direct determinants of sea level. Three-quarters of **Earth**'s freshwater is locked in ice, which covers about 10 percent of the land surface.

Cold temperatures allow water to freeze and this phase change is caused by energy loss from strings of liquid water molecules to rigid crystalline configurations. Pure water freezes at 0°C but seawater of average salinity does not freeze until it reaches –2°C. Over much of the world, precipitation starts in clouds as snow. If temperatures below the clouds are below freezing the precipitation is received as snow on the surface. There are several major forms that snow and ice can take on Earth's surface.

Snow cover is present in some places year round and seasonally in many places, including middle latitude lowlands and tropical mountains. The ground surface is blanketed and, ironically, this greatly moderates ground temperatures so that life can survive underneath. Snow can be considered to be a reservoir of fresh water. For instance, human activities in the U.S. Southwest are dependent on stream flow that emanates from the summer melt of the snowpack in the mountains.

Glaciers are large bodies of ice that have formed because season after season of snowfall has been greater than the melting and sublimation subtracting from the mass. The snow becomes denser over time, translating into glacial ice. Ice is interesting because it seems so brittle in our common experience, but it is capable of bending and flowing when subjected to slow, steady pressure. Once ice is more than a few tens of meters thick it will move under its own weight and gouges and deposits materials on a grand scale.

Earth "recently" emerged from the Pleistocene Epoch (the "Ice Age"), which ended 12,000 years ago. Winters were noticeably longer and colder than today and about a third of Earth's continental surface was covered by ice. A concomitant effect was the dramatic lowering of sea level on the order of 100 m caused by the storage of the water in glaciers. The glaciers that remain provide excellent guidance to glacial land shaping processes more widespread during the Pleistocene.

Ice caps are large areas hosting thick ice coverings sometimes in excess of 3 km. Ice caps are not limited to land and where the glacial ice extends over the ocean that part is known as an ice shelf. Today, major glacial ice caps cover Antarctica and Greenland but sometimes the sea ice of the Arctic and Antarctic are included in the discussion of ice caps. Another distinction is that over land, ice sheets are masses of ice covering less than 50,000 km^2.

Sea ice is a common feature of the world's cold oceans. Sea ice is found in the north and south polar seas and fluctuates seasonally. Averaging 25 million square miles in extent (the approximate size of North America), sea ice expands and contracts according to season. Scientists divide sea ice into first-year and multiyear varieties. Multiyear ice is 2- to 4-m thick and has survived at least one summer melt. Rather than being smooth, sea ice frequently has significant creases, cracks, and mounds due to drift caused by currents and winds.

Approximately equal in area to sea ice is frozen ground also known as permafrost. Permafrost is more widespread in the Northern Hemisphere than in the Southern Hemisphere because of the much greater area of unglaciated higher latitudes in the former **region**. The parts of the planet subject to a cold winter frequently experience frozen ground. However, permafrost is soil material frozen for more than a year at a time. Permafrost has variants with more than 30 percent of the volume in ice to practically no ice at all; the defining characteristic of permafrost is the temperature of the soil material rather than the amount of water. Depending on local climates, permafrost can occur in continuous, discontinuous, and sporadic concentrations. In North America, some small pockets of permafrost occur as far south as the mountains of New Mexico. Permafrost may be quite deep with record depths approaching 1,500 m in northern Siberia. Of great importance is that the large majority of the area underlain by permafrost undergoes thaw in the surface materials during the cool summers. The top of the permafrost is, therefore, known as the active layer and is usually thawed to depths ranging between a half-meter to 4 m. This thawing allows the existence of plant life, but the plant life is limited because of the short duration of summer, the poor drainage of moisture out of the active layer, and the rock-hard barrier of still-frozen permafrost inhibiting rooting underneath the active layer.

The resulting biome is tundra and human use of the landscape is severely restricted. The activities of humans must take into account the permafrost's seasonal transformation from a frozen surface to a series of mushy bogs. Telephone poles, road surfaces, and house foundations can be ruined. Witness the extraordinary engineering measures of the Trans-Alaskan Pipeline stretching 1,300 km southward from the Arctic Ocean. The pipeline is carried on a series of supports

The area covered by ice and snow is declining on our warming Earth. The Mendanhall Glacier near Juneau, Alaska, has retreated 4 km since 1500. (iStockPhoto)

that prevent the warm oil from melting the permafrost and damage from seasonal heave of the tundra. Currently, the cryosphere is noticeably shrinking. **Global warming** is causing ice caps, glaciers, seasonal and permanent snow, and seasonal and permanent sea ice to decrease in area. For instance, Arctic Sea ice is decreasing in both area and thickness with the maximum winter extent lessening by more than 1.5 percent per decade. Similarly, the ice and snow on Earth's surface modifies temperatures.

Cultivation Regions

Sometimes called agricultural regions, cultivation regions are units of territory associated with a specific type of agricultural activity. This activity can involve the actual tilling of the land or can be some type of animal husbandry. Cultivation regions emerge due to both cultural and environmental factors.

A region of *shifting cultivation*, often labeled "slash and burn" agriculture, is found in tropical and subtropical climates in Latin America, Africa, and Southeast Asia. Fields are prepared here by killing the natural vegetation and then burning away the dead material. The ash residue from the burnt vegetation helps to enrich the tropical soil, which is typically tilled and planted by hand with food crops. One type of crop is often planted between the rows of another crop, in a system of *intertillage*. Once the soil is exhausted farmers "shift" to another location and repeat the process.

Plantation cultivation is also associated with tropical or subtropical climates. This system is focused on the production of a single commercial crop, such as coffee, tea, palm oil, cotton, tobacco, or others. Historically, this form of cultivation relied on abundant, cheap labor, but many modern plantations have incorporated machinery into the production of their crops. Plantations are usually located close to road, rail, or water transport, because much of the production is typically grown for export. Crops may be refined or processed on the plantation before being sent to markets.

Yet a third cultivation region found primarily in the tropics is *rice paddy cultivation*. This is a labor-intensive style of agriculture, and has been widely mechanized only in areas lying outside the tropics, like Japan and the lower Mississippi River valley in the United States. In the tropical climates of South and Southeast Asia, fields are small and prepared using draft animals. Most rice production is for subsistence in the tropics, and typically two crops are produced in a single year, a practice called *double-cropping*.

In many developing areas outside the tropics, a system of subsistence production called *peasant grain, roots, and livestock farming* is found. Also labor intensive, the grains produced by farmers in this system typically are wheat, barley, oats, millet, and corn. Animals are kept for a variety of purposes—they may provide milk, meat, and in the case of cattle and horses, draft power. In some regions, tubers such as yams, potatoes, and cassava are an important component of the local diet. In well-watered areas with a longer growing season, a cash crop may also be grown.

Mediterranean agriculture is encountered around the littoral of the Mediterranean Sea, but also in some parts of the west coast of South America. Because of the marked seasonality of rainfall in this region, hardy grains like barley and wheat are raised in the winter, along with vineyards and olive, fruit, and nut orchards that produce in the dry summer months. Livestock, primarily goats and sheep, are also part of the agricultural landscape here, and provide wool, meat, and, in the case of goats, milk. Both subsistence and commercial cultivation may be encountered in this region.

The mid-latitudes are the site of large stretches of *commercial grain farming*. Much of the Interior Plains of North America, the pampas of South America, Western Europe, and interior Asia is occupied by this region. Grains produced in these regions include primarily wheat, but also expanses of corn and other grains as well. The production system is highly mechanized, scientifically produced hybrids and commercial fertilizers are widely employed, and yields per farmer are among the highest in the world. In North America and Western Europe, **agribusiness** has become dominant in the agricultural economy.

In developed countries located in the mid-latitudes, a system of *truck farming* or *market gardening* has evolved to service the large urbanized populations. Cultivation is intensive and highly specialized, with one or two crops grown. The focus is typically on high-value agricultural products like vegetables, nuts, and fruits, and virtually the entire output is directed to local or national markets. Warm to moderate climates are necessary, and in the United States, Florida, southern California, and the Rio Grande valley of Texas are particularly important for this type of agricultural production.

In the colder reaches of North America and Eurasia, as well as stretches of Australia and New Zealand, zones of *dairy farming* are found, often adjacent to the regions of commercial grain farming. Here commercial production of milk and associated dairy products are commonplace, involving a high level of mechanization. Farms producing liquid milk typically are located close to urban markets, while those making cheese and butter can be situated further from the main markets. Corporate dairy farming is becoming more common, and some operations use hundreds or even thousands of cows.

Transitory livestock herding, although once a widespread agricultural activity, is now confined to relatively few regions on the globe. Primarily found in Africa, the Middle East, and the central section of Asia, regions where this activity predominates are marked by the migration of both humans and large herds of livestock, which may consist of sheep, cattle, goats, horses, or even camels. Due to government-enforced settlement policies of the last century, especially in North Africa and Central Asia, much of the world's nomadic herding has disappeared.

A sizable region of *mixed farming and livestock* stretches across much of the eastern United States and Central Eurasia. Farms in these belts produce grain and other feed for livestock, which are then sold. In the United States, the feed is often corn, which has a high caloric content, and soybeans, which are high in protein. *Feedlots* are numerous, which specialize in fattening livestock for slaughter by purchasing feed from local farms. Much of the world's meat, especially beef, is produced in these regions, and most of the grain and soybeans grown there is for livestock consumption.

The final major region of cultivation is *ranching*. Most of the United States west of the Rocky Mountains, almost all of Australia, and large tracts of territory in southern South America are devoted to rangeland for cattle and sheep. These are arid regions, and the livestock is allowed to forage on the natural grasses present, although some ranchers will grow supplemental fodder crops and other types of feed. With the exception of Argentina, sheep predominate in the Southern Hemisphere, and beef cattle are the favored ranch animal in most of the Northern Hemisphere.

Cultural Diffusion

The dispersion of ideas, practices, technologies, techniques, language, or other cultural attributes throughout space. The process of cultural diffusion has occurred as long as modern humans have interacted and has perhaps reached its ultimate expression in **globalization**. Many technological innovations and ideas may have arisen independently, but a very large number of cultural attitudes and characteristics originated in a single location and spread to others. Diffusion from **culture hearths** was particularly important in the historical development of civilization. The movement of cultural traits through space can lead to other cultural processes. For example, **religious syncretism** is a direct result of cultural diffusion and typically results in a new cultural expression of faith. Indeed, diffusion is the mechanism whereby culture evolves and adapts. In the modern world most cultural diffusion is voluntary, but historically a great deal of cultural transfer and change was effected by cultural imperialism, when cultural traits from one society were imposed by force onto other societies. This was a common practice during the age of **imperialism**, when Western, European-based culture was forcefully, and frequently violently, carried to most of the world's land masses. But involuntary, compulsory diffusion has also occurred in many other eras and been employed by many other cultures. Cultural diffusion may be inhibited by many factors. Simple distance may prevent the transfer of culture, as a result of the influence of **distance decay**. Physical features, especially mountain ranges and large bodies of water, historically blocked the movement of people, and therefore stood as barriers to the exchange of elements of culture. Political **boundaries** frequently stop diffusion by prohibiting the passage of people and their **cultural identity**. And of course, some cultural ideas or practices may not diffuse simply because they are rejected or ignored by other cultures.

There are two general types of diffusion. *Expansion diffusion* happens when a cultural component moves through a given population or society, gradually being adopted by a larger number of people as the area of diffusion increases. Conceptually, this process works through a **core and periphery** relationship, with the point of origin of the cultural component represented by the core, and the diffusion area represented by the periphery. An example might be the diffusion of jazz music in the United States in the 1920s. Jazz was a musical style that originated in New Orleans in the first decades of the 20th century, and by the 1920s began spreading to other major metropolitan locations, as players from New Orleans carried their unique sound to new venues. One of the first cities outside the American South to host a thriving jazz scene was Chicago, and by the end of the decade most big cities in the United States featured numerous dance clubs playing jazz music. Jazz itself was a syncretic product of cultural diffusion, as the music incorporated many diverse elements, ranging from Negro spirituals to French Creole folk music. Jazz also gave rise to still more new innovations in musical style, including swing music, and ultimately the rock and hip hop music of today. These various kinds of music all have their point of origin in the music halls of New Orleans.

Relocation diffusion occurs when a group migrates to a new location and brings a new aspect of culture with them that is adopted by those they settle among. Religions and languages often spread in this fashion, and a good example is the adoption of English in many parts of the world. As British settlers relocated to colonies in the 18th and 19th centuries, their language became the language of administration and higher education, which compelled a sizable portion of the people they controlled to learn English as a second language. The British Empire was so far-flung that today the effect of this relocation diffusion is that English is a truly globalized tongue, spoken as a first or second language from North America to India, and from Nigeria to Singapore. The spread of Christianity followed a similar pattern, except that the area of diffusion was much broader, because several European countries spread their versions of the faith to every corner of the colonial realm.

Expansion diffusion may be further divided into three specific types. *Hierarchical diffusion* involves the spreading of ideas or attitudes from urban centers in one culture to those in another, or in rare instances, even from a single person to others. Over time, the new cultural elements then diffuse from the urban locations into the remainder of society, or from the initially "select" group into the larger population. This type of diffusion is quite common in the spreading and adoption of aspects of popular culture, like clothing styles, new trends in dance or art, esoteric religious movements, or in some instances, diseases may spread in this way. Hierarchical diffusion is characteristic of the international fashion industry, where the latest clothing styles typically appear in fashion shows in Paris and Milan and then are adopted by designers and consumers in New York, Tokyo, Rio de Janeiro, and other large metropolitan centers around the world. Even gestures may diffuse in a hierarchical manner. During World War II, Winston Churchill, Great Britain's prime minister, began flashing a "V" symbol with his fingers when he was photographed or when he spoke in public. This gesture stood for "victory," and the practice quickly caught on among American politicians once the United States entered the war. The gesture became widely diffused in the United Kingdom and the United States, and by the end of the war Americans from all walks of life understood the meaning of this symbol and used it as a common reminder of the ultimate goal of the war.

Stimulus diffusion is a second type of expansion diffusion. With this kind of diffusion, a culture may not adopt a practice or technology completely, but may adopt the concept and apply it to their own cultural conditions. Scholars believe that the transfer of writing from the culture hearth of Mesopotamia to Egypt around 3000 BCE is one of the earliest examples of stimulus diffusion. The Sumerian civilization in Mesopotamia had developed a pictographic writing system some time earlier, and it appears that knowledge of this system eventually reached the emerging culture of Egypt. The Egyptians did not adopt the Sumerian system, a script called cuneiform, but rather were stimulated by this invention to develop their own pictographic script, which is recognized today as ancient Egyptian hieroglyphics. It is likely that many early writing systems were constructed as a result of stimulus diffusion. Probably the most famous example of stimulus diffusion is the construction of the written script for the Cherokee language in 1821 by Sequoia, who had seen written English but did not understand it. Showing remarkable ingenuity, he devised an alphabet to represent spoken sounds in his native Cherokee, and in only a few months, many of his people had adopted the writing system.

A third pattern of expansion diffusion is represented in *contagious diffusion*. Here new cultural elements are adopted rapidly and en masse, spreading through the population like a highly infectious virus. New movements in popular culture often spread contagiously, especially in the era of visual media and the Internet, when millions of people have almost instantaneous access to outlets that follow and promote the latest trends. In 1964, the Beatles first appeared on the Ed Sullivan show, an immensely popular variety show watched by millions of Americans, and their longer haircuts immediately caused controversy. But within a few months, thousands of young American men were wearing their hair in a similar style and donning "Nehru jackets" that the Beatles often wore while performing. The British quartet had changed American cultural standards and behavior almost overnight. Before the advent of mass visual media, contagious diffusion of ideas or practices generally required person-to-person contact. A second example of this type of diffusion is the spreading of proselytizing religions that seek mass conversion of a population. Both Islam and Christianity historically diffused when a few believers from these faiths encountered nonbelievers, converted them, and they in turn converted additional followers in a classic sequence of contagious diffusion based on personal contact.

One of the most influential theoreticians in cultural geography regarding the concept of diffusion was the late Torsten Hagerstrand, who developed statistical models that illustrate how diffusion operates in space. His theory rests heavily on a core and periphery framework, which he employs to explain why new ideas and cultural trends, which could possibly arise at any point in space, tend to concentrate and thereby create core areas and corresponding peripheral regions. A *neighborhood effect* also influences the movement of cultural traits through space at the local scale, when an individual acquires the new trait and then influences those near him, including his family and immediate neighbors. One of the shortcomings of the Hagerstrand approach is its failure to explain why many cultural attributes *fail* to diffuse, especially in an era of global communication and cultural exchange. Nevertheless, Hagerstrand's ideas continue to be the focus of much discussion and debate, and provide a conceptual foundation for further study of diffusion and its effects.

Cultural Ecology

A theoretical perspective common to cultural geography and derived from anthropology that focuses on the relationship between human activity and environmental conditions. Early theory in cultural ecology evolved from anthropological studies that were concerned with how human societies are changed due to alterations of the natural environment. Thus, early on, cultural ecology as a theoretical approach shared *some* common elements with **environmental determinism**, although the cultural ecologists as a whole rejected the idea that environmental

Carl Sauer (1889–1975)

Carl Sauer was one of the most influential thinkers in cultural geography in the 20th century. Sauer earned a doctorate from the department of geography at the University of Chicago in 1915, where he was initially influenced by the theory of **environmental determinism**, a perspective that many faculty at Chicago were promoting in their research and publications. A few years after graduating, Sauer took a position at the University of California at Berkeley, and met the eminent cultural anthropologist Alfred Kroeber, who fundamentally changed his perspective on the human-environment relationship. Sauer abandoned his deterministic approach, and began laying the theoretical groundwork for the developmental perspective that would eventually emerge as **possibilism**. Sauer's benchmark work, *The Morphology of Landscape*, argued for a more sophisticated approach to the concepts of the cultural landscape and **cultural ecology**, and the impact of humans on the environment. Sauer was a strong proponent of the notion of **cultural diffusion**, and was considered one of the world's foremost experts on the diffusion of domesticated species in the New World. He was the founder of the so-called Berkeley School in American geographic thought and trained many doctoral students who continued to develop his ideas.

agents shaped the nature of cultures into some predictable pattern of characteristics. To some degree, the emergence of cultural ecology in academic geography was an effort to apply systems theory, social Darwinism, and other more "scientific" frameworks to the cultural landscape concept promoted by the so-called Berkeley School, and championed by Carl Sauer, which itself was a reaction to the dominance of deterministic assumptions in the philosophical structure of geography and other social sciences. Although more rooted in the perspective of **possibilism** rather than the rigid **ethnocentrism** of the determinists, cultural ecology recognizes that the impact of human activity on the landscape, and the influence of the physical environment on human culture are both factors that dynamically shape the cultural geography of a region. In general, cultural ecologists study how nonindustrialized societies interact with their physical environment, especially through the development of strategies that allow them to exploit their physical surroundings. Cultural ecology therefore falls within the "man-land" tradition of the discipline of geography, which is concerned with the relationship between human activity and the environment.

In the first part of the 20th century, anthropologists and cultural geographers had constructed their analysis of human societies around the nature of group interaction and relationships, or common features and beliefs that defined the culture, i.e., customs, religious practices, kinship, or marriage structures, etc. Much of the research produced was highly descriptive, and typically considered to be specific to the culture under examination—few scholars attempted to find commonalities that would allow for the generalization from one society or cultural group to another, or that would lead to broader theoretical understanding of the concept of culture. Partially this was due to the influence of the eminent anthropologist Franz Boas, who had made a strong case for the application of "historical particularism," itself a reaction to the "unilateralist" theories of his predecessors. Boas vigorously disputed the notion that human culture followed a single, sequential formula in its development, and that the historical circumstances of societies largely shaped their path of evolution.

In the 1950s, the work of a number of scholars once again became directed at how cultures evolve over time, and especially toward the factors that drive such change. Leslie White and Julian Steward were two of the most influential thinkers of this era, and Steward is generally credited with formulating the theoretical basis for "cultural ecology." Steward followed in large part the "neoevolutionary" perspective developed by White, but offered what he termed a "multilinear" explanation of cultural evolution. Steward argued that while the environment did not *determine* the nature of social organization and cultural phenomena, it did confine such evolution to a finite cluster of alternatives, any one of which might be the path taken by any specific society, making Steward an early proponent of possibilism. This theoretical perspective quickly became the foundation of cultural ecology, and Steward is often credited with founding the subdiscipline. Steward's simple definition of his theory of cultural ecology encapsulates the essence of his broader philosophy: "the adaptive processes by which the nature of society and an *unpredictable* number of features of culture are affected by the basic adjustment through which man utilizes a given environment" (emphasis added).

The cultural ecological approach quickly expanded to other social sciences, among them cultural geography. To cultural geographers, a consideration of the ecological factors affecting the development of culture effectively combined the two schools of philosophy that had seriously divided the discipline throughout the first 50 years of the 20th century: the determinists, who argued that culture was entirely dependent on the ambient physical ecology; and those following Sauer and the Berkeley School, who in response offered theoretical alternatives in the form of **sequent occupance** and related notions. Theoretically, cultural ecology offered to marry these two opposites by promoting the view that neither the environment nor humanity were independent actors. Rather, each influenced and shaped the other within a range of possible relationships, and what ultimately emerged was in large part the product of the society's responsive strategy to the challenges presented by the physical surroundings—neither was absolutely deterministic, and both played a pivotal role in the shaping of culture. This integrative approach was highly useful in explaining the spatial diversity of culture, the presence of a great variety of culture **regions** and cultural **landscapes**, elements affecting **cultural identity** and the emergence of **cultural hearths**, and for articulating adaptive strategies that influence **carrying capacity** and **sustainable development**.

The work of the cultural ecology school has become quite influential in the larger context of cultural geography, with several branches of thought emerging from the basic concepts of cultural ecology. Possibilists continue to view the environment as playing a key part in the evolution and character of culture, although many acknowledge that technological innovation has in many cases come to overshadow the impact of the physical conditions of a location. Some scholars, while acknowledging the importance of the dynamic between environment and culture, emphasize that the way humans perceive their environment may be as important as the actual conditions they encounter, and that the adaptations they engage in are strongly shaped by these perceptions. Moreover, human perceptions of the environment are not entirely accurate and contain distortions and misconceptions, frequently in regard to casual agents in the environment. Cultural adaptation is therefore conditioned not simply by the environmental factors that exist, but also, and perhaps more so, by how those factors are viewed in the society and such forces originate. Such perception may be an essential part of how religious beliefs form and evolve, for example.

Cultural Identity

Almost all human beings live in groups. What identifies one as a member of a group may be a physical characteristic such as skin color or gender, or the criteria may consist of shared attitudes, rituals, language, religion, and values. These are all elements of culture, and a group membership derived from these elements is said to be one's cultural identity. Unlike physical qualities, cultural identity is an acquired identity, and a child born in one culture but raised in a separate culture will have the identity of the latter. This is because he or she has gone through *acculturation* in the process of acquiring the cultural identity he or she claims, and all other members of the culture have also experienced the same training, so to speak. That is, the individual has acquired a command of the group's common language or dialect, and typically speaks it as a first language. He or she also follows the religious perspective that members of the group support. But knowledge of the language and shared faith are seldom sufficient to qualify one as sharing a cultural identity, because languages can be learned, sometimes to a level of native fluency, by outsiders, and many religions allow for conversion from other faiths. Shared values, customs, attitudes, and mores are also imprinted on each member, further solidifying the identity.

"The Golden Man," a set of golden armor discovered in southern Kazakhstan, has become a symbol of the Kazakh people. Here the "man" is perched atop a column in downtown Almaty, the former capital city. (Photo courtesy of Reuel R. Hanks)

In some cases outsiders may join the group, but such a newcomer must indicate that he or she has absorbed all of the characteristics necessary to qualify as possessing the cultural identity of the larger collective.

In reality, cultural identity operates at multiple levels, and rarely does an individual identify exclusively with only one group. A certain resident of Miami, Florida, may be on one level Cuban, on another Floridian, on another Hispanic, on another American, and on yet another level, Roman Catholic. All of these are cultural groups, some admittedly overlapping, that such an individual feels loyalty to and identifies with. That is, she believes that she "belongs" to each of these collectives, and that each group in turn represents a component of her individual cultural identity. But this is not to suggest that cultural identity may be clearly marked off by definite **boundaries**, or that such identity is static. Cultural identity is a fluid concept, the building blocks of identity evolve over time, and new identities emerge. Furthermore, there is often a strong geographical basis to cultural identity, typically expressed through **territoriality**. A specific piece of land can itself come to play a central role in cultural identity. This may be due to an historical event that occurred in that place, or may appear because of a special significance assigned by the cultural group to a location.

Under these conditions, the geography so recognized may become a **sacred space**, and members of the group may feel compelled to defend, control, or occupy the land. The economic value of the land may be minimal, but the cultural value is immense, at least in the view of the people who confer such value. An excellent example of this is the region of Kosovo, now an independent country on the Balkan Peninsula. Although the territory of Kosovo has been occupied by a majority population of Albanians for several centuries, the land has played an extremely important emotional role in the cultural identity of the Serbs. In 1389, a Serbian army was defeated by the Turks in Kosovo, and although the battle was lost, Serbs view this event as a defining moment in the history of their nation. To many Serbs, the sacrifice at Kosovo represents the seminal event in shaping the Serbian character.

Some commentators have proposed that the environment in which a group originates crafts the cultural identity of the group, a view known as **environmental determinism**. The Turner Thesis, purporting to outline the elements of a unique American identity, is one such perspective. This view holds that the qualities of the North American **landscape** engendered specific character traits in forming an American identity, especially the concept of the frontier in American history. While many would contend that this represents an oversimplification of identity formation, it certainly is the case that concepts of the "land" and its characteristics often figure in the construction of group identity, especially the notion of a specific territory being "god-given" to a particular group. This can be seen in the identity of many peoples. The Jewish people, for example, have historically linked their group identity to the land of Israel, in spite of being driven from that land in the early Christian era.

Cultural identity in a **nation-state** plays a vital role in shaping the **centrifugal and centripetal forces** present in the state. The failure to craft a cultural identity that evolves into a unifying national identity can lead to the collapse of state cohesion, and the spatial disintegration and **balkanization** of the country. A case in point is the Soviet Union, a state that devolved into 15 independent countries in 1991. Officially founded in 1922, the Union of Soviet Socialist Republics (USSR) attempted for most of its short history to establish a Soviet cultural identity in the country. Soviet leaders recognized that with more than 100 ethnic groups in the country who spoke dozens of languages, it was vital to find a common cultural denominator that would serve as the basis for a shared identity. Soviet propaganda featured the personification of this concept in the "New Soviet Man," an idealized and fictitious "person" who would be devoid of all cultural loyalties except those to the Soviet state. Russian language became the center piece of this effort, and Russian became the tongue of inter-ethnic communication, most of the official media, and higher education. Yet this effort ultimately failed because many non-Russians maintained the use of their native languages, while using Russian to communicate with others outside their ethnic group. Thus, non-Russians in many cases maintained a competing identity that undermined, and eventually displaced, the state-sponsored identity. Many countries today face challenges in forming a centralizing, coalescent cultural identity that are similar to those experienced in the Soviet Union.

Culture Hearth

One of several locations where major cultural and technological advancements appeared in the early development of human civilization. These places emerged as centers of innovation and artistic and scientific achievement, with complex social orders and sophisticated political systems. Moreover, each hearth benefited from regularly cultivated and proficient agricultural systems, often based on extensive irrigation. The latter feature generally enabled the production of a surplus of food. Most of the hearths developed on the banks of major rivers, or at least in close proximity to streams, allowing for a regular supply of water, and also benefited from the rich alluvial soils laid down by flooding.

Scholars generally recognize four major cultural hearths, and some include three secondary hearths as well. The four major hearths are the Tigris-Euphrates River Valley located in modern Iraq; the Nile River Valley in Egypt; the Indus River Valley situated in modern Pakistan; and China's Huang Ho River Valley. The three secondary hearths are located in western Africa, especially Ghana and its environs; Central America, especially southern Mexico, Belize, and northern Guatemala; and the Gangetic Plain of northern India. Each of these locations witnessed the emergence of a common set of advancements, either entirely in isolation from the others or at least partially independent of external influences. The common denominators in each included the organization of society into hierarchical classes, a division of labor ranging from manual labor to higher intellectual pursuits, a high degree of urbanization, a relatively advanced transportation and communication system, and a high level of skill in the physical and literary arts. These characteristics appeared in part because of the food surplus made possible by the rich soils and abundant supply of water. A significant percentage of the workforce was freed from the task of cultivating crops, allowing them to develop technical and scientific skills in areas as diverse as metallurgy, astronomy, and philosophy. Religious belief and language both acquired more organization, complexity, and sophistication. Religion became more interpretive and ritualistic, both characteristics that were secured by the flowering of religious literature and a class of religious officials in the form of priests or others who functioned as intermediaries between humanity and the god(s). Skills in architecture and construction were refined, as monuments either to the gods or to rulers (sometimes they were one and the same) were raised in religious tribute. As new technology appeared and the literary arts flourished, the vocabulary of language became richer and its discourse deepened, enabling speakers to express themselves in new and innovative ways.

Writing systems emerged independently in the hearth regions as early as 3000–4000 BCE. Probably the first was early cuneiform in the Sumerian city-states located in Mesopotamia, followed within a few centuries by an early form of hieroglyphics in the lower Nile region in Egypt and in the Indus Valley in South Asia. Writing systems were necessary for record-keeping and may have evolved because of the increased commerce brought on by specialization of labor and crafts and the need to record the quantities of grain produced in surplus years.

As such systems became more complex over time, a class of specialists in the art of writing, typically working as scribes, record-keepers, poets, and others, parlayed their advanced skills into a new, elite class in society. Writing and reading would remain the purview of specialists for several thousand years, and it would not be until the 19th century that any society would claim a literacy rate exceeding 50 percent. Many modern states reached this milestone only in the 20th century. But the birth of this element of culture in the hearth regions would completely transform human interaction, and language would ultimately become one of the most important factors distinguishing one culture from another.

The culture hearths also allowed for more detailed, stylized, and elaborate religious expression. There is no question that humans possessed certain beliefs about their surroundings that were "religious" in character prior to the establishment of the hearth regions. These beliefs were not expressed through the construction of permanent religious structures in most instances, however. Natural features like mountains, rivers, etc., might be assigned supernatural qualities, but because humans were mostly nomadic prior to the foundation of the hearths, there was little utility in building large monuments, temples, shrines or similar symbols to deities. But in the hearth regions, humans felt compelled to raise buildings that reflected the power of the gods, the might of those ruling the societies, and the wealth and skill of the culture itself. Religious officials also derived much of their authority from the creation of a religious landscape, and since in most hearth areas the heavens played a central role in the local religious system, monuments that approached the sky in the form of pyramids or great temples often appeared. This was especially the case in the hearth in Mesopotamia in the form of the ziggurat, and the huge pyramids built by the Egyptians in the Nile Valley and the Maya in Central America.

Over a long period of time, the many innovations in culture and technology spread outward from the hearths into surrounding regions, influencing the peoples who resided there. This process is known as **cultural diffusion**. As these groups encountered the new influences, they of course modified them to meet the standards, requirements, and expectations of their own societies, resulting in new variations of language, religion, and other aspects of culture. This in turn led to the development of the various expressions of **cultural identity** found across the spectrum of human existence today. In today's era of **globalization**, one may identify new cultural hearths, especially those of Western culture, which has diffused to every part of the Earth over the last century. It was in the original handful of hearths, however, that we find the first evidence of human beings starkly defining and differentiating themselves in terms of the set of attributes scholars collectively describe as culture. These locations set the basic pattern that subsequent civilizations would follow in molding the global cultural landscape.

D

Deforestation

Over most of human history, the majority of the **population** has populated forested portions of the planet. The forest vegetated realms evolved in several climate types (see **Climate**) and they range from the tropical rainforests to the taiga (see **Biomes**). *De*forestation is usually defined to mean the reduction of a forest's canopy cover to below 10 percent of an area.

Over the extreme arcs of **Earth**'s history, areas have deforested because continental drift has moved regions into climate types not supportive of forests. So, too, ancient deforestations can be recognized as being caused by volcanic eruptions and explosions and, more extensively, by basalt flows covering hundreds of thousands of square kilometers. It is manifest that natural climate change causes some deforestation. The cold of the Pleistocene Epoch caused equatorward shifts in climate types, and forests died in places far from the actual continental ice sheets. In Sumatra and Borneo, the Pleistocene lowering of sea level by many tens of meters caused shallow seas to disappear to create a large peninsula connected to the Asian mainland and this peninsula became heavily forested. The rise of sea level after the Pleistocene flooded and deforested the area and it became sea bottom again. Thus, deforestation was ongoing far before the times dominated by humans.

Humans have become inexorably intertwined with deforestation. With today's rapid increase in world population, deforestation is of increasing concern. Earth is losing forest at the estimated rate of 24 hectares (60 acres) *per minute*. Geographically, there are significant differences in deforestation. South America has the greatest amount of deforestation per year while Africa has the greatest rate. Brazil and Indonesia have the greatest total loss of forest while the central African country of Burundi has had recent deforestation estimated at about 9 percent per year. In some developed places such as the United States and the European Union there have been increases in forest cover. However, this does little to staunch the world trend. The largest remaining forested tracts are those associated with low densities of human habitation and inaccessibility. They include the boreal (northern) forests of Alaska, Canada, and Russia, and the remote tropical forests of the northwestern Amazon basin and the eastern Congo basin.

Trees are large and obvious with their species and diversity emblematic of related **landscape** conditions. For humans, trees are useful. Humans use wood for fuel, fiber, shelter, and numerous other purposes. The advent of agriculture supplanted trees with crops. Other human deforestations include creating a defensive "field of fire" around habitations and fortifications. Historically, many more trees have been cut than replanted so forests have become depleted over time. Deforestation implies that the change is for long time periods in which forest has been converted to another land use (e.g., urban or agriculture).

Anthropogenic deforestation has been a significant imprint on landscapes for a very long time. Here, four cases may be considered: First, the cedars of Lebanon (*Cedrus libani*) are referenced in the Bible. Found in mountainous areas near the Mediterranean, these 40 m tall trees were renowned in ancient times for ship masts, building timbers, and resin for mummification. Wealthy, but tree-poor Phoenician, Egyptian, and Mesopotamian civilizations cut and transported the wood hundreds of kilometers and this led to depletion. By modern times the wood was used for railroad ties and the species is now all but extinct except in protected reserves. Second, Easter Island (Rapa Nui) in the southeastern Pacific Ocean has undergone extensive deforestation in the last three-and-a-half centuries. This has been variously attributed to cooling during the Little Ice Age, the cutting of trees by its residents, and the deleterious effects of introduced animals on the native tree species. Third, Great Britain once had an extensive forest cover. The British dominance of the seas and dependence on trade led to deforestation; much timber was used in the construction of thousands of ships. There was such concern about the depletion of the forests that protection laws were passed in the 1600s. Rapidly increasing population of the Industrial Revolution in the 18th and 19th centuries exacerbated matters creating additional needed fuel and railroad ties. The result is that less than 1 percent of Great Britain has been continuously forested for more than four centuries. Much of the bushy landscape now known as "heath" and "moor" represents the result of human-induced deforestation. Fourth, the United States was created land wealthy and the forest resource seen as virtually unlimited. The widespread destruction of the unbroken forests east of the Mississippi River for agricultural purposes changed both the forest mass and species composition.

The deforestation of land surface for human needs has had a number of unintended consequences. For instance, wholesale tree removal has rendered large drainage basins vulnerable to erosion; the deposition of silt at downstream locations has rendered useless many ports on all the inhabited continents. The biodiversity of plant and animal life invariably decreases with forest removal as many ecological niches caused by the presence of the trees no longer exist. The spatial arrangement of remaining forest patches can be crucial as there are many forest plant and animal species that do not survive unless there are suitably large expanses of forest canopy.

Just as climate impacts forest cover, forest changes have been shown to drive some climate changes. Forests provide significant sequestration of the greenhouse gas carbon dioxide. As deforestation proceeds, it is estimated to be responsible for over a fifth of the world's carbon dioxide emissions. The burning of trees significantly decreases atmospheric visibility and local energy balances. Land cover changes provide other feedbacks into the climate system. The loss of trees usually decreases the amount of oxygen produced by photosynthesis. Because the main source of atmospheric oxygen is photosynthesis, the loss of forest by deforestation interferes with the planet's oxygen cycle. The loss of trees lessens the amount of energy associated with water vapor (latent heat), resulting in a local rise of temperature. Finally, deforestation changing forest canopy to bare soil and incomplete crop canopies changes local energy balances and tends to warm the atmosphere.

Demographic Transition Model

A model that defines four progressive stages of demographic development, in an attempt to elucidate and analyze the various conditions and factors that influence population growth. Some sources use the term "demographic transformation" to refer to this model. The key relationship at the center of the model is the dynamic between death rates and birth rates, and how these are altered by changing socioeconomic conditions over time. In particular, the model relates increasing rates of economic advancement and higher levels of urbanization to a general, and at least in theory, long-term, reduction in population growth. The model is based on the experience of economically advanced countries that underwent industrialization in the late 19th and early 20th centuries.

The first stage of the model, typically called the "preindustrial" stage or "high stable" stage, is marked by both high birth rates and high death rates. The total fertility rate and infant mortality rate are also both quite high, compared to later stages of the model. Life expectancy for both genders is also typically reduced in comparison to subsequent stages, due to the influence of food shortages, disease, and frequent violent conflict. **Population** growth in this stage is generally low, because while the birth rate may exceed the death rate for a short interval, it is also just as likely for the death rate to be higher, and neither is marked by a sustained trend upward. The age structure of the population at this stage is typically youthful, with large numbers of individuals in the lower-age cohorts, and few in the older cohorts, reflecting the high rates of birth, fertility, and mortality, as few people survive into "old age." Represented by a population pyramid, a country or region in this stage would show a pyramid with a broad base (high birth rate) tapering steeply to a sharp pinnacle (high death rate). Few countries are considered to be currently at this stage of the model.

The second stage of the model, often called the "early expanding" or "early industrial" stage, reflects the impact of advancements in food production and medical science on the death rate. Technological achievements in science and the mechanization of agriculture have the result of dramatically reducing the death rate and expanding life expectancy. This results in more people living longer, as well as lower infant mortality, meaning that more individuals survive to adulthood. The birth rate, however, remains unchanged, or perhaps may even decline slightly during this stage. The plunging death rate, however, causes a disparity with the relatively constant birth rate, resulting in rapid population increases. The death rate declines throughout this entire stage, although toward the end of the period it begins to level off, as the impact of technological advancement on life expectancy and mortality diminishes. A population pyramid of this stage indicates much less tapering toward the apex of the pyramid (a much larger number of people living into old age), but retaining its broad base, because birth rates have not significantly changed from the first stage. Many countries of the "developing world" are considered by proponents of the model to be currently experiencing this stage.

In stage three, or the "late expanding" portion of the model, the death rate has stabilized at a low level, and the birth rate now declines rapidly. Population growth continues at almost the same rate as in the early expanding period however, because although the birth rate is falling, the death rate has dropped to such a low point that births still exceed deaths by a sizable margin, resulting in overall population growth. The decline in the birth rate during this stage is generally caused not by scientific or technological progress as was the case in the second stage, but by shifts in social values and traditions, often having to do with the role of women in society. For example, during stage three, women may enter the work force in proportionally large numbers compared to stage two, or larger numbers of women may pursue higher education, the widespread use of contraception and abortion may become more socially acceptable, both women and men may delay marriage until later in life, and average family size will typically decline along with the total fertility rate. All of these changes in fact occurred in the United States during the 1960s and 1970s, and supporters of the model generally place the United States in the later phases of stage three. A pyramid representing this stage would not preserve much of a traditional shape; rather, the base of the structure would be only slightly broader than the apex, especially a pyramid illustrating the latter part of stage three.

The fourth and final stage of the model, the "postindustrial" stage, is characterized by both low birth and death rates in the population. Life expectancies remain high, but the population in question has achieved a zero growth rate and possibly even is reproducing at a rate below replacement value. The total fertility rate, in other words, has declined to about two or even lower, meaning that couples are not replacing themselves in the population. Average family size is four or less. A country at this stage can grow substantially only through immigration and may experience shortages in the labor supply, particularly in lower-paying jobs. A number of countries may be considered to have reached this stage of the model, including many of the countries of Western Europe and Japan. The population pyramid representing this stage is not a pyramid at all, but resembles a column, due to the nearly equal numbers of people in most age categories, except for those at the very limits of the human life span.

The Demographic Transition Model has many detractors, and indeed the model has shortcomings when applied to individual countries or regions. For example, critics of the model argue that it fails to take into sufficient account the complexity of factors influencing birth and fertility rates in developing countries, and also point out that the model is based exclusively on patterns observed in the westernized, developed world, and therefore is plagued by **ethnocentrism**. Nevertheless, the model remains widely employed and debated among population geographers.

Desert

The term derives from the Latin word *desertum* for "abandoned." Ironically, the Sahara Desert and the Gobi Desert are redundant names because "Sahara" and "Gobi" both translate as "desert" in their original languages. During the settlement of the trans-Mississippi west of North America in the 1800s, the term "Great American Desert" was commonly applied to semiarid areas thought not capable of supporting agriculture and significant habitation. Occasionally, current writers will refer to "polar deserts" because polar climates are some of the driest on the planet. For instance, places in the interior of Antarctica receive less than 25 mm of **precipitation** a year; moreover, the precipitated water is in frozen form not directly usable by life.

How many deserts does the world contain? Although there are various ways to define desert, the widely used Köppen classification system defines desert climates as those averaging at least twice as much potential evapotranspiration as precipitation during the year. This definition fits large stretches of the planet with adjacent land considered to be "semiarid" with substantial problems for their use. The largest desert is the Sahara, which occupies 9 million sq km of northern Africa. The arid **region** of which it is a part extends through southwestern and central Asia. All continents save Antarctica contain dry, hot deserts. Australia is usually reckoned to be the driest continent in that somewhat over 40 percent of its surface is desert with much other adjacent semiarid land. The long, narrow strip of the South

American Atacama is taken to be the driest land region on the planet containing **locations** where it is many *years* between rainfall events.

Three types of deserts are generally delineated according to their causes. Subtropical deserts inhabit the latitudes between 20° and 30° north and south of the equator. They are caused by the year-round presence of subtropical highs in the Hadley Circulation. The strong descent of air in the subtropical highs precludes significant cloudiness and these are latitudes in which the solar angles are relatively high year round. Thus, the subtropical deserts are the hottest places on **Earth**. The Sahara is this type of desert.

A second type of desert is the middle latitude deserts. They are caused by the summertime influence of the subtropical highs and also lack precipitation because of topographic blockage and/or great distance inland from an oceanic source of precipitable water. The middle latitude locations mean that summer **temperatures** are not as high so that the imbalance between precipitation and evapotranspiration is not nearly as great and the aridity not as great as in the subtropical deserts. A noteworthy aspect of such regions is that winter brings cold temperatures and some snow with the passage of **middle latitude cyclones**. The Gobi is an outstanding example of a middle latitude desert.

The third type of desert is coastal desert. Whereas one might suppose the presence of a coast would make for a water source and a wetter environment, there are certain coasts that are impressively dry. Coastal deserts are found near the Tropics of Cancer and Capricorn and, ironically, the aridity is amplified by the nearby oceans. This desert type is found along coasts with cold currents bolstered by upwelling. Air passing over these near-shore waters is cooled and stabilized, sapping any chance to rise to initiate the precipitation processes. Accordingly, these coastal deserts are quite a bit cooler than their inland counterparts. The Atacama and eastern African Namib deserts are examples of this extraordinary dryness.

The first of several common misconceptions about the desert environment is that they are covered by sand. Dry places have much slower rates of weathering of rocks into the tiny pieces of clay so common in moist areas, so deserts have greater amounts of sand at their surfaces. Yet, sand dunes are not the "average" desert surface. It is believed that sand covers only about 15 percent of desert surfaces. More common are rocky surfaces with high percentages of sand in the soil material. Of course, sand dunes are the iconic symbols of deserts. Several types of sand dunes are classified based on the local amount of sand supply and the seasonal nature of direction and strength of wind. The world's tallest dunes approach 500 m (Algeria). The largest stretch of sand surface is the Rub' al Kali (Empty Quarter), covering about 650,000 sq km of southern Saudi Arabia, Yemen, Oman, and the United Arab Emirates.

The fact that wind-crafted sand dunes are associated with deserts does not mean that they are windier places than their moister counterparts. Wind is simply more

Meager life is supported in Death Valley, California, which is the hottest place in the Western Hemisphere. (Photo courtesy of Steve Stadler)

effective in deserts because of the lack of vegetation to hold soils and small pieces of weathered rock in place. Although sand blown by the wind polishes some desert surfaces, water is a far greater shaper of desert landscapes. In the driest deserts the plentitude of dry stream channels is ample evidence that water erodes and deposits much more total material than wind.

Desert precipitation is noticeably scarce. How dry is a desert? Quantitative definitions such as that of Köppen are based on precipitation amounts that change with temperature. For instance, Albuquerque, New Mexico garners a meager precipitation of about 300 mm per year. This renders Albuquerque quite dry with a natural cover that was scattered bushes and grasses in response to the large evapotranspiration of the summer. The northeastern portion of the province of Alberta, Canada has similar yearly precipitation, but it is a land of forest and swamp because of considerably lower evapotranspiration. A hallmark of deserts is the unreliability of precipitation. There is huge year-to-year variability in the amount and timing of precipitation. In both the subtropical and middle latitude deserts precipitation is intense when it does come because a great percentage of it is from convectional storms.

The popular image of a desert is a hot, dry area devoid of vegetation, but this stereotype is not uniformly true. Net primary productivities are lower than any land regions but for those covered by ice. Both plants and animals are scarcer and less diverse than in other realms but they do exist. Plant life is adapted to extreme dryness through the development of deep or wide root systems or forms that lessen transpiration such as leaves modified into thorns. In short, desert plants are very conservative of water. Most desert animals are nocturnal and so not obvious to the casual observer.

An oasis is a desert locale in which there is verdant life clustered around springs, lakes, ponds, or streams emanating from **groundwater** sources. The water table is close to the surface and artesian water emerges under its own pressure. The vegetation around oases provides food for limited numbers of grazing animals so for thousands of years nomadic peoples have traveled from oasis to oasis to allow their animals to eat and drink.

Deserts make up about a third of Earth's lands and, so, have shaped the habitation patterns of the planet. Perhaps 5 percent of humans live in deserts. There are large cities (e.g., Phoenix and Cairo) in some deserts, but these places are able to obtain water not dependent on local rainfall. In the case of Phoenix, it uses groundwater and an aqueduct that brings water from the Colorado River almost 300 km away. Cairo is situated on the Nile River, which flows without significant tributaries, bringing water from wet tropics near the equator thousands of kilometers distant.

Desertification

Much of the world's land surface is arid or semiarid (see **Deserts**). First used by the French scientist Aubreville in 1949, the concept of desertification has maintained its perceived importance in an increasingly populated world. Desertification is land degradation in arid and semiarid areas and has a negative connotation for sustaining human **populations**. It is a transformation that includes the lessening of water resources, decreased fertility of soil, and disappearance of most of the biomass. This is the result of interrelated processes that are difficult to sort. Some desertification is natural and some a complicated interplay of natural and human causes. Desertified land has a larger yearly moisture deficit (**precipitation** minus evapotranspiration), more modest **humidity**, and higher **temperatures** than land that has not undergone desertification.

Desertification can be identified using land resources satellites (see **Remote sensing**). The assessment of biomass status is now common and has monitoring potential. There are no worldwide assessments but case studies have been conducted in several areas. How much land has undergone desertification? There are several varying estimates depending on particular definitions of desertification; however, it is apparent that the total exceeds a billion hectares (about two-and-a-half billion acres).

Clearly, natural **climate** change can lead to desertification. The Sahara of northern Africa is a significant case in point. There is good proxy evidence from pollen and lake levels that the Sahara once harbored a much greener environment. All evidence points to two drying periods: 6,700 to 5,500 years ago and a more abrupt drying from 4,000 to 3,600 years ago. The drying climate eradicated widespread low bushes and annual grasses and led to the present-day stark covering of bare rock and sand. This result was devastating to nascent agriculture and led to its disappearance from the area, destroying some civilizations. There is abundant evidence that the spreading sands encroached on the monumental works of the Old Kingdom Egyptians. Another natural climate change can be sensed in the 13th century Mongol outpourings from central Asia. Progressive desiccation of already marginal lands placed pressure on the remaining grasslands. Some authors have attributed the ferocity of the empire-building of Genghis Khan as borne of changing climate. Although this single cause is not sufficient to explain Mongolian expansion, it certainly must have been a contributing factor. These two cases are cautionary and hint at deep societal problems when we ponder the possibility of global warming.

The Sahel, the transition between the Sahara and the **forests** of central Africa at, roughly, 15° to 18°N across the entire longitudinal extent of Africa has been pointed to as a pernicious sort of desertification affected by humans, first noticeable in the droughty late 1960s and early 1970s. There are several features of Sahelian societies that deepened the practical severity of the droughts and have lessened the carrying capacity of the Sahel to the point it has not yet recovered. There has been a long-standing shift from nomadic to sedentary societies since the beginning of European colonization. This shift forced **populations** to stay put and intensively use the **landscape** resources around them. Significant population growth has exacerbated the demand on lands; instead of "resting" in between nomadic visits, the land is now continuously exploited. In many areas the exploitation is nonsustainable. Expanded domestic herding promoted overgrazing and subsequent killing of grasses, especially around water sources. Many trees and shrubs were removed for use as fuel and wood. Others were cleared for agriculture. The result was the exposure of dry soil to the **atmosphere** which is readily eroded by wind and water. Poorly managed irrigated areas underwent salinization, killing crops and precluding natural revegetation. Soil fertility is invariably lowered by the removal of most all of the biomass. This created barren landscapes much more difficult to exploit and thus having a much lower **carrying capacity** for animals and humans. This landscape change has all sorts of ramifications such as increasing temperatures and suppression of rain because of increased dust in the air. The re-establishment of previous grass and tree species is problematic because desertification has altered the water balance in a marginal zone. In all respects, much of the Sahel has become the desert.

How permanent is desertification? The term usually implies finality to the landscape's status but this is not certain. This is a question of immense practical importance. Over long spans of time such as hundreds and thousands of years, there are many examples of regions undergoing desertification and then becoming greener. However, can desertification be reversed in shorter time frames? This has been debated under the name of "permanent" versus "temporary" desertification and is in no way resolved.

The plight of the Sahel has led to the investigation of desertification in other places such as parts of the U.S. Great Plains (the "Dust Bowl"), southern Africa, Australia, and Central Asia. Large amounts of the world's desert fringes have been identified as having moderate to very high potential for desertification. The specter of CO_2-induced global warming holds the threat of increased desertification if the number and intensity of droughts increase.

In the 1970s it was posited that the degradation of land causes desertification because of increases in albedo. This simplistic explanation has been discarded. The bulk of the research suggests that desertification does not *cause* dryness, but it is a factor exacerbating the impacts of natural droughts. The human/environment connection is undoubtedly present but it is not a convincing climate control. The geographic community has come to understand that although desertification exists, the human imprint does not cause desert climate. This does not mean that the topic of desertification will be ignored. The Food and Agricultural Organization of the United Nations houses a large amount of information about desertification. Many developing countries have formulated plans for slowing and preventing desertification. There have been limited successes in local cases, but there is no template that can be used worldwide. The desertification of previously usable land is real and of concern for the fifth of humanity living in arid and semiarid lands.

Devolution

A process signifying the transfer of political power from a larger geographical entity to smaller, regional units. A reflection of **core-periphery** relations, devolution often occurs in response to **centrifugal forces** within a governed political space and is the opposite of **supranationalism**. Demands for devolution of political power often are the result of the coalescence of a national consciousness on the part of a minority group within a **nation-state**. Regional constituencies, sometimes based solely on location, but more frequently based on distinctions of history, language, religion, or economic interests (or perhaps all of these), may call for devolution to increase their political and legal autonomy over their "home" region. Devolution in the extreme can lead to the collapse of a larger state into smaller, sovereign countries, as the degree of political control transferred to the smaller units so weakens the authority of the central government that it can no longer maintain its political integrity, and it disintegrates along the borders of the autonomous regions.

Almost all nation-states evince devolutionary tendencies to some degree, but for some, devolution is seen as a means of appeasing regional demands for greater influence in the political process, while simultaneously avoiding the exacerbation of centrifugal forces. Ironically, in recent decades calls for greater devolution in many European countries have increased along with the rise of supranationalism, in the form of the European Union. Two countries in Western Europe in particular represent examples of recent devolutionary trends: the United Kingdom and Spain. In the case of the United Kingdom, strong regional identities began emerging in Wales and Scotland in the 1970s, followed by a similar movement even within the borders of England itself—for example, in the region of Cornwall in western England. In 1997 plebiscites were held in both Scotland and Wales, where public surveys showed strong support for "home rule," or more precisely, the devolution of local political decision-making to legislatures in both regions. In 1999, for the first time since 1707, a Scottish Parliament was convened, charged with making law on local issues for Scots, as separate from the British Parliament. In Wales, a governing assembly was assigned the authority to determine budgetary allotments for the region, although in both Scotland and Wales the British national government retained authority in matters of foreign policy, national defense, and other matters. Spain is a country that has overseen even greater devolution of political power to its constituent regions. Since the death of Francisco Franco in 1975, the Spanish national government has steadily relinquished authority to 17 autonomous regions and 2 autonomous cities. Critics of devolution argue that certain regions of the country, particularly Catalonia and the Basque region, function almost independently of the Spanish federal government, and that such a high degree of devolution of power to the provinces threatens to weaken the Spanish state, as well as leading to a redundant and inefficient administrative structure.

In Spain and the United Kingdom devolution has taken place in a nonviolent context for the most part, but in the case of the former Yugoslavia, the process ultimately led to state collapse and the most destructive episode in Europe since the end of World War II. The legal basis for the implosion of the Yugoslav state lay in devolutionary measures contained in the Yugoslavian constitution promulgated in 1974, in response to calls for more autonomy from a number of the country's provinces, especially Croatia and Slovenia. The constitution allowed for the devolution of political decision-making to the provinces, and established two "autonomous regions" within Serbia, the politically dominant region of the country. The constitution even contained provisions that would, at least in theory, allow individual provinces to secede from the Yugoslav Federation. Serbian leaders were largely opposed to these reforms. Josef Tito, the dictator of Yugoslavia, held the country together in the 1970s through the force of his will and by keeping the aspirations of nationalists in the provinces in check, but on his death in 1980 strong devolutionary pressures once again emerged. Kosovo, one of the autonomous regions in Serbia, was demographically dominated by ethnic Albanians, many of whom demanded recognition of the region as a province of Yugoslavia, a status that would grant Kosovo the right to leave the country. Tensions over this issue and the Serbian administration's response to the Albanian demands led to Slovenia and other republics eventually declaring independence by the early 1990s, and subsequently to bloody civil wars that killed tens of thousands of people. By 2008 Kosovo itself had attained independence. Many would suggest that devolution in the 1970s in Yugoslavia set in motion the downward spiral of federal authority, and the eventual collapse of the Yugoslav state.

Canada is a final example of a state that has used devolution in response to demands for autonomy or outright independence. A number of regions in Canada, a federative state, have in recent decades called for increased political devolution. The most obvious case is that of Quebec, Canada's French-speaking province that is culturally distinct from the remainder of the country. The Parti Quebecois, a French Canadian nationalist party, has taken the concept of devolution to the point of advocating Quebec's complete political independence. The Canadian government responded in 2006 by declaring that Quebec is a separate nation, but holds that status "within united Canada." But Quebec represents only the best-known controversy over devolution in Canada. The "territories" located in the northern reaches of the country have in recent decades demanded increased home rule and control of the natural resources within their boundaries. In response, the Canadian government in 1999 established a new territorial unit, Nunavut, which represented a large-scale devolution of authority to the native peoples of the region by creating an enormous territory under their jurisdiction. The Canadian federal government will eventually grant control over all of the territory's natural resources to the native people living there, an example of both political and economic devolution. The crucial challenge that countries like Spain, Canada, and others face when devolving political power is determining the critical limit between successfully decentralizing authority, or undermining the geographic integrity of the state.

Distance Decay

Distance decay is a concept linking geography and the occurrence or frequency of a pattern of activity. Some literature uses the term "distance lapse rate" to identify this connection. The underlying principle is that any phenomenon or influence diminishes in intensity as the distance from the point of origin increases. This is due to the effect of **friction of distance**. The concept is enshrined in the geographer Waldo Tobler's famous dictum that all things are related, but those that are nearest to one another are more related than those located at a greater distance. Distance decay is an important precept of spatial analysis, especially for **spatial interaction models** and notions of **cultural diffusion**. Distance decay is inherent in various applications of gravity models in economic and urban geography and underlies some of the fundamental theoretical models in these subfields of geography, including Christaller's **Central Place Theory** and the **Von Thunen model** of agricultural development. Gravity model theorists, for example, suggest in their simpler propositions that a precise mathematical relationship exists between the strength of a spatially expressed phenomenon and the distance over which it is distributed, in a formula closely akin to Isaac Newton's equation defining gravitational force. Whereas Newton proposed that the force of gravity is directly proportional to the mass of the objects in question and a mathematical constant, divided by the square of the distance between them, geographers using gravity theory in their work attempt to construct a similar model analyzing labor or trade flows, transportation, shifts in population, diffusion of languages, religions, technologies, etc. Just as Newton's theorem postulates that the force of gravity between two objects declines with distance, so too do gravity models in geography, based on the concept of distance decay. A simple example would be a model that attempts to gauge the influence of German as a first language used in business transactions. Such a model would likely show a high degree of use in Germany (center of gravity), with moderate use in surrounding countries in Europe (weakening gravity with distance from the center), and little use in North America or Africa (decay of gravitational influence with distance). More sophisticated approaches attempt to integrate a geometric component into the model, because in real world applications the rate of decay with distance is dependent to some extent on the geometry of the distance covered, not simply the absolute distance intervening between the two interacting points. A simple illustration of the influence of geometry (or in more accurate terms, topography) would be a mountain range lying between two towns that trade with one another but are separated by a distance of only 5 miles, and the same two towns located on a level plain but separated by a distance of 10 miles. The distance decay rate (as expressed by a mathematical formula or a curve) in the first relationship could likely be higher (a steeper curve when graphed) than in the latter case, due to the physical barrier presented by the mountain range, in spite of the fact that the towns are separated by only half the distance. Some geographers suggest that the impact of **globalization** and mass electronic communication are weakening the relevance of the distance decay concept.

E

Earth

Earth is the formal name for the planet on which we live. Although we know of other planets, including some outside of our own solar system, there is not another known planet with the combination of physical characteristics as ours. Earth is large enough to support several billion human inhabitants and one-and-a-half million species of plants and animals. It is small enough to present a finite set of resources in terms of long-term sustainability of life as we know it. Moreover, it is subject to all sorts of natural and human caused catastrophes, some of which could possibly destroy the viability of life on the planet.

Seemingly large to humankind, Earth is incredibly small when considering the size of the known universe. The third planet from its local star, the sun, Earth orbits the sun at an average distance of 150,000,000 kilometers while rotating on its axis (see **Seasons**). Earth was created approximately 4.6 billion years ago out of the conglomeration of materials from a disk-shaped concentration of gas and dust widely surrounding the sun.

The planet seems massive by human standards. Earth contains about 5.97×10^{24} kg of mass, although it is not nearly the most massive planet in our own solar system. Mass is added on a daily basis as dust and meteorites enter the atmosphere. Conversely, some mass is lost by gases escaping the microgravity at the top of the **atmosphere**. At present, gains and losses do not seem to have an appreciable imbalance. Something of Earth's size can be gleaned from watching a ship disappear over the horizon at sea or observing the planet from a commercial jet aircraft: it is clear the Earth has a rounded shape. The circumference of the planet approximates 40,000 km at the equator.

The shape is roundish, but Earth is not a perfect sphere on two counts. First, the radius—the distance from mean sea level to Earth's center—is not consistent in all **latitudes**. At the equator, the radius is 6,378 km while at the North and South Poles, the radius is only 6,357 km. Earth belongs to a class of shapes called oblate spheroids. Earth's spheroid is "fat" in its tropical latitudes and this is caused by shape deformation of Earth's non-solid interior by virtue of outward forces caused by rotation. Notice the polar and equatorial radii differ by only three-tenths of a percent. For all but the most esoteric mapping functions and to the astronaut observing from orbit, this difference is negligible. In practicality, the radius of Earth is usually rounded and taken to be 6,400 km. The second nonspherical component of the planet is its topography above sea level. The oceans that constitute most of the planetary surface present an almost-spherical shape to the atmosphere. The largest exceptions are two lunar-caused tidal bulges rhythmically increasing and decreasing sea level approximately twice every 24 hours. Most land on the planet is within a thousand meters of sea level, but some very high elevations exist. The tallest is Mt. Everest (Sagamartha) in Nepal and is 8.8 km above sea level and the lowest is the Dead Sea in Israel at minus 4 km. Underneath the ocean there is an average depth of 3.8 km with the deepest being 10.9 km at the Challenger Deep in the long, narrow extent of the Mariana Trench in the southwest Pacific Ocean. The difference in elevation from highest to lowest points on Earth approximates 13.6 km and is dwarfed by the radii of the Earth. This is not to say that these topographic variations are inconsequential. For instance, altitude above sea level is a major control of climate experienced at a location. So, too, varying physical and biological features can be found with depth into the oceans.

The interior of Earth is of intense interest to the geographer because interior processes have a profound effect on exterior landforms and the human geography of the surface. Far from being a solid mass, Earth's interior is differentiated over depth and is quite active. How do we know this is the case? First, there are dramatic surface illustrations of activity through volcanoes and earthquakes. Second,

Terrestrial Magnetism

Earth has a magnetic field and it has been studied scientifically since the early 19th century. Terrestrial magnetism (geomagnetism) is caused by giant, sluggish convection currents in the molten, iron-rich outer core of Earth. This magnetism has some irregularities so it is not equally strong everywhere. Nonetheless, it surrounds Earth out to several thousand kilometers and is known as the magnetosphere. The magnetosphere is "teardrop" shaped with its long end away from the sun shielding the planet from the particles and energy of the solar wind. The magnetic field is a dipole because it intersects in two places with Earth's surface. One place is near the magnetic north pole near the northern end of Earth's axis; it has a counterpart in the Southern Hemisphere. Compasses point to magnetic north pole rather than the geographic North Pole, which is the north end axis of Earth's rotation. The angular difference between a magnetic pole and a geographic pole is called declination, and it can be over 40° in some places. The magnetic poles systematically shift. For instance, the north magnetic pole is currently moving at 60 kilometers per year (36 mi per year) from northern Canada toward Russia. Additionally, the North and South poles reverse with a periodicity averaging 300,000 years.

Earth has a magnetic field that is thought to be caused by the circulation of interior fluids combined with Earth's rotation producing a "dynamo" effect.

The deepest borehole ever drilled was to a depth of about 12.3 km and this leaves humans without direct sampling of most of Earth's interior. Of crucial value to us are the interior motions of the planet we detect as earthquakes. There is a global system of seismometers measuring location and strength of the thousands of major and minor earthquakes occurring each year. Through careful study of the timing of the passage of various energy wave types through the interior, the basic nature of the interior has been revealed. Near the surface, explosives are used to create small crustal motions to discern rock configuration that could point to the presence of exploitable oil and natural gas.

The inner core extends outward 1,450 km from Earth's center. It is solid and dominated by iron and nickel or iron and silicate. At temperatures between 5,000°C and 6,000°C, the core is kept hot by nuclear reactions under tremendous pressures. Although the core is cooling, it is cooling exceedingly slowly. The outer core is probably made of the same materials and extends from 1,450 to 3,500 km from the center, but it is liquid and the source of Earth's magnetism.

Above the outer core is the mantle extending to the crust within a few tens of kilometers of the surface. It is less dense than the deeper layers. Yet, because it is farther from the center, it has the greatest volume and mass of any layer. It is composed mainly of iron and magnesium and is hotter with depth. The top of the mantle down to 100 km under continents seems to be fairly rigid. Below, down to 350 km, the rocks are heated so much that they are capable of great deformation and slow, plastic flow. This zone is the asthenosphere and forms the basis for the motion of crustal plates collectively known as **plate tectonics**.

The solid surface of Earth is known as the crust and it has been likened to a thin, brittle eggshell surrounding a more pliable interior. The crust contains an impressive mixture of rock types with increasing density of materials with depth. The boundary of the crust with the upper mantle is at markedly different depths beneath continents and oceans. When studying the respective suites of rocks it has become clear that continental crust reaches greater depths (40 km) than oceanic crust (8 km). This configuration makes for continents and ocean basins and was imparted to the planet early on as lighter-density liquids were extracted from the deepest parts of the mantle and cooled. The resulting continental rocks are less dense than oceanic rocks and, so, do not press as much on the asthenosphere and maintain higher elevations. The upper part of continental crust is sometimes referred to as "sial" from the relative abundance of silica and aluminum; a typical rock type is granite. Oceanic crust is referred to as "sima" from the combination of silica and magnesium; the signature rock type is basalt.

Earthquakes

The study of **plate tectonics** makes it clear that the crust of the **Earth** is in long-term motion being driven by giant convective currents in the materials of the mantle. The average rate of motion is small but able to account for massive crustal change on the order of tens of millions of years. Rather than moving slowly and smoothly in all places, the crust is subject to differential motions because of irregularities in the convection of the flow of plastic materials in the mantle and variations in resistance of rock formations to pressures. In many instances rock is able to accommodate the pressures and will warp and **fold** over long amounts of time. Sometimes, rocks will suddenly snap and crack propagating underground motion in the form of seismic waves. Such events can be very large. In 2004 the western edge of the Burma Plate jerked to the west and upward tearing some 400 km of crust and causing massive destruction via tsunamis. In other circumstances small pieces of the crust are rearranged a few millimeters along previously existing **faults**. Sometimes, ground motion can be felt hundreds of kilometers away.

Earthquakes are measured using devices called seismometers, which record time series of data called seismographs. Seismometers measure motion in the crust. In the simplest incarnation, the frame of the device is anchored to a hanging pendulum. As earthquake waves move the crust underneath, the pendulum maintains its position by inertia and the relative motion of the crust is traced onto a seismograph.

The center of earthquake motion is the focus while the epicenter is the surface position directly about the focus. Most earthquake energy is used in the deformation of rock material, but up to 10 percent of the energy is released as shock waves. These seismic waves emanate from the focus with their speed governed by the rock materials in which they are working. The fastest waves are known as P (primary) waves and propagate at speeds between roughly 1.8 and 8.0 km per second. S (secondary) waves include side-to-side and up-and-down motions and travel at 60 percent to 70 percent of the speed of P waves. P and S waves both register on seismographs with P waves arriving before S waves. Their speeds of travel can range widely, but their ratio of speeds is always within a few percent and this allows the establishment of the distance of the focus from a seismometer. So, distance to the epicenter is readily found with greater distances being associated with the greater difference in arrival times between S and P waves. Exact epicenter location is unknown until it is "triangulated" from the data of at least three seismographs. The Global Seismographic Network encompasses over 150 stations capable of rapid exchange of standardized data so that events are very well monitored. In addition there are country-based and local networks in operation.

Energy released by earthquakes varies of orders of magnitude. There are a couple of dozen major earthquakes (Richter magnitude 7.0 or greater) expected each year and over a million smaller ones. Most earthquakes are not discernible without the aid of instrumentation. Seismograph observations are converted in various ways to measures of earthquake intensities.

The Richter scale and its modification is the earthquake measurement scheme most familiar to the public. This scale is one of energy release and has no theoretical upper limit. What is usually unappreciated is that the magnitudes are rated on a logarithmic scale and so every increase of 1.0 represents an energy increase of 32 times. So, a major earthquake of magnitude 7.0 releases more than a magnitude 5.0 event. The Richter scale has been replaced by the moment magnitude scale, which gives somewhat better estimates of the strengths of large quakes. The modified Mercalli scale is based on observable damage rated from I to XII. An earthquake of magnitude I will not be felt by most of the population except, for instance, if one is on top of a tall building. Magnitude I earthquakes do not leave behind damage. Magnitude V events are felt by nearly everyone with broken windows and dishes falling from shelves. Magnitude XII quakes are unimaginably damaging but, fortunately, rare. In a magnitude XII event, the destruction is complete with rolling waves observed on the landscape and objects thrown in the air.

There is a definite geography to the large earthquakes. They tend to occur on tectonic plate boundaries although they occasionally occur associated with hot spots and other weaknesses in the crust. The ten largest quakes of recorded history have occurred on the plate boundaries ringing the Pacific Ocean. In the United States, Alaska and California have the greatest propensity for experiencing large earthquakes. This is because of their proximity to ocean bottom where the Pacific Plate is subducting underneath the North American Plate.

Deaths are common in large earthquakes and there are reliable written records pointing to several million deaths within historic times. Some earthquakes have killed more than three-quarters of a million people. In the United States, the most fatal event was the San Francisco temblor of 1906 in which 3,000 people were lost. Earthquake deaths are not limited to the land close to the focus but can also be thousands of kilometers away in the case of earthquake-generated tsunamis. In 2011, an earthquake of 9.1 magnitude rocked the eastern coast of Japan's Honshu Island. Most of the thousands of deaths were caused by the ensuing tsunami, which devastated areas near the coast.

The largest earthquake in recorded history in the United States occurred off of southern Alaska in 1964. Its estimated moment magnitude was 9.2. The onshore areas were not heavily populated and the death toll was only 131. However, some towns were destroyed through the resulting tsunamis (tidal waves). Anchorage was heavily damaged because of the differential rise and fall of the land and the resulting rending of gas lines, water mains, and sewer lines. Parts of southern Alaska's surface rose by 10 meters while other portions sunk by 3 m.

The world's largest earthquake recorded by instrumentation was the Chilean quake of 1960 measured at 9.5. There were over 2,000 fatalities and 2,000,000 homeless and related tsunamis caused loss of life in Hawaii, Japan, and the Philippines.

Economic Development, Geography of

Economic development, like all human social processes, has a geographic dimension. Geographers study the various spatial aspects of economic development at every **scale**. At the local level, geographers studying the developing world are interested in what factors may lead to a **population** exceeding its **carrying capacity**, potentially leading to **desertification** or other problems. **Sustainable development** is a key concept that many geographers now emphasize in their work. In developed economies, the influence of **break-of-bulk points** and **agglomeration** may be the focus of inquiry, along with many other spatial elements that play a role in development. At the scale of the **region**, economic geographers may study the effects of **capital leakage**, a problem that is especially acute for developing countries. Global economic relationships have been the concern of geographers for centuries, and geographers have offered conceptual approaches to international trade flows with theories like **World Systems Theory, comparative advantage**, and many others. Geographers are of course interested in the effects of **globalization** on the spatial distribution of development and wealth, and use a spectrum of methodologies and theoretical tools to comparatively assess both the positive and negative impact of globalization on the spatial relationships between developed and lesser-developed countries. It is not just economic geographers who contribute to the investigation of these quite complex relationships, but political geographers also work on policy issues that often influence economic development, like trade policy or the distribution of foreign aid. Cultural geographers may study the **push-pull concept** driving **migration**, yet another consideration in a quite complex and multifaceted process.

A good deal of the geographic literature on economic development is directed at explaining the spatial disparity between the developed world and lesser-developed countries, or LDCs. The gulf in the quality of life between these two regions is large and cannot be explained through uneven distribution of resources, climate, or other simplistic notions. A majority of the human family lives on an income of only about one U.S. dollar per day, but in developed countries average

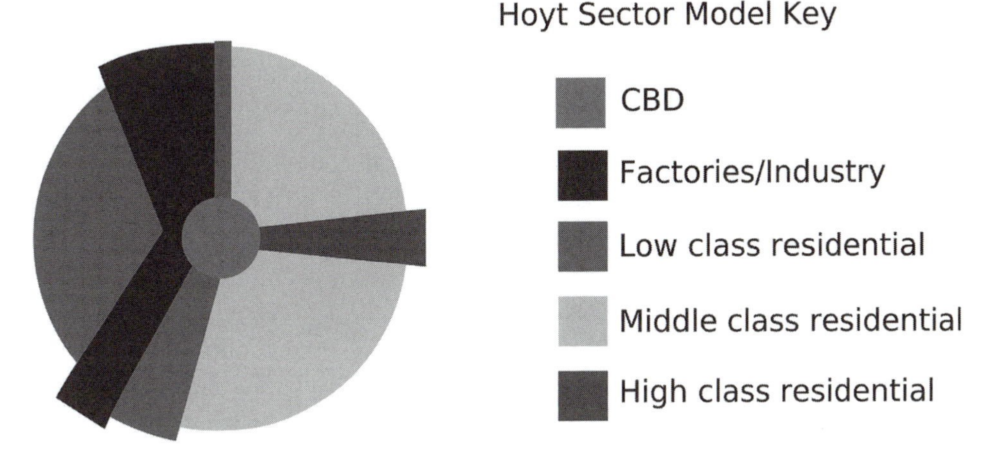

The Sector Model, also called the Hoyt Model, represents an attempt to conceptualize the spatial differentiation of economic development in the urban environment. (Cieran 91/ABC-CLIO)

annual income per capita is typically several tens of thousands of dollars per year, often more than one hundred times greater than in the world's poor countries. The measure of **Purchasing Power Parity (PPP)** indicates that same magnitude of disparity. Some economic geographers have attempted to explain at least a portion of this gap by holding that it represents the residue of **imperialism** and mercantilism, as the exploitative economic relations of the 18th and 19th centuries have not really ended. But this theory fails to adequately articulate why some regions that were exploited by colonialism, like Latin America, have fared much better than others by most measures, such as Africa. In other words, there is a regionalization to the underdeveloped world that presents a much more complex geography than a theory of post-colonial, capitalistic exploitation can explain. Indeed, countries like Argentina and Chile can hardly be described as underdeveloped at the beginning of the 21st century, and to include them in a spatial analysis along with the poorer states of Africa or South Asia no longer seems appropriate.

A common regional scheme encountered in the geography of economic development is the so-called "North-South" divide, which splits the world into a well-developed northern tier of countries that enjoy high standards of living, relatively low rates of population growth, low **infant mortality rates**, and other positive indicators, while the Southern Hemisphere (the "South"), with the exception of Australia and New Zealand, is a zone of much lower personal income, higher mortality rates, shorter life expectancies for both genders, etc. The North is dominated by the Group of Eight (G-8), representing eight of the most influential economies in the world. However, this geographic pattern may be changing, as the G-8 has now been replaced as the main forum for the discussion of economic policy by the Group of 20 (G-20). The G-20 contains a more diverse membership, with several countries from the "South" that are rapidly developing, experiencing substantial increases in wealth and an emerging middle class, and gaining global economic influence. China, for example, as of early 2010, is considered by many economists to be the world's second-largest economy, and living standards there have been increasing dramatically since the 1980s. India, to take another example, now boasts a middle class of some 300 million people, or about a quarter of the total population. Brazil, Argentina, and South Africa, while all holding significant pockets of poverty and **spatial inequality** of wealth, nevertheless are advancing economically and are members of the G-20 forum. The dichotomy represented by the "North-South" regionalization between the hemispheres appears to be collapsing, and new conceptualizations may soon be needed. The geography of economic development is a broad, dynamic, and complex field that will provide scholars with many challenges for many decades.

Electoral Regions

Spatial units designed to provide representation to the electorate, based on a set number or percentage of the voters. These may be found at various scales. In the United States they range from local city precincts and wards, to state senate and house districts, as well as states themselves for state-wide offices; to congressional house districts, and finally to the individual states, which function as electoral regions for U.S. senators, and finally to all 50 states collectively—the electoral region for U.S. presidents. Some informal groupings have emerged in the discussion of electoral regions in recent years, primarily based on the popular notion of "red" (Republican) states and "blue" (Democrat) states, and associated clusters of these states based on voting patterns. Some electoral regions are dynamic and change shape on a regular basis in the United States. One example is congressional districts, which are redrawn every 10 years, based on the results of the U.S. census. These electoral regions change shape, and typically change the number of people they represent over time. This is because the number of seats in the U.S. House of Representatives is set at 435, and the number of representatives a state receives is based on the state's population. If a state gains population from one census to another in proportion to other states, it will gain additional representatives, and therefore must add new electoral regions in the form of congressional districts. If it loses population vis-à-vis other states, it loses one or more seats in the House, and still must redraw the **boundaries** of its congressional districts. Districts with controversial boundaries are often labeled **gerrymandered** districts.

The electoral regions for U.S. senators do not change over time, because these regions are the states themselves, since each state is guaranteed in the U.S. Constitution representation by two senators in the United States Senate. Likewise, in presidential elections, the electoral region is all 50 states collectively and the District of Columbia as represented through the Electoral College, in which 570 total electoral votes are available. However, the number of electoral votes assigned to each state changes every 10 years with new census results, although the number of electoral votes assigned to the District of Columbia remains constant. Moreover, this is not to suggest that the geography of presidential elections is somehow static—regions of the country tend to change their political tendencies over time, a fact that makes the element of spatial strategy in presidential campaigns of paramount importance. For example, in the early 20th century, states in New England and the upper Midwest often supported Republican presidential candidates, while the American South consistently and solidly backed Democratic candidates. In the elections of 1920, 1924, and 1928 the Republican candidates (who all won their respective elections) carried every state in New England and the upper Midwest with the exception of Wisconsin in 1924, and Massachusetts in 1928. States in the Deep South almost uniformly supported the Democratic candidate in each election. This had been the case as well in the elections of 1900, 1904, and 1908, again all won by the Republican candidate. At the beginning of the 21st century, this geographical pattern has almost completely reversed, with the South now a Republican bastion in most elections, and New England and the upper Midwest solidly in the Democratic column in most elections. The shape, size, and location of electoral regions and their political characteristics change as often as attitudes and allegiances do in politics.

Entrepôt

Also known as a free port, an entrepôt is a duty-free port where high-value commodities are traded. Such goods often are transported long distances, and the entrepôt functions as an exchange point between carriers. Commodities may be stockpiled at an entrepôt, allowing for a more consistent supply to the market to be maintained, and thereby avoiding large fluctuations in price that destabilize demand. Goods exchanged at an entrepôt typically are in high demand, and are produced in monopolistic or near-monopolistic conditions. Such exchanges were common under the mercantilism systems of the colonial era. Entrepôt trade developed primarily due to the long distances involved in moving commodities from the point of origin to the final market.

For example, spices in the 1600s were frequently transported along a series of entrepôt ports of call between Southeast Asia, where nutmeg, cinnamon, and other high-value goods were purchased, and the major market area for the spice trade, Europe. Singapore and Penang, Malaysia both developed into major economic centers through entrepôt trade. By employing a system of entrepôt ports, companies in the spice trade, at the time primarily the Dutch East India Company, could stabilize supply and also reap additional benefits, such as replenishing damaged stock, exchanging information concerning fluctuations in production or the status of suppliers, and other non-tangible business advantages that collectively helped to maximize profits and limit risk. Managing the supply of spices was essential, because the demand for these commodities was stable, and an unexpected glut in supply could result in a dramatic drop in price. If too many ships carrying cargoes of spices arrived at Dutch ports unexpectedly, the surplus supply could result in the entire expedition failing to turn any profit, a serious failure indeed since the Dutch East India Company raised revenue to finance its ventures by selling stock. Too many failures to make a profit would not only result in financial losses, but undermine the faith of investors and damage the company's ability to raise future investment capital. Under the monopolistic circumstances of the spice trade, the entrepôt system allowed adjustments to be made in deliveries that virtually eliminated the dangers of oversupply.

The system also took root in Europe itself, with first Antwerp and later Amsterdam functioning as thriving centers of entrepôt trade. Antwerp was the leading entrepôt port in central Europe until 1585, when the city was badly damaged by military action as a result of the Dutch Revolt against Spain. Ironically it was the connection to Spain that had enabled Antwerp to dominate other ports in the region, but by the early 1600s Amsterdam had replaced Antwerp as the leading free port in the Netherlands. After the establishment of the Dutch Republic in 1581, Amsterdam became the home port of the Dutch East India Company, and the profits from the high-value commodities that the port traded, like spices, silk, and porcelain, led to the accumulation of enormous wealth. Eventually competition from British merchants undermined Amsterdam's dominance, but for a century and a half its trading status made the Netherlands the leading maritime power in the world.

Environmental Determinism

A highly controversial theory, environmental determinism has influenced geographical thought for many centuries, and became a dominant philosophical approach in American geography in the late 19th and early 20th centuries.

Environmental determinism, in its simplest form, argues that human cultural traits are either determined, or at least strongly influenced, by the physical environment in which culture originates. It is a theory with strong elements of **ethnocentrism**. The foundations of the theory may be found in the writings of many ancient thinkers, dating back at least to the ancient Greeks. Herodotus, for example, linked the physical environment of some of the peoples he described to cultural characteristics he (allegedly) observed, and centuries later, Aristotle argued that **climate** had a strong influence on human character and behavior. The Greek geographer Strabo, who lived at the beginning of the Christian era, continued this tradition, suggesting that peoples who lived in warmer climates lacked energy and initiative, while those who lived in cold latitudes were less intelligent. Nor was the deterministic view confined to European thinking—the medieval Muslim geographer Ibn Khaldun (see sidebar) was convinced that the differing mental and cultural qualities between urban and rural dwellers were due to the "fact" that living in a city made one more intelligent than living a nomadic existence. During the Enlightenment, perhaps the most influential proponent of environmental determinism was the French philosopher Charles Montesquieu, who believed that climate greatly shapes the nature of human society and even proposed that various political systems were the product of the climate in which they developed. Tropical climates, in the view of Montesquieu, invariably led to the emergence of despotic regimes and spawned slavery, while more moderate climates (meaning the climate of Europe) gave rise to democratic systems. Montesquieu's ideas were widely accepted by European scholars of the time and would subsequently influence some of the most important social scientists of the 19th century.

In the late 19th century, the impact of Charles Darwin's theories on evolution and natural adaptation shifted from the natural sciences to the social sciences, resulting in the collection of theory generally termed Social Darwinism. Leading proponents of this perspective in history, anthropology, geography, and other social sciences grounded their thinking in Darwin's concept of "natural selection." These scholars reasoned that if the ability of a species of plant or animal to survive was determined by that species' capacity to adapt to changes in the environment, then the same must hold for human societies. Indeed, according to many of this school of thought, the very characteristics of human civilizations must be shaped largely by the natural environment, just as the physical attributes of an animal were determined by its surroundings, according to Darwinian principles. Examples of the application of Social Darwinism to geographical scholarship at this time abound. Rudolf Kjellen, a Swedish political geographer, argued in his most influential work, *The State as an Organism*, that the law of natural selection applies to political states as well as to biological species, and that countries typically engage in a competition of the "survival of the fittest" in which one state "naturally" conquers and absorbs weaker neighbors. Kjellen's thinking was deeply influenced by the work of Friedrich Ratzel, a German Social Darwinist, who developed a life-cycle theory of the political state. Such theories were of course used to justify the expansionistic, aggressive foreign policies that underlay European **imperialism**, and rested on the notion that some states or cultures were "inferior" due to the natural environment that surrounded them; therefore, their conquest by stronger states was simply part of a "natural" process. By the early 20th century the theory of environmental determinism had acquired overtly racist and chauvinistic elements, a trend that would continue until World War II.

One of Friedrich Ratzel's students in the late 1800s was a remarkable young American woman named Ellen Churchill Semple. Probably more than any other individual, Semple was responsible for bringing Ratzel's ideas regarding the life cycle of states and the influence of geography on civilization to the American geographical community. Semple taught at both the University of Chicago and Clark University, and her writings shaped geographical discourse in the United States for several decades. Many of her contemporaries were already examining the concept of "geographic influence" on human activity and history, and Semple's work provided significant momentum to this effort. One of Semple's more controversial positions held that religion is an outgrowth of natural forces and phenomena, a notion that had its origins in the philosophy of both Montesquieu and the French scholar Ernest Renan. Other American geographers took environmental determinism considerably further, most notably Ellsworth Huntington, a well-known scholar at Yale University who published numerous articles and books on the subject. Huntington's book *Climate and Civilization* was read not only by academics but also by the general public, and he possibly did more to popularize the concept of determinism than any other American scholar. By the 1920s the "scientific" basis of environmental determinism was widely accepted by many in the United States. At about the same time, in Germany ideas associated with environmental determinism emerged in the new field of *Geopolitik*, or geopolitics. The main proponent of this new approach was Karl Haushofer, a professor at the University of Munich. A German nationalist, Haushofer's publications and lectures may have reinforced the racial and ultranationalist views of some of the leaders of the National Socialist (Nazi) Party. Rudolf Hess, one of Adolf Hitler's close associates, studied under Haushofer. Moreover, the fundamentals of Geopolitik were well known to many who passed through the halls of the University of Munich in the 1920s and 1930s.

By the 1930s and 1940s philosophical responses to environmental determinism had begun to appear in some academic circles on both sides of the Atlantic. In the United States, the influence of Carl Sauer (see sidebar) and the Berkeley School of geographers offered a counterview to the determinists. While acknowledging

Ellsworth Huntington (1876–1947)

Ellsworth Huntington was a professor at Yale and one of the early giants of the discipline of geography in the United States. He was a student of William Morris Davis at Yale and conducted Asian fieldwork in the early 20th century. He was one of the first scholars to recognize the imprint of climate change in the ruins of ancient societies and became the leading proponent of **environmental determinism**. His premise was that climate was the root cause of the rise and fall of civilizations. Far from being known only in academic circles, Huntington's engaging writing style and prodigious production made him a popular author. His seminal books included *The Pulse of Asia, Civilization and Climate* and *Mainsprings of Civilization*. In retrospect, they were based on some scholarship and many grand generalizations that are simplistic and sometimes racist. For instance, he believed that middle latitude storminess "energized" populations and led to the progress of civilizations in Western Europe and North America while the lack of such storminess inhibited African development. Huntington's popularity instigated a backlash among geographers to the extent that his ideas were castigated and provided fuel for the development of social and cultural geography, which took a more complex view of the interrelations between nature and humans.

that the physical conditions a group faced imposed certain limits, Sauer and his supporters argued that environmental determinism placed too much emphasis on the inevitability of natural influence. At the same time, the determinists failed to recognize the enormous capacity of human beings to not only adapt to environmental challenges, but also human ability to alter the impact of the environment through technological advancement. Most of the Berkeley School scholars were advocates of the theory of **possibilism** to one extent or another, a movement that also had support in some corners of Europe, especially after World War II. Possibilism holds that there are many possible responses that a culture may make to the limitations the natural environment may impose, and that cultural attributes are not "determined" by physical geography, only shaped by such forces. In this context, many of the adherents of the Berkeley perspective stressed the approaches and methodologies of **cultural ecology**. Somewhat later, another theory emerged that represented something of a compromise between these two perspectives. *Probabilism* contends that while the environment does not determine the qualities of human cultures, it nevertheless makes some outcomes more likely than others. Although environmental determinism has lost much of its influence in geography over the last 50 years, and few geographers today would claim to completely support the theory, the debate continues to rage in academic geography over the degree that physical conditions may direct the development of human activity.

Ethnocentrism

The point of view that one's own culture, ethnic group, religion, or race is fundamentally superior to others. Such attitudes can lead to distortions, misconceptions, and prejudice in disciplines such as cultural geography, ethnography, anthropology, and sociology, among others. Ethnocentric bias is difficult to completely avoid when studying human cultures different from one's own, because all individuals experience the process of enculturation. Enculturation is the wide range of practices and values that each person in a society acquires, as he or she matures and becomes an active member of the group. One is taught that there are "right" (using tableware) and "wrong" (using one's fingers) ways to eat; for example, to be accepted by others in the group, one must consume food in the "right" fashion. When encountering different cultures where the "wrong" way of doing things is acceptable, it then is easy to conclude that such groups are "crude," "primitive," or "uneducated." Scholars of other cultures must constantly guard against the trap of assigning ethnocentric value judgments to the behavior of the people they study.

Ethnocentric attitudes are found in every culture, and in fact have had a marked impact on the development of world geography. In the West, notions of cultural and moral superiority helped drive the expansion of **imperialism**. "Christian" civilization was seen as more intellectually advanced and principled than "primitive" cultures in the regions where European countries wished to establish colonies. There were, to be sure, economic motivations for imperialism as well, but the "civilizing mission" of spreading the Christian faith was commonly used as justification for acquiring an empire during the colonial age. Rudyard Kipling's poem *The White Man's Burden*, although published toward the end of the colonial era, is frequently cited as an anthem overtly expressing a Western ethnocentric bias. In the poem Kipling characterizes colonial peoples as "half devil, half child," a phrase that many commentators view as indicative of Kipling's, and many others', ethnocentric chauvinism and sense of "natural" superiority. At the same time, this perspective also led to the global spread of English-speaking culture, dramatically shaping the modern political and cultural landscape.

Other cultures and civilizations have also adopted ethnocentric prejudices. Early in the history of China, the Chinese elite acquired the view that outsiders were inferior, and used the blanket term "barbarian" to refer to any foreigner. They named their country the "Middle Kingdom," meaning that it was below heaven but superior to the remainder of Earth, an opinion that persisted among many in the upper classes of China up to recent time. According to Herodotus, the ancient Persians regarded their culture and empire as the center of the world, and themselves as the most advanced people on Earth. Some writers living in the urbanized portions of the Islamic realm in the Middle Ages held quite negative attitudes toward nomadic peoples, especially those in Central Asia, considering them little better than "savages" who lacked intellect and motivation. In some instances, Muslim commentators held that the peoples of sub-Saharan Africa were mentally inferior to other people, and lacked any concept of civilization. Ethnocentric bias is a trait that most, and perhaps all, societies share to some degree, a fact that compounds the difficulties in building cross-cultural communication and understanding.

The influences of ethnocentrism in geography may be found throughout the discipline's history. Studies using the philosophical framework of **environmental determinism** were especially susceptible to such bias, and one can detect elements of implied cultural and racial superiority in the thinking of many scholars of this school, as evidenced in some of geography's earliest writings. Textbooks on geography also were not immune to ethnocentric notions. Roswell Smith, the author of a geography textbook published in the United States in the mid-1800s, offered a typology of "civilization" that utilized the classifications of "barbarous," "half-civilized," "civilized," and "enlightened." He meticulously placed countries into the categories based on the characteristics of their inhabitants, and as to be expected, his assessment of his own country was that it fell into the "enlightened" column. In fact, most of the regions lying outside of Europe and North America, according to Smith, were "barbarous." More than 60 years later, Hendrick Van Loon published *Van Loon's Geography: The Story of the World We Live In*, a book widely distributed in American schools and libraries. Van Loon's portrayal of non-Western cultures was strongly ethnocentric, nationalistic, and, in some cases, overtly insulting, but his writing shaped the perspectives on the rest of the world for a generation of Americans. Arnold Toynbee, an imminent historian who had an enormous influence on social sciences outside his immediate discipline, including cultural geography, argued that regions where the climate or environment present few challenges had little incentive to develop "civilization." According to Toynbee, "the greater the ease of the environment, the weaker the stimulus towards civilization." The tropical latitudes, in Toynbee's analysis, were regions that lacked such stimulation.

Indeed, even in the 20th century some of the most influential writers and thinkers in geography evinced an ethnocentric bent. Ellen Churchill Semple, one of the foremost proponents of the environmental determinist perspective, argued that residents of the highland regions of Europe lacked artistic talent, while those who lived in river valleys and lowlands were superior writers, artists, and intellectuals. But Semple failed to recognize that her analysis was laden with imposed and biased cultural and personal value judgments, including her own definition of "art" and "talent." Many other geographers of her time made similar errors when describing other peoples and places. A lesson that may be taken from the ethnocentrism of the past is that geography is a discipline that lends itself readily to oversimplification and prejudice, especially when evaluating other ethnic groups and cultures, and geographers must be aware of the inherent bias their observations may contain. It is unlikely that ethnocentric tendencies can be completely purged from scholarship, but their intrusiveness can be limited by a guarded and careful approach to other groups and their ways of life.

F

Federation

A means of ordering political space, consisting of a cluster of states or territories with limited sovereignty, joined together by a central, or *federal*, governmental authority. A condition of federalism usually implies a sharing of power between the central authority and the constituent units. The federation is typically organized by a constitution or treaty that defines the relative assignment of political power. In general, responsibilities for national defense, treaty agreements with foreign powers, and the regulation of trade with other countries, along with other related tasks, fall to the federal government. The individual states are allowed to establish statutes governing both tort and criminal law, but this legislation cannot contravene law at the federal level, and in particular, cannot violate conditions and stipulations set forth in the constitution.

In general, the distinction between a federation and a *confederation* is that the political units that compose the latter are typically joined together on a voluntary basis to pursue mutual interests, or to increase their collective security against external threats. The Confederate States of America, otherwise known as the "South" during the American Civil War, was one such confederation. Slavery was the issue that triggered the crisis, but the war was actually fought over the question of whether the federation represented by the United States of America was a voluntary affiliation. In other words, did individual states have the authority to secede from the federation, or were they compelled to remain a part of it? The question was settled only after the deaths of more than 600,000 Americans.

The actual balance of authority between the states and federal authority is rarely so starkly resolved, however, and the matter of states' rights versus federal power and even which level of government has jurisdiction is one that is frequently in question, and is often adjudicated in the federal courts, in many cases in front of the U.S. Supreme Court. For example, at the time of this writing (2011), two major controversial issues with constitutional ramifications have appeared on the American political landscape and, once they are resolved, will likely affect the balance of power in the federation. These issues are addressed in the Patient Protection and Affordable Care Act (PPACA), referred to in the media as "Obamacare"; and the matter of illegal **migration** into the southwestern states of the United States. In the case of the PPACA, the statute was passed into law at the federal level in

March 2010, but by the end of the year 18 states had filed or joined lawsuits challenging the authority of the federal government to enforce many of the provisions of the law, claiming that the law violates the sovereignty of the states. Also in 2010, the state of Arizona passed a controversial law allowing the detention of illegal immigrants by local law enforcement. In this case, the federal government has sued the state of Arizona, arguing that the state has no authority to pass and enforce laws governing immigration, which lies in the realm of federal authority. Both of these cases will likely be resolved by the Supreme Court, illustrating that in most federations, the power dynamic between centralized government and individual territorial entities is often a matter of both debate and litigation.

Folding and Faulting

Plate Tectonics is explained later. It is clear by all evidence that Earth's crust is not permanently configured. Instead, the surface is moving with a shuffling of continents and creation and destruction of the crust over millions upon millions of years. Within the panoply of effects resulting from plate tectonics is the striking evidence that "solid rock" is not everlasting in its structure or position. Movements within the crust are usually slow and hardly consistent. With careful study of **geomorphology**, some localized evidence of crustal rearrangement can be related to larger scales while others seem to have very local sources. The crust is capable of various contortions and two of the most important are folding and faulting where the former refers to rocks bent from their original position and the latter refers to rocks that have broken. Both folding and faulting are difficult to comprehend from usual human experience. Without machines, a human cannot apply the correct pressure over a long enough time to fold rock. Similarly, humans perceive rocks as hard substances that can be made to break if force is sharply applied with a tool or an explosive. The natural processes apply huge forces over vast time and rocks are quite capable of being bent or broken under these conditions. The term diastrophism applies to this rearrangement of solid crust. Whereas some diastrophism can be explained by intrusion of molten material to cause surface distortion, it is evident that most is not.

All sorts of forces might be applied in three dimensions to rock. Stress is force applied over area. Stresses can be uniform from all directions and the stressed rock will not permanently deform. However, there is often differential stress applied from different directions and differential stress from a single direction. Three types of stress are delineated. The first is shear stress in which forces are simultaneously applied from varying directions. The second is tensional stress that pulls in opposite directions and lengthens a section of crust. The third type of stress is compressional stress that shortens a section of crust. In reacting to stress, rocks might undergo elastic deformation, which is reversible if the stress is removed. However, with the long-lasting stresses experienced within the crust the rocks commonly undergo plastic deformation in which they bend or crack and are so changed permanently. It is the interplay between rock type and amount and direction of the stresses that governs the result and the stress is applied slowly enough not to break the rock.

Several different types of folds have been distinguished. Sedimentary rock is always laid down horizontally, and it is crustal stress that brings it out of horizontal and this angular displacement is easily observed in nature. The stresses folding rocks are compressional. Individual folds can vary in scale from tens of kilometers down to a very few centimeters.

A simple, common fold connecting horizontal layers is a monocline. A second type of fold is a syncline. It is a concave compressional downfold having symmetric limbs extending upward from the center. Sedimentary rock layers are progressively younger with distance through the limbs toward the surface. An anticline is a third type of fold. It is a convex compressional upfold of rock with a horizontal geometry similar to an arch. Older rocks are found with increasing depth through the limbs and anticlines served as traps for enormous reservoirs of oil and natural gas. A fourth kind of fold is an overturned fold, where the axis has been pushed past horizontal because of greater compression stress in one horizontal direction than the other. A fifth kind of fold is a recumbent fold, where the forces have pushed the center of the fold from vertical to horizontal. A fifth type of fold is an overthrust fold, which is much like a recumbent fold except the extreme folding has caused the rocks to fault and one side rides up over the other side, sometimes for long distances. This produces the situation where one could drill a well and go through a sequence of rock, pass through the fault, and penetrate the same sequence of rock further down. Notable overthrusts are located in Glacier National Park USA (80 km of overthrust), the Glarus region of eastern Switzerland (35 km), and in the Himalayas (100 km).

There are myriad examples of major folding around the world. The ridges and valleys of the Allegheny Mountains in the eastern United States have hundreds of thousands of square kilometers of landscape created by folding. Other famous examples occur in the Andes, Himalayas, and Urals. Far from having simple, symmetrical folds, many areas have incredible complexity with several types of folds and faults found in association with each other.

When the stresses are greater than the internal strength of the rock, the rock fractures and these fractures are known as faults. A fault is the displacement of rocks that were connected to one another. The displacement is along a fault plane that might be oriented horizontally, vertically, or in between. Faults are of various sizes and observed displacements can range from a few centimeters to hundreds of kilometers. The landscape rearrangement might be minor and noticeable only by close examination of a road cut or so major that a range of mountains is formed. In some faults the motion is imperceptible, and in others, displacement of tens of meters can happen virtually instantaneously, in **earthquakes**. Displacements seldom occur at even rates over any amounts of time and, in many places, can be thought of moving in "fits and starts."

There are many varieties of faults, but they are generally classified into four main types. The first is the strike-slip fault. This is the situation in which the fault plane is nearly vertical and the main forces are those of horizontal shear so that the sides of the fault slide past each other and there is little vertical displacement. The largest of this type are the transform faults separating crustal plates. The San Andreas Fault of California is a prime example and its lateral displacement has shown to be over 300 km. A second fault type is a normal fault. Tensional forces stretch and break the rock along a mainly vertical fault plane and one side of the fault is upthrown compared to the other. Vertical movement along some very old faults has been shown to be over 15 km but, of course, the topographic expression of this displacement is much softened by erosional forces. The reverse fault is the third type and results from compressional forces displacing the one side of the fault up and somewhat over the other side. The fourth fault type is the overthrust characterized by compressional forces along a low angle fault plane over long distances described above as resulting from folding.

Landscapes of faulting can be quite dramatic. In the European Alps and the American West are complex mountain ranges caused by faulting. When tensional forces pull apart the crust, large blocks of many square kilometers and surrounded by normal faults can be thrust upward or downward considerable elevations. The former situation makes horsts (fault mountains) and the latter grabens (fault valleys). The altitude difference between the bottom of Death Valley, California, United States (graben), and the abutting Panamint Range (horst) is somewhat over 3 km. In larger form this also appears as the Rift Valley system in eastern Africa.

Folk Culture

A folk culture is composed of a traditional, nonurbanized group, who adhere strongly to an established pattern of life with few cultural intrusions or innovations from outside. Frequently, rules and rituals are in place in folk cultures that prohibit adoption of many aspects of modernization. Members of folk cultures are often self-sufficient, making many of the items required for their work and pleasure themselves by hand, so unique crafts are often characteristic of such groups. In some folk cultures, it is not necessarily the *use* of modern conveniences that is forbidden, but the *location* of such conveniences is restricted. For example, one of the most studied folk cultures in the United States is the Amish, or "plain people," as they refer to themselves. In many Amish sects, telephones in the home are forbidden, but the use of a telephone, especially for outgoing calls, is not. Thus in some Amish communities, it is common to see a rather incongruous telephone booth located on the corner of a country lane, several hundred yards from the nearest home. Members of the community who wish to make a call may use this telephone, but rarely are incoming calls received. Likewise, most Amish are prohibited from owning an automobile, but are allowed to travel in motorized vehicles when necessary.

Folk cultures may be regionally defined. Such regions may be identified on the basis of material objects, such as architectural styles, farming techniques or practices, cuisine type, or other qualities that distinguish a group. For example, in western Ukraine several folk regions may be defined based on their traditional costumes, handicrafts, dialect, and other features. The Hutsuls, Lemkos, and Boykos are all mountain-dwelling peoples who produce unique clothing, and the traditional decorated eggs known as *pysanky*. The areas each of these groups occupies can be defined based on the differences in the styles of these cultural attributes, as well as others. In the United States, various folk regions may be delineated on the basis of a characteristic as specific as agricultural barn architecture—traditional barns in the upper Midwest, where many Scandinavians and Germans settled, are quite distinctive when compared to barns in the South or southern California, where other ethnic groups dominated the cultural landscape. Nonmaterial folk culture consists of music, dialects, folk tales, and other intangible manifestations of culture. Although geographers who study folk culture distinguish between material and nonmaterial folk culture, in reality both are typically found together. Indeed, material folk regions often spatially correlate closely with the **linguistic geography** of a location. Folk tales, legends, and folk beliefs are also a central part of nonmaterial folk culture, and these too may be correlated to a specific folk region. The Ozarks of southern Missouri, eastern Oklahoma, and northern Arkansas represent a distinctive folk region in the central United States, and one expression of the region's nonmaterial folk culture are numerous unique folk tales and legends. Music is another important form of folk culture. Certain styles of folk music are associated with specific regions. Bluegrass music, for example, although now an internationally recognized genre of music, originated from so-called hillbilly music played and sung mostly by Scots-Irish settlers in the Appalachian highlands of the eastern United States. Bluegrass music also represents material folk culture, because it features distinctive folk instruments like the dulcimer and banjo. This style of music also provides a good example of a component of folk culture that has entered popular culture. Once confined to the American southeast, bluegrass music is now played internationally and has a large following in Europe and elsewhere.

The popularization of bluegrass music shows that folk culture, like all other aspects of culture, has the potential to spread to other regions via the process of **cultural diffusion**. This happens frequently in the case of unique folk cuisines, because styles of food can be transplanted rapidly to other locations via restaurants that specialize in a specific kind of cuisine. Some cultural geographers study the spatial distribution of beverages and food styles by examining **foodways**, although this does not always include folk culture. An example is the diffusion of "soul food," a cuisine that is unique to African American folk culture, and which originated in the American South. This style of food is distinctive for the ingredients used, especially vegetables and greens that are not widely used outside the American South such as pokeweed and collard greens; and the types of meats employed, such as chitlins and beef or pork tongue. Over the past several decades, soul food cuisine has diffused across the United States and one may find several soul food restaurants in any large American city. Likewise, the popularity of "Tex-Mex" cooking, a folk cuisine originally found only in the American southwest that combines preparation and ingredients from both traditional Mexican cuisine and white southern traditional recipes, has diffused into the popular cuisine of American culture. Tex-Mex differs from traditional Mexican cuisine in its heavy use of cheese in many dishes and grilled preparation. The several varieties of *fajitas*, a dish made by grilling thin slices of meat with chopped peppers and onions, and then serving with flour tortillas is a classic example of Tex-Mex cooking that has diffused globally. Music, cuisine, or any aspect of folk culture has the potential to diffuse and subsequently to enter popular culture.

Those who live in folk cultures are usually farmers, ranchers, or others who make a living from the land they occupy. This close connection with the local environment sometimes is illustrated in the emergence of aspects of folk culture like folk medicine, or in other variations where the surrounding physical geography plays an important role in the culture. Folk medicine utilizes treatments and methods supplied by the local environment, usually employing plants and animals that folk healers identify as beneficial. Even in many regions where modern, technological medical care is available, folk methods remain popular and widely used. In the Appalachian highlands of the eastern United States many traditional cures are still employed by rural residents. These include tea or broth made from the sassafras root for a cold or sore throat, wild ginger root for a variety of ailments, sumac as a treatment for allergies, and many others. Techniques common to folk medicine systems include bleeding the patient, either directly or through the use of leeches, an approach common in many folk medicine systems; the use of compresses containing herbs or other substances to heal wounds or broken bones; or the application of pressure to certain parts of the body to relieve pain. The best known example of the latter is acupuncture, a traditional form of medical treatment derived in China centuries ago that now is used globally by thousands of practitioners. Geophagy, or the eating of earth, is also a form of folk medicine that is a custom found in several widely dispersed parts of the world. Those who engage in geophagy believe that the practice will result in improved health, and it may be done out of religious motivation as well.

Folk culture may be connected to the environment in ways that go beyond the use of folk medicine. In some sections of the United States *dowsing* or *water witching* is a common practice. This activity requires an experienced individual who uses a specially prepared stick, often in the form of a wishbone, to identify where an underground source of water may be found and accessed. Dowsers, or those who claim skill in this practice, are still in demand in some arid western states, where land may be uninhabitable without a functioning well or other water supply. In Chinese folk culture, geomancy, or *feng shui*, a complex system that takes into account local topography along with astrological features and other factors, is used to situate structures on the landscape. Feng shui has been a part of Chinese folk culture for many centuries and is deeply incorporated into social belief and behavior. Many residents of China and Hong Kong consult a specialist in this art before starting construction on a new home or building, and some

Geomancy

For centuries, the Chinese have believed that mythological forces are present in the **landscape**, or in other spaces occupied by humans. It is vital that these forces, often taking the form of beasts, be placated when any changes are made to the local geography, including constructing a new building, or even arranging the furniture in one's home. In Mandarin, the art of determining the location of these influences, and how to best appease them, is called feng shui. Many residents of China take this part of their folk culture quite seriously, and will hire a master of feng shui to evaluate the natural forces in their surroundings before making any significant changes. A feng shui master might be called in to advise someone on the best location for a new home, for example, or on the proper design for the structure. In some cases, an individual or family that is experiencing what appears to be unusual misfortune will consult an expert on feng shui to determine if by chance they have offended or disturbed the forces connected to their property. In traditional Chinese culture, the proper orientation and arrangement of one's living space is vital to success and harmony with nature.

employ a feng shui expert even in the placement of furniture and use of the floor plan within a home. To the cultural geographer studying these behaviors, it is not important that they lack a scientific basis. Rather, what is of interest is the fact that so many people continue to have faith that such practices are beneficial to their lives. In the age of **globalization**, folk cultures often struggle to maintain their identity in the face of assimilation by popular culture, and the traditions and values of folk cultures may appear quaint or old fashioned to those outside the group. Yet these cultures remind us of the character and context of the human experience, and offer a spectrum of perspectives on society and the place of humanity in the world that makes "modern" culture seem uniform and uninspired. By studying and cataloging these endangered folk cultures, cultural geographers can help preserve and promote the richness of human cultural landscapes.

Foodways

Foodways is a term that has numerous connotations in modern scholarship. In general, the concept of foodways identifies specific types of food associated with certain cultural, ethnic or regional groups, and the manner in which such food shapes the lives and identity of the people who prepare and consume it. Foodways are often seen therefore as a component of **folk culture**, and food as much more than just a source of nutrition and energy. Rather, the qualities of food represent a distinct cultural element that contributes to an ethnic, regional, or **cultural identity**. In the study of foodways, geographers are primarily interested in the spatial variations of the way food is produced, prepared, and consumed, and how these in combination contribute to the notion of place. The cultural geographers Pete and Barbara Shortridge highlight this spatial approach to foodways in their book, *The Taste of American Place: A Reader on Regional and Ethnic Foods*. As they note in the introduction to this work, "Geographers also integrate information on food habits with other aspects of cultural variation to create a more complete profile of the people who live in a given area."

Patterns of food production and consumption (and for that matter, beverage production as well) clearly have a spatial dimension that may be expressed at almost any **scale**. It is not necessarily the case that regions may be distinguished on the basis of *distinctive* kinds of food or food preparation—spatial differentiation may occur even within a specific food group or type. Those who are familiar with these distinctions are able to immediately identify the origin of the food item, or at least the origin of those who prepared it. In Uzbekistan, *non*, or flatbread, frequently shows such distinctions. The bread is traditionally baked in an outdoor oven called a *tandir*, but local cooks often prepare the bread using techniques unique to the local area. For example, the kind of bread produced in the city of Samarkand is thicker than other variations in the region, and the top of the bread is elaborately decorated with abstract patterns using indentations and sesame seeds. Other types of bread are unique to cities in the Fergana Valley, western Uzbekistan, and other locations in the country. The same regionalization occurs with the Uzbek national dish, *plov*. The basic ingredients of plov are rice topped with chunks of beef or mutton, but various vegetables or spices are sliced or chopped and included with the rice, and these are distinctive to a specific region or even a particular city. If one is served plov that contains chopped quince, for example, it is a certainty that the chef is from the city of Dzhizak in central Uzbekistan, as this recipe is unique to the residents of that city.

Ethnicity is frequently connected to foodways. Various groups have specific foods or dishes that identify them, or perhaps possess certain cultural or religious restrictions pertaining to food that serve as distinguishing characteristics. In some cases it is not the nature of the food item itself that carries the greatest importance, but rather how it is prepared or served. Such dietary codes may be quite extensive and complex, as in the case of the food restrictions among the Jewish people commonly known as *kosher*. The rules governing what is determined to be kosher (permitted) and what is forbidden for consumption are primarily derived from Jewish scripture, but there is not universal agreement even among the Jewish people on what foods are kosher and which are forbidden, as Orthodox Jews view some foods as non-kosher that other Jewish groups consider acceptable. Generally, not only are certain foods considered "unclean" and therefore unfit for consumption by many devout Jews, but the manner in which some foods are prepared is also quite important. Some examples are that meat and dairy products should not be prepared together in the same dish, and that fish and meat must not be eaten together at the same meal. Some foods or ways of preparing foods are prohibited only during specific times. Probably the best known example of such a restriction is the rule against consumption of leavened bread during the period known as Passover. *Matzo*, a type of flat unleavened cracker, is commonly eaten during the Passover holiday. At other times of the year leavened bread is an acceptable part of the diet of the Jewish people.

Religious dietary codes are not the only manner in which foodways can be associated with ethnicity. In some cases the two are connected through historical development, as certain foods became a common part of the diet of a minority group, especially those groups that were discriminated against or socially underprivileged. A case in point is *soul food*, a cuisine style typically tied to African Americans in the United States. Soul foods are a composite of Native American, African, and Southern food items and practices, and originated as traditional foods among slaves and poor whites in the "Deep South" of the United States. Corn (maize) was a staple among Native American groups in the southeastern United States, and many items of soul food are made from this grain, such as grits and cornbread. Meat protein was often limited in the diet of enslaved blacks in the southern United States, and this is reflected in an emphasis on sources of vegetable proteins like black-eyed peas, and on the use of many animal parts that whites typically eschewed, such as entrails, organs, or fatty tissue. Some well-known examples are chitlins, the intestines of pigs; and fatback, a layer of hard fat taken from the upper back of swine. When pigs were slaughtered by slave owners, the better cuts of meat were reserved for whites, while the intestines, fatty cuts, and other less desirable portions were given to the slaves. Out of this discriminatory behavior the uniquely ethnic cuisine of soul food evolved, and as blacks migrated to other regions in the United States, their foodways traveled with them. Soul food restaurants may now be found in any large American city.

Regions may also become tied to foodways. This may initially be due to a specific ethnic group that occupies the region, but over time the region itself becomes associated with distinctive food types, spices, and cooking techniques. Tex-Mex, a style of cooking that distinctively combines traditional Mexican spices and techniques with ingredients from Anglo cuisine, is a good example. Cattle ranching in the northern Mexico and southern Texas regions led to the emergence of a local cuisine characterized by beef preparations flavored with spices and frequently broiled or fried. *Fajitas*, now a popular "Mexican" dish across North America, is in fact part of Tex-Mex cooking. A popular snack, nachos and cheese, is also derived directly from the Tex-Mex tradition. In this dish, the use of melted cheese is an Anglo (or "Tex") contribution, and its combination with chilies or peppers for flavor is derived from the Mexican tradition. Cultural geographer Daniel Arreola in his book *Tejano South Texas* identifies regions in Texas where the same food may be identified by different names, based on the influence of Tex-Mex cuisine. A second example is the Creole/Cajun cuisine of Louisiana in the United States, which developed over two centuries, based on a mixture of unique local ingredients, traditional French and Spanish cooking, and special spicing techniques. In southern Louisiana, the traditional cuisine remains highly variegated, and one may encounter numerous local specialties in almost any town in the region, depending on local preferences and techniques. Even a common dish like gumbo, a stew usually made with seafood and sausage, will often be prepared and flavored differently, depending on the location.

In the era of **globalization**, some foodways have become internationalized. A classic example is the hamburger, originally a local preparation from the city of Hamburg in northern Germany called "hamburger steak." In the United States in the 19th century this cut of meat evolved into a ground beef patty placed between two slices of bread. Although there is great controversy and conflicting claims over where the American hamburger was first served, it is a food item that now indelibly belongs to "American" cuisine. Since the 1960s, the hamburger as a common meal has spread to every corner of the world, primarily via American fast food chains. But it is not just American food that has been globalized; the Japanese rice delicacy called *sushi* has also spread internationally over the past 20 years, and has become a popular alternative to fast food even in the United States. The great diversity of foodways reminds us that food and the ways it is prepared vary geographically, and it is these spatial patterns that give the peoples and cultures who produce and consume it a unique character that enhances and enriches the human experience.

Forests

Forest is, most technically, a term naming a particular floristic association on **Earth**. A forest is an extensive area or region dominated by trees and their canopy of leaves that shade Earth's surface. Usually, the trees are accompanied by bushes and other plants of lesser mass. One such definition cannot do the term justice because many, many definitions have been offered. This is a testament to the importance forests have held for various purposes. Of greatest importance is that trees are the sole provider of wood—cellulose fibers embedded in lignin—which has great strength and can be used in a myriad of applications.

A tree is usually the largest single piece of biomass in the vicinity and quite obvious to the viewer. Less obvious is the solar energy stored during photosynthesis, the water and nutrients taken up from the soil, and the web of plant and animal life predicated on the existence of the tree. How did much of the planet's surface come to be covered with trees? Land life came from the ocean somewhat over a half a billion years ago with land plants becoming common about 460 million years ago in the Ordovician Period. Trees appeared and began to cover the land some 370 million years ago toward the end of the Devonian period. The evolution of trees included development of their ability to help alter the upper rock crust of Earth into soils and allowing the further evolution of more tree species, other plants, and animals. Early trees were gymnosperms (conifers), carrying their seeds in cones; it was this type of tree that dominated the first forests. The angiosperm trees have been the more prolific over the last 50 million years. The angiosperms dominate with a now-diminished number of gymnosperm species. Approximately one hundred thousand tree species are alive on Earth with many more having been lost during our evolutionary history.

Forests are widespread over the tropical and middle latitude regions of the globe. Humans have actively utilized and cleared this resource causing marked lessening of areal coverage in the last few hundred years and a striking decrease in the last few decades (see **Deforestation**). The net effect is that global forest cover has decreased by about one-fifth since prehistoric times. As is true with other physical elements of Earth, forest cover is unequally distributed. According to the Food and Agricultural Organization of the United Nation, about two-thirds of Earth's forest cover is split among ten countries, with the largest amount of forest in the Russian Federation. The forested area of Earth is roughly 38 million sq km (14.7 million sq mi). This constitutes about 30 percent of Earth's land area. About half the forests are in the tropics, a third in upper middle latitudes and lower polar latitudes of the Northern Hemisphere, a tenth in temperate climates, and the rest in the subtropics.

Forests require a growing season long enough to add significant biomass via photosynthesis and conditions in which precipitation is greater than evapotranspiration to ensure a reasonable supply of soil moisture. Forests were able to evolve in various climates so there are various types and subtypes of forests. At the greatest level of abstraction, three forest types are usually delineated: tropical, temperature, and boreal.

Tropical forests are associated with the wetter tropical places. The tropical forest has some subtypes. The *rainforest* has the most superlatives attached to it. Found associated with the rainforest climate, this subtype is associated with the greatest number of species (both trees and other life), the tallest species (some 50 m and more), and the greatest biomass per hectare on Earth. Some rainforests receive more than 7,500 mm (250 in) of rain per year; there is no dry season. The trees are evergreen, losing and gaining leaves gradually rather than in anticipation of seasonal dryness or freezing temperatures. The *monsoon forest* (moist/dry deciduous forest) has a considerable dry season in the low-sun part of the year, huge precipitation totals in the high-sun period, and no freezing temperatures. These tropical trees are broadleaf species and seasonally drop their leaves in anticipation of winter dryness. The *tropical deciduous forest* is found in climates in which the precipitation totals are not as great as in the first two subtypes and the drier season longer. These forests are sometimes known as semi-evergreen forests because the taller trees are deciduous while the understory trees are evergreen.

Temperate forests are found in central and western Europe, the eastern half of North America, northeastern Asia, and in middle latitudes along the west coasts of continents. Growing seasons are up to six months long with annual precipitation greater than evapotranspiration. Winter and a concomitant lowering/stoppage of photosynthesis are limitations to growth. Temperate types are delineated primarily on the basis of seasonality and amount of precipitation. Mountains and high

In places where there are long growing seasons and adequate water, forests are the most common vegetated landscape. There are many varieties of forest; this one is a temperate rainforest near Ketchikan, Alaska. (Photo courtesy of Steve Stadler)

plateaus (in much of the American West, for instance) produce *dry conifer forests* with relatively low average annual precipitation. *Mediterranean forests* have peaking of precipitation during mild winters and droughty summers. Broadleaf trees such as small oaks and some evergreen, thick-leaved, water-conserving species are common with winter being the "green" season. Many of the trees are virtually dormant in parts of the summer. In selected areas on upper middle latitude coasts (for example, Washington state's Olympic peninsula) are *temperate rainforests* that have in excess of 2,000 mm of precipitation per year (much less than tropical rainforest but much more than any other temperate forest). Infrequent freezes and a year-round excess of precipitation over evapotranspiration make this type fairly diverse and high in biomass. *Deciduous forests* have rainy summers and cold, snowy winters for which the trees prepare by losing their leaves. This kind of forest, topped with tall trees, is ecologically diverse and covers the greatest area of all the temperate forest subtypes.

The boreal forest (taiga) represents the largest forest on Earth. Dominating the northern continental surfaces from 50° to 60° latitude (there is little land at accordant latitudes in the Southern Hemisphere), the boreal forest has developed in climates that have short, cool summers and long, cold winters. Growing seasons average about four months so that accumulated biomass is low. Here, there is a relative lack of numbers of species and ecological diversification compared to the tropical forests. Soils tend to be poor and acidic (spodosols). The coniferous trees are needle-leafed and close enough together to prevent full solar penetration to the ground and this limits the biomass of understory plants and animals.

Friction of Distance

A spatial concept that relates distance to interaction, and a key concept in the **geography of economic development**. Interaction in this case may take many forms, such as economic exchange or transfer of cultural attributes. Simply put, the greater the distance involved the greater the resistance to movement across space, resulting in a lower degree of contact. A closely related geographical notion is **distance decay**. The notion of friction of distance plays a role in a number of geographical theories, including the **von Thunen model**. Friction of distance affects an array of processes: **cultural diffusion, migration**, trade, and many others show the correlation between increased distance and a diminution of activity. This typically happens because both actual costs and time invested increase as the space between two points grows, and thus the incentive to interact is inversely proportional to increases in distance. The degree of "friction" may be influenced by several factors, including physical barriers (mountain ranges, deserts, bodies of water), political **boundaries**, and the sophistication of the communications and transportation systems involved.

In societies that enjoy highly mechanized transport opportunities, for example, the friction of distance is generally reduced in comparison to those in which transport is not technologically as advanced, because travel by automobile, train or aircraft will efficiently cover more distance than travel on foot or on the back of an animal. Many people in **rural settlements** in developing **regions** never travel more than 50 miles from their place of birth, while it is common for residents of economically advanced countries to cover this distance on a daily basis as they commute to and from their place of work. Even in regions with well-developed transport systems, however, political boundaries can greatly increase the friction of distance by increasing costs of transport, both in terms of time (time spent for customs inspections, etc.) and actual costs (tariffs on imported goods).

In practical terms, the *type* of exchange may affect the degree of friction between two **locations**. In an economic exchange, low cost goods will have a high degree of friction over distance. Consumers will not travel more than a short distance to secure common staples like bread and milk, as the cost and character of these items creates a great deal of friction beyond a relatively short distance, typically only a few miles. On the other hand, goods that are more expensive and rarely purchased, such as automobiles, have a much lower level of friction in proportion to distance. A customer may travel a considerable ways to procure such high-value items, because the additional investments in travel cost and time are viewed as acceptable due to the qualities of the good being sought. For trade, migration, or tourism to occur between two points, transport systems must overcome the friction of distance. By increasing the efficiency of movement between locations, the expense and time of transport are lowered, reducing the friction as well.

Fronts

A front is the boundary between air masses and a linear zone of low pressure that can extend for thousands of kilometers. Fronts usually represent sharp horizontal divisions between weather types as they move at varying speeds over the **landscape**. During frontal passage, a weather observation station will experience changes in air **temperature**, air pressure, **humidity**, cloudiness, wind speed, wind direction, and might receive **precipitation**. How different must **air masses** be for a front to be defined? The answer is "not much." Sometimes, there are only small differences in temperature and moisture on either side of a front. How cold must it be to define an air mass change as a cold front? The answer is "not very." Summer temperatures might be in excess of 38°C before frontal passage and a toasty 35°C after the front—there is no absolute rule for the temperatures involved on either side of the fronts explained below.

The concept of fronts is attributable to Norwegian meteorologists who developed the concept of air masses and **middle latitude cyclone** theory. They recognized that air mass boundaries were the locations of the most vigorous weather activity and made the analogy between these boundaries and the military fronts of recently completed World War I. Essentially, air masses were likened to opposing armies engaged in trench warfare. At various times the military fronts were stationary and, at other times, one army gained the upper hand and pushed the enemy out of the way. Air masses, energized by the push of the strong flows of air in the middle and upper troposphere, are able to mechanically force other air masses out of the way. In the case of the atmosphere of the middle latitudes, the air masses "fight" back and forth over time because the polar front jet stream is arranged in a series of immense moving waves. The jet stream separates warm, tropical air masses from cold, polar air masses in each hemisphere. Thus, no one air mass permanently holds a single position and daily weather is variable in most middle latitude locations.

Even though the width of a front occupies a very narrow distance in contact with Earth's surface, it is important that the frontal zone extends up several kilometers into the atmosphere. Air masses of different densities push on each other, so there is mechanical lift along the front aloft and this means that significant cloudiness and precipitation might extend far away from its surface position.

Weathering mapping has internationally standardized symbols indicating position of fronts and their directions of movement. A cold front is indicated by a series of triangles attached to a line (the frontal position) and pointing in the direction of motion of the advancing cold air. A cold front is the situation where cold air is pushing on the warm air. Because cold air is denser than warm air, cold air is able to wedge in under the warm air and the warm air is forced to rise. A cold front extends into the air, a typical rise over run is 1:50; that is, for each 50 m along Earth's surface, the front is another meter away from the surface. Compared to other frontal types, cold fronts are usually faster (45 to 75 kph) and have steeper rises over runs resulting in more dramatic uplifts of air, intense precipitation, and severe weather. As a cold front approaches a weather station in the Northern Hemisphere, the typical weather would be warm with high dewpoint temperatures, lowering barometric pressure, winds from the southwest or west, and showery precipitation in unstable air. As the front passes, the barometric pressure would steadily rise with temperatures cooling and the wind changing to northwest or west, and the atmosphere stabilizing. The heaviest precipitation would be close to the frontal zone and be over in a short amount of time.

A warm front represents the advance of warm air from lower right to upper left. Because the warm air is less dense than the air it is replacing, the warm air will overrun the cold air at the surface and rise. In this case, the surface speed of the front is typically 15 to 40 kph. A warm front extends into the air at a gentler ratio of 1:80. This means that there is rising air and precipitation considerably *ahead* of the surface position of the front. Precipitation tends to be steady and gentle for up to a day before the passage of the front at the surface. Additionally, as the warm front approaches, temperatures are cool, the barometric pressure drops, the winds are southeast to south, and low clouds are common. As the front passes, the air warms, the barometric pressure rises, the winds shift to the southwest or south, and precipitation ceases.

An occluded front usually results from the interaction of cold and warm fronts. Cold and warm fronts are attached to a common, large disturbance known as a middle latitude cyclone. Over a number of days, the cold front progresses more quickly than the warm front and all the warm air mass is raised from the surface.

The sharpness of surface air mass differences fuels middle latitude cyclones so occlusion represents the dying out of a cyclone. The expected weather with an occluded frontal passage is cool air followed by cold air with cloudy, showery weather. The occluded front lengthens by "zippering closed" the surface cold and warm fronts.

A dryline may be present in selected areas of the middle latitudes such as the southern Great Plains of the United States. The dryline is a front-like feature representing the boundary between hot, dry continental tropical (cT) air and warm, moist maritime tropical (mT) air. In some cases, the dryline tends to move in one direction during the day, and in the opposite direction at night over a several hundred kilometer span. In fact, in certain times of the year at some locations, the passage of the same dryline might occur for several days in a row. Wind converges from both sides of the dryline. The typical directions are from the southeast in the mT air side and from the southwest on the cT air side. The strongly converging air streams are forced to rise in the dryline. The mT air is so unstable that there are typically huge and sometimes tornadic thunderstorms in front of the daytime advance of the dryline. The cT air is much warmer and without clouds.

G

Gentrification

Gentrification is the process whereby urban neighborhoods are revitalized through the use of economic incentives, designed to induce residents or businesses to acquire degraded property and restore it. Examples of economic incentives are reduced property taxes, or even the complete suspension of such taxes for a set period of time; the sale of condemned property for a reduced price, with the agreement that such property will be improved and utilized for a specific purpose; or in the case of residential neighborhoods, the refurbishment of housing either through local government incentives or strictly by private investment. Gentrification not only has obvious economic effects on neighborhoods, but also brings substantial demographic and cultural changes to an area, and therefore is a controversial method of urban renewal.

Gentrification is a response to the decline of inner-city neighborhoods and districts. Beginning in the 1970s, cities found that they could attract relatively wealthy buyers to declining urban housing, especially if the housing had some historic or cultural interest, by offering the real estate at bargain prices. They were able to attach certain restrictions on the use of the property as well, and in this way were able to greatly increase property values. A municipal government or a local neighborhood association might buy houses that are in dilapidated condition at a low price, and then offer the property to qualified buyers, who must agree to improve the property's value to a certain level. The property may have conditions and covenants that restrict its use. For example, a common requirement for gentrified property is that it must be occupied by the property owner, he or she may not rent the property, and the property must be used solely for residential purposes. The great advantage of this process is that low-value housing is converted, with minimal investment on the part of the city government or local entity, into high-value housing that raises property values in the area (a major incentive for neighborhood associations and local residents to promote gentrification), and generates much greater revenue in the form of higher property taxes (a major incentive for local governments to promote gentrification). There are additional benefits as well. Gentrified areas typically attract more businesses, due to the more affluent consumer base that results from the addition of higher-income residents, which in turn results in greater revenue for the city in the form of increased sales taxes.

Many cities in North America and Europe now feature gentrified districts. Some of the better known are Georgetown in Washington, D.C., and Barnsbury in London. Those who have been most drawn to such neighborhoods are often so-called yuppies (young, upwardly mobile professionals), who typically have the financial means and motivation to purchase housing close to the center of the city, that also has the potential to be gentrified. In the United Kingdom government grants have assisted buyers in the process of acquiring housing, while in the United States gentrification has been fueled by local government incentives and favorable prices. There are several theories that attempt to explain the motivation behind gentrification. In the 1980s and 1990s, a so-called back to the city movement arose, which was in response to the negative aspects of suburban living: sprawl, long commutes to work, the uniformity and conformity of suburban housing, the expenses of suburban property, and other characteristics. Many who led the "back to the city" effort were encouraged by municipal urban renewal projects that combined gentrification strategies with investment in projects that integrated commercial and residential development. Those providing momentum to the "back to the city" approach are typically single professionals, married professionals without children, gay couples, and retirees. Some scholars ascribe an economic, rather than cultural or aesthetic rationale to gentrification. This explanation is often labeled the "rent gap hypothesis." According to the urban geographer Neil Smith and other proponents of this theory, the "rent gap" occurs in declining residential areas when the value of a substandard structure actually is below that of the vacant property. The difference in values is the "gap," and this differential is sufficient, according to the hypothesis, to induce outside investment. In the view of Smith, gentrification then represents not so much a shift in demographic and social patterns, as a shift of economic capital and investment patterns.

The phenomenon of gentrification cannot be explained by a single theory or a single factor, however. Chris Hamnett, a British geographer, has identified three factors that appear to be essential in fostering gentrification. First, potential gentrifiers must be present in an urban space. Such groups are often a by-product of the local economic structure: cities that have a large segment of the local economy in high-paying service and professional jobs typically have many residents who potentially may engage in gentrification. Second, gentrification requires a sufficient amount of property that may be gentrified. Such property is more commonly associated with former working-class neighborhoods in old industrial cities, meaning that cities that lack this history may offer little opportunity for gentrification. Phoenix, Salt Lake City, Dallas, and a few other cities in the western United States have fewer gentrified neighborhoods than New York, Chicago, Boston, and Philadelphia, for example. Third, a demand for residential property in the inner-city area must exist. This is a function of the local demographic, cultural, and economic environments.

An urban area that holds a large number of single professionals, married couples without children, and other groups who make up much of the average gentrifying community is likely to have several gentrified neighborhoods, if the first two criteria have been met.

Many social geographers have pointed out that gentrification, in addition to the economic benefits it brings, also has negative effects. One of the most obvious negative influences of gentrification is that the rise in property values that gentrification invariably generates often results in the original, lower-income residents of the neighborhood moving to another location, as they cannot afford higher rents and/or taxes. The displacement of residents, who in the United States are often African Americans or other ethnic minorities, carries racial overtones, and sometimes results in tension between the gentrifying residents and those who lived in the area prior to gentrification. Poorer residents, who are more likely to be renters than property owners, often find that their rents are steadily increasing as a result of gentrification in the local community. Moreover, local businesses may be driven out as well, as the new consumer base does not patronize stores that do not offer either the type or range of products they wish to buy. New businesses may locate to the gentrified area to take advantage of the business opportunities offered by the "gentry" now living there, and "Mom and Pop" style stores, small neighborhood restaurants, and other establishments may be forced out of business. Surrounding neighborhoods, even those lying outside the immediate zone of gentrification, may be adversely affected, as their property values, and therefore property taxes, may increase simply by being located adjacent to gentrified housing. In areas where many of the gentrifying groups are gay, conflict over social values with the local community may erupt, leading to friction. In spite of the drawbacks to gentrification, it appears highly likely that the trend will continue well into the future.

Geographic Information Systems

Cartography, the science of making **maps**, has long been the province of the geographer. Modern cartography has largely transitioned to the use of the computer. In the 1960s, computer mapping required mainframe computers to produce simple and poor-quality maps. Today, the advent of computing thousands of times faster with high-quality output available at the desktop allows manipulation and modeling of geographically referenced data into output maps that would be difficult if not impossible to achieve with manual cartography. This type of environment is known as geographic information systems (GIS), and its academic pursuit is sometimes called geographic information science. The power behind

GIS is the ability to conceptualize the geographic relationships between multiple data sets. GIS is very much to the point of answering the essential geographic question of "What is where?" Also, it can aid in exploring the relative location of features and model the likelihood of encountering a particular feature (e.g., suitable bird habitat) using a combination of spatial data layers.

The idea behind GIS is simple enough. Suppose a developer wishes to build a large group of single family houses on a large tract. It is known that some of the land is boggy, some on the floodplain, and some within areas a county has zoned for commercial development. The developer would like to avoid each of these types of areas in situating his/her houses. Whereas the developer could examine paper maps of each of these three patterns and attempt to associate them by flipping between the maps, GIS provides a rational manner of simultaneously excluding unsuitable areas and highlighting the suitable areas. Mathematically, it is a spatial version of Venn diagrams in which the three spatial layers are laid over the tract and the areas not exhibiting one of the three limitations are deemed suitable and automatically mapped.

The applications of GIS have become quite diverse. For instance, GIS has been applied to modeling of runoff and evapotranspiration, the solution of least-cost transportation problems, decisions as to which agricultural land is to be idled, the siting of wind farms, realignment of census boundaries, the exploration for oil and natural gas, city evacuation plans, location of new retail outlets, and construction of ecoregions. Most cities and agencies of state and federal government now have GIS operations. Thousands of private companies have followed along. The opportunity of employment using GIS has exponentially increased and has dramatically altered the employment landscape within the discipline of geography. There are many opportunities for programmers, but *geographers* are useful because of their knowledge of cartography plus their knowledge of the spatial patterns of the human and physical landscapes. This employability has changed undergraduate and graduate geographic education to incorporate formalized courses and degrees.

Although the premise of GIS seems simple, the implementation of GIS has some profound implications. The inputs to GIS are electronic, spatially referenced data sets. That is, the data set is divided into areas and each area is identified by coordinates so that all areas in the data set can be portrayed on a map. The areas (in raster or vector styles as defined below) are assigned one or more attributes. Examples of attributes are elevation, population, land parcels, type of vegetative cover, presence of a road, or average temperature. The type and number of attributes assigned are a function of the purpose of that particular GIS.

The design of any major project must be done with care. Choices made early on can limit the utility of the products and not readily rectified. The sources of GIS data

Geographic Information Systems (GIS) have multiple applications. Here a researcher helps build a GIS that will serve as an early warning system for tsunamis, potentially saving thousands of lives. (AFP/Getty Images)

layers include satellite data, paper maps, aerial photographs, electronic maps, and locations geopositioned by satellite (GPS). Many times, a GIS worker must build spatial datasets particular to a project. One common data input is the use of a digitizer to capture data from a paper map into electronic form with spatial coordinates. This sort of data layer, of course, can be only as good as the original map. Frequently, GIS incorporates GPS locations that have been collected by the data set builder.

Private companies have created many data sets using their own inputs meshed with government sources. These GIS projects are sometimes proprietary and other times can be purchased. Fortunately, there are many electronic data sets freely available for potential use in GIS analysis. In the United States examples include atmospheric, land cover, population, wildlife, and elevation. It must be noted, however, that much of the coverage of spatial data is uneven on a state-by-state basis.

No matter how the input data are constructed, careful consideration must be given to the traditional cartographic problems of **scale** and **projection**. These are both important in the final output. Yet, GIS work has a different spin in that the input data layers are from sources of varying scales and projections. The ideal situation is to have the data layers geographically matched. This process is time-consuming and all-important. The degree of accuracy of the output cannot be better than the accuracy of the least accurate data layer. Frequently, the GIS user must decide on the degree of inaccuracy acceptable for a project. For a project involving parcels of land, high-orders of accuracy (to millimeters if possible) might be warranted. For estimation of runoff from a large drainage basin, lesser accuracy might be acceptable because the soils data layer includes boundaries that are lines attempting to make sharp differentiation in what is usually a continuum of soil characteristics transitioning over many meters. The production of spatially accurate geographic data is not a trivial matter.

There are two types of geographically referenced data models attached to GIS attribute data. They are raster and vector data. Raster data are conceived of as in a regular grid of cells of definite size; sizes vary considerably among GIS projects. The cells usually range from meters to many kilometers on a side. They are useful with spatially continuous data such as elevation. For example, an elevation can be assigned as a cell attribute (the elevation could be a spatial average of all elevations in the cell or the elevation at the center of the cell). Vector data are composed of polygons, lines, and points that represent discrete objects and features. Today's vector data models include topology in which a polygon, line, or point is associated with its neighboring vectors. This allows spatial associations of adjacency and connectivity to be maintained no matter which scales or projections are used. In complex GIS work, the GIS analyst frequently makes conversions from raster to vector and vice versa and this involves conscious trade-offs of characteristics to obtain the necessary accuracy of results.

Geomorphology

Geomorphology is the science of the description, analysis, and the specification of processes causing landforms. The term was coined in the U.S. Geological Survey in the 1880s, although its scientific roots were much older because previous scientists had been studying the erosion of sculpting landscapes. The geographer William Morris Davis was a forceful early proponent of geomorphology and his considerable influence within geography helped it become the intellectual linchpin of the fledging discipline of geography in the early 20th century. Academically, geomorphology became part of both geography and geology curricula with some consternation among geographers and geologists continuing to this day. If one considers the scope of geography, geology, and geomorphology, it is clear that geomorphology occupies the nexus between the disciplines and contributes to both.

Nascent geomorphology has been around for two millennia. Classical Greek, Roman, and Arab scholars wrote numerous travel geographies that have survived.

William Morris Davis (1850–1934)

Davis was a giant in early academic geography in the United States. He is remembered as a geographer, geomorphologist, and a founder of the Association of American Geographers and the National Geographic Society. He is considered one of the fathers of modern American geography. Never finishing a PhD, Davis became a physical geography instructor at Harvard in 1884 as geography was emerging as a discipline. Davis' supreme insight was the application of evolution into the landscape to the effect that he was first to devise the concept of landscape evolution through development of drainage systems. First published in 1889, it is usually called the "geographic cycle" or "cycle of erosion." Starting with a flat surface above sea level, streams will erode into the landscape in a series of V-shaped valleys (youthful landscape). Stream erosion over eons will result in a landscape all in hill and valley (mature landscape). Finally, erosion will subdue the hills and the landscape will become a virtually flat surface known as a peneplain (landscape of old age). Today, this model has been superseded by equilibrium theory, which deals with forces and mass. Yet, Davis' contributions still stand as seminal to those trying to understand landscapes.

In these geographies there are plentiful descriptions of landforms and landscapes along the travel routes. In modern times this tradition has been expanded into descriptions of regional geomorphology.

In geographic terms, geomorphology answers "what is where" on the physical landscape and then "when" and "how." Early on, geomorphic inquiry was intellectually handcuffed by the notion that the landforms of Earth had all been impacted by the biblical flood of Noah. In the 1700s, scientists started to question a flood origin of landforms and began to use Hutton's concept of uniformitarianism that espoused series of processes acting over long times to create landforms. In that it uses uniformitarianism to deduce the past, geomorphology can also ascend to the pinnacle of science by providing forecasts of the evolution of future landscapes.

Since the 1960s, concepts of **plate tectonics** have revolutionized geomorphology. The realization that there are crustal plates and they move has caused an epiphany in geomorphology. While geomorphologists have long realized that parts of Earth's crust rise and fall, plate tectonics have provided a mega-overview that has been used to great advantage. Although not all the details are known, the parts that are certain are of immense scientific value. Why do most large volcanoes cluster around the rim of the Pacific Ocean? Why do oceanic ridges and trenches exist? Why are the oldest rocks exposed at Earth's surface younger than Earth by, perhaps, a half a billion years? These are question for which there are now firm answers.

It should be noted that while the tectonic forces inside Earth are a major key to understanding landforms, the planet's atmosphere has much input to the shape of the landscape. Most continental surface landforms are largely the result of destructive processes such as **weathering and mass wasting**, and erosion. Weathering loosens materials, mass wasting brings the materials downslope because of gravity, and erosion moves the materials away (sometimes far away) from their origins. There are constructional landforms—notably the floodplains built by stream flooding, deposition by glaciers, and deposition by wind—but the resulting landforms are not as widespread as the landforms of destruction.

Although previous scientists had made contributions to the corpus of knowledge of geomorphology, it was the geomorphologist Davis who summarized the science of geomorphology as rooted in structure, process, and time. This succinct characterization has proven useful for over a century.

Structure is a term applied to an inherent quality of rock. "Structure" implies several things. It refers to the minerals composing the rocks because minerals can be quite varied as to the amounts and nature of possible deformations. Although rock seems solid and immutable under normal circumstances, there is no doubt that deformation occurs over long times when considerable force is applied. Structure is concerned with the physical arrangement of rock masses and layers. Rocks can be warped, folded, or broken (see **Folding and faulting**). A horizontal set of rock layers, for instance, will produce a markedly different set of landforms than rock, which are arranged in anticlines. An examination of structure implies that the geomorphologist is querying about deformation of the rocks over past time.

"Process" refers to any of the multitudinous array of internal and external factors combining to produce the observed landscape. These include erosion and deposition by liquid water, ice, and wind. Also, physical and mechanical weathering and mass wasting are important players. Internally, the tectonic processes and related volcanism have observable impacts. Yet, these factors have definite geographies and can be more or less important in different regions. For instance, weathering by solution is a dominant process in precipitation-rich regions underlain by limestone; solution is responsible for the distinctive **karst** topography. However, limestone weathers more slowly with less solution in desert area so that karst is not very common.

"Time" is an exorable characteristic of the physical universe and immense amounts of time are available in which the processes can shape landforms. Some landscapes are rearranged in a matter of moments—the explosion of a stratovolcano for instance. Other landscapes are shaped bit by bit, obvious only to the keenest of observers. Geomorphologists have traditionally written about "stages" of landscape development in which sequences of events take place over thousands or millions of years. Davis, for instance, invoked the notion of peneplains, which are low-lying surfaces of modest topography developed over extremely long time sequences (presumably millions of years). Modern rock dating techniques developed in the last few decades have eased some time problems by fixing sequences of events in time. Still, it is a form of mental gymnastics to conceptualize landscapes evolving over times many multiples of the human life span.

Geopolitics

According to political geographer Saul Cohen, geopolitics "is defined as . . . the analysis of the interaction between . . . geographical settings and perspectives and . . . political processes." As a field of modern academic inquiry, geopolitics emerged from the era of **imperialism**, when European countries set about establishing political control over large sections of Africa, Latin America, Asia, Australia, and North America. In competition through the system of mercantilism, these imperialistic powers sought advantage over one another by controlling **straits, passages and canals** of the world's shipping lanes, acquiring **natural resources** to support large-**scale** industrialization, establishing monopolistic overseas markets, and building powerful militaries. Early modern geopolitical thinkers attempted to develop theories about the nature of the **nation-state** and the relationships between states. As new powers arose during the late colonial era (Germany, Japan, the United States, Russia) to challenge the supremacy of Great Britain and France, an emphasis on power and strategy tended to dominate the thinking and writing of scholars and policymakers. The strategic advantage some countries appeared to enjoy over others, and methods of obtaining such advantage, became a focus of geopolitical thinking. Geostrategic notions of buffer states, **shatterbelts**, and other concepts entered the lexicon of geopoliticians between the world wars, as scholars grappled with the changes wrought by the Treaty of Versailles, the collapse of the Ottoman Empire, and the emergence of the Soviet Union. After World War II, American foreign policy was profoundly shaped by geopolitics, especially by the strategy of containment and the "domino theory," both of which became the foundation for increased American involvement in Vietnam after 1954.

The nature of the state and the functioning of its institutions has been the subject of scholarly inquiry and theory for centuries. The roots of geopolitical investigation may be traced to ancient Greece. Plato is likely the first to offer an **organic theory** of the state, and his famous protégé, Aristotle, built further on the concept. The Muslim philosopher Ibn Kaldun, writing in the 1300s, also argued that cities and civilizations behave much like animate creatures, experiencing a cycle that takes them from "youth" to "old age" and eventually collapse. During the 19th century, this concept was revived by German geographers, who laid the foundation for modern geopolitics. Carl Ritter, a contemporary of Alexander von Humboldt, suggested that states experience a life cycle similar to living organisms, an argument that was further developed by Freidrich Ratzel, a scholar frequently labeled the "father of modern political geography." Ratzel is credited with first using the term *lebensraum* ["living room"] as applied to the growth stage of a state. Essentially, he proposed that states naturally seek to expand their geographic boundaries in competition with other states, resulting in "stronger" states enlarging at the expense of weaker neighbors. Although Ratzel sought to understand the dynamics of interstate relations, his theories provided the basis for Germany's pursuit of a colonial empire in the second half of the 19th century, as well as the aggressive policies of Adolf Hitler and the Nazi party. His ideas had a profound influence on later geopolitical theorists.

Ratzel's student, Rudolf Kjellen, is generally given credit for first using the term *geopolitics*. Kjellen was a Swedish scholar who carried the organic theory of the state to an extreme in his book *The State as an Organism*. In this work Kjellen identified five viscera of the state and held that the ideal condition for longevity was for the state to achieve "autarky," or complete economic self-sufficiency. States could typically achieve this condition only by acquiring resources through the conquest and absorption of other, weaker states. Kjellen's book, originally published in Swedish but almost immediately translated into German in 1917, had a profound impact on scholars in Germany.

Simultaneously with Ratzel and Kjellen, strategists outside of Germany were also beginning to think in geopolitical terms. In the United States, the most influential of these was Alfred Thayer Mahan, a naval officer and educator who argued persuasively that naval power was the key to world power, and that control of the global shipping lanes would provide a crucial advantage in time of war. Mahan's numerous writings not only influenced American policy and the expansion of a global naval presence, but he was widely translated and read in Germany and Japan, both of which were engaged in building colonial empires by the beginning of the 20th century. Although he never employed the term "geopolitics" in his books, papers, or lectures, Mahan nevertheless was a pioneering theorist who almost single-handedly initiated the study of global military strategy in the United States. But the most influential thinker among the geopolitical scholars of the late 19th century was Halford J. MacKinder, a professor and member of the British parliament. MacKinder appeared before the Royal Geographical Society in 1904, and in a famous lecture laid the foundation of the **Heartland theory**, a concept that influenced British and American foreign policy throughout much of the 20th century. MacKinder suggested that the key to global domination was controlling the resources and territory of the heart of Eurasia, a region encompassing Eastern Europe and the western reaches of Russia and Central Asia. Unlike

Mahan, MacKinder held that the domination of the Asian landmass was essential to geopolitical strategy, not sea power.

Germany's defeat in World War I spurred a reexamination among German scholars of power and policy issues. Central to this effort was a desire to explain why Germany had lost the war, and how the country could regain status as a "great power." Karl Haushofer, a professor and former military officer, emerged as the chief figure behind a resurgence of German interest in geopolitics. Haushofer founded the Institut fur Geopolitik [Institute for Geopolitics] at the University of Munich, established a journal devoted to the subject, and attracted a number of students, one of whom was Rudolf Hess, a close associate of Adolf Hitler. In his work Haushofer drew heavily on the organic theories of Ratzel and Kjellen and was familiar with the writings of both Mahan and MacKinder. Unfortunately the Munich School of geopolitics, as it came to be called, effectively became a propaganda outlet for German nationalism and aggression. Haushofer attempted to rationalize German expansion as "natural" and "scientific," and it is likely Hitler took some of his ideology from Haushofer's publications, especially the necessity of *lebensraum* as a requirement for Germany's advancement as a power. Hitler's articulation of lebensraum in *Mein Kampf*, his twisted philosophical treatise, closely parallels that promulgated by Haushofer. The apparent connections between the Munich School and Nazism damaged the reputation of geopolitics as an academic field of inquiry, but the aftermath of World War II and the threat of Soviet aggression once again forced both policymakers and scholars to confront the world in geopolitical terms.

Writing during the last years of the Second World War, the American scholar Nicholas Spykman reformulated MacKinder's heartland theory into the **Rimland theory**. Spykman agreed with MacKinder's basic concept of the need to control strategic territory, but argued that it was not the heartland that was vital, but rather what MacKinder had called the "inner crescent," and what Spykman labeled as the "Rimland," a region that surrounded the heartland. At the end of the war, the Rimland corresponded closely to the zone immediately adjacent to the USSR on the Eurasian landmass. Spykman suggested that holding this region was the key to limiting the power of a state occupying the heartland. This concept was most eloquently and forcefully put forth by George Kennan, a diplomat working in the U.S. State Department, in the journal *Foreign Affairs* in 1947. Although Kennan does not refer to Spykman in his article, his suggested policy of "containment" reflected the fundamentals of the Rimland theory and laid the groundwork for the Truman Doctrine and subsequent American foreign policy during the Cold War. The "domino theory," which motivated American policy in Southeast Asia in the 1960s, also may be partially traced to the Rimland theory and Spykman's writings. Spykman's concepts on securing a "balance of power" and forming alliances among Rimland countries to restrict the influence of the heartland have also been echoed in the international strategy of the United States since 1945, and continue to inform American foreign policy.

The end of the Cold War has not meant a reduction in the relevance of geopolitics; if anything, the collapse of Soviet communism and the rise of radical Islam has magnified the importance of the field. Influential geopolitical commentators include Jakub Grygiel, the late Samuel Huntington, Robert Kaplan, Fareed Zakaria, Zbigniew Brzezinski, and many others. Within the discipline of geography, some scholars have promoted the rise of a "critical geopolitics" that challenges the foundations of what its proponents designate as "classical geopolitics." Critical geopolitics is rooted in post-structuralist and postmodern theory, and claims to offer a perspective on geopolitics that avoids the "binary" and "post-colonial" distortions toward interstate relationships that characterize the "classical" approach. The leading advocates of the "critical" school are John Agnew and Gerald Toal.

Gerrymandering

Gerrymandering is the deliberate setting of political **boundaries** to provide unfair electoral advantage. The resulting political district is said to be a "gerrymander," or to be "gerrymandered." The term originated in the United States in the early 1800s, and is formed partially from the surname of the governor of Massachusetts, Elbridge Gerry. In 1812, Gerry presided over a session of the state legislature that was charged with re-drawing the state's congressional districts. In one specific instance, the borders of the resulting district were so convoluted that a political cartoonist for the *Boston Weekly Messenger*, a local newspaper, represented the region as an animal. According to legend, he described it as a salamander, but his boss remarked that it was more appropriately called a "gerrymander." The newspaper used the term in an editorial, and it soon entered the political vocabulary of the United States. The United States employs a system of proportional representation for seats in the House of Representatives, and individual states are given the authority to redistrict after every census is conducted. The party in control of the legislature therefore has the ability to shape districts in a way that gives candidates from its side an advantage. Gerrymandering is a technique that can create "secure" seats for a party for the subsequent decade, thus greatly affecting the political landscape at the national level, especially if the state in question has a large population and therefore many seats in the House of Representatives. In many cases, gerrymandered districts have been legally challenged, and a 1985 Supreme Court ruling held that such districts were unconstitutional and unfair.

Yet, with every census and subsequent redistricting, districts that are clearly gerrymandered are created in some states.

There are numerous ways a political region may be gerrymandered. A common technique to dilute the voting power of a specific political, ethnic, or racial group is to divide them among various districts, thereby making them a minority in all. This is called the *wasted vote strategy*, because it renders the votes of this specific bloc effectively meaningless, as they are too dispersed to collectively elect a candidate. A second means of limiting the voting power of a specific group is to concentrate most of the members into a single district. This political geography enables the group to elect a candidate they favor, but also limits their representation to a single district, allowing candidates of the opposition party to win seats in the remaining districts, where competition has been minimized by the gerrymander. This is called the *excess vote strategy*. Finally, *stacking* is a strategy that relies on the creation of unusual, distorted boundaries that are drawn in an effort to concentrate voters based on party affiliation, race, or economic interests. Such districts may not even be contiguous, but rather may be broken into segments. There are many examples of such gerrymanders, but some of the more notorious resulting from re-districting after the census of 2000 are Illinois Congressional District 4, California Congressional District 38, and North Carolina Congressional District 12.

Globalization

A broad process transcending physical and political barriers that results in greater connectivity, integration, and homogenization of human systems. These systems may be economic, cultural, social, or political and are influenced by the **cultural diffusion** of characteristics and qualities at the international **scale**. Globalization has occurred at some level for centuries, but over the past several decades appears to have accelerated, as new mechanisms for the process have emerged. For example, the advent of advanced telecommunications technology over the past 50 years, represented by innovations like satellite television, cellular telephones, and the Internet, has globalized the flow of information and dramatically enhanced connections across cultures. This in turn has made it possible for trends and practices from around the world to influence the tastes and behavior of individuals without those people even leaving their places of residence, a condition that has never held before in human relations. Changes in global **linguistic geography** have also assisted the process of globalization. The emergence of a truly global language, English, partially as a result of the age of **imperialism** when first British and later American **cultural identity** was spread internationally, has provided a common medium for cultural and economic exchange. Additional factors fueling globalization are greater levels of **migration** and international tourism, increasing contact between peoples of different cultures. In 2008, there were more than 900,000,000 international tourist arrivals recorded around the world. The great majority of these travelers originate from regions that are driving globalization, i.e., North America, Europe, and Japan.

Examples of globalization may be found everywhere. One of the most frequently cited instances of globalization is the American restaurant franchise McDonald's, which opened a single facility in California in 1940, but by the year 2000 had more than 25,000 locations in 118 countries and territories. The company claims to serve an average of 58 million people per day worldwide, a figure roughly equivalent to the entire population of France. Indeed, many commentators view multinational companies as the primary instigators of globalization since the end of World War II. They argue that corporations like McDonald's fueled globalization by pursuing overseas markets and overwhelming local competitors through advantages in economies of scale and competitive advantage. The convenience and availability of "fast food," certainly an American innovation, thereby became a global phenomenon and American **foodways** became the norm for millions of people living outside the United States. But it is not only Western, and especially American, corporations and their products that have acquired a global reach—it is also Western popular culture. Western popular music, movies, and modes of dress now are found in every corner of the planet. But perhaps even more interesting is the adoption by local performers of foreign styles of music and dress, and even the methods of promoting their music. An example of the latter is the near universal adoption of the music video as a mechanism for publicity by performers worldwide, albeit with some modification to suit local standards. But globalization does not occur in only one direction—the popularity of sushi and karaoke outside of Japan, along with the global emergence of "Bollywood" cinema and stars shows that popular culture flows in all directions across the global **landscape**.

The effects of globalization permeate into all aspects of modern life, but two broad categories of the phenomenon may be identified—economic globalization and cultural globalization. Broadly put, economic globalization signifies the emergence of a unified, interdependent economic system that transcends international **boundaries**. The global economic system is characterized by a greater level of mobility of capital, labor, and information than ever before in history. The decline of international socialism in the 1980s and 1990s, represented by broad capitalistic market reforms in the People's Republic of China and the collapse of Soviet communism, fueled the development of increased levels of international trade, larger levels of capital investment in foreign markets, and transfers of information, technology, and marketing strategies. Pools of highly skilled labor in emerging markets like India and Southeast Asia, available at lower cost than domestic sources, induced numerous companies in the developed world to "outsource" many of their labor requirements. Production facilities that contribute to the manufacturing sector are frequently located abroad, but also facilities that provide services are now situated outside the market service area. Examples are the technical support services offered by cellular phone companies, computer hardware and software makers, and other businesses. An American customer may contact a call center for assistance and speak to a technician who is actually located in India or Latin America.

Even major retail outlets, most from the developed world, have become globalized—the American corporation Walmart, the world's largest retailer, now has stores (although not all operate under the "Walmart" name) in more than a dozen countries located on five continents.

Cultural globalization is the outgrowth of cultural diffusion on a scale never before seen in human experience. The Internet has clearly been an important medium for cultural transfer and exchange, but the process of globalization began long before the personal computer became part of millions of lives. Cultural globalization began to accelerate during the era of European **imperialism**, when languages, religious faiths, social and behavioral standards, and many other dimensions of European culture spread from the "home" countries to the colonies. The use of English in the British Empire among elites, as well as its continuance as

Group of Twenty (G-20)

The "Group of 20" formed from the G-8, or Group of 8, which was composed of eight of the most influential economic powers in the world. The G-8 consists of the United States, United Kingdom, France, Canada, Germany, Italy, Japan, and Russia. The organization began holding annual summit meetings to discuss policy as the G-6 in the 1970s. The G-20 is an effort to include more emerging countries in the annual meetings, because a number of important states were not part of the G-8, like China and India. Some countries in the G-20 are important suppliers of resources, like Saudi Arabia and South Africa, while others play a vital regional role in economic development, like Brazil and Argentina in South America. The 20 countries represent a disproportional amount of the world's total economic output and volume of trade, at more than 80 percent of each, and also account for a large percentage of the world's population. The most influential and traded currencies are also included in the group, primarily the U.S. dollar, the yen, and the euro. The rationale for the group's conferences, to be held yearly beginning in 2011, is that the era of economic globalization makes it imperative that the most important players in the global market coordinate policy to avoid a worldwide economic downturn.

a national tongue in many former colonies, helped to elevate the language to a global **lingua franca** even before it emerged as the primary language of the "information superhighway." But additionally, there is no question that new forms of communications technology have exponentially increased the level of contact between people from around the world. Social networking sites like Facebook and MySpace, easily accessible media sites like YouTube, the now-ubiquitous cell phone with its option of text messaging, and the ability to connect to individuals and businesses across the globe virtually free of charge via email have all contributed to the emergence of a "global culture" by dramatically increasing the degree of communication between individuals living in diverse, geographically dispersed **regions**. Western popular culture, such as movies and music, also has become disseminated globally, both through traditional means like theaters, as well as via the World Wide Web. The American film *Avatar* opened on more than 14,000 screens in over 80 countries in 2009, and with distribution on DVD and the Internet it may have been viewed by almost half-a-billion people worldwide since its release.

While globalization has resulted in increased contact between people and brought investment capital and economic development to many places, there are negative consequences as well. Some advocates of **World Systems Theory** and its derivatives view globalization as a continuation of the drive by the developed world to dominate international relations and control the production and distribution of wealth. In addition, many see the spread of Western cultural attributes, both in terms of physical culture and social values and mores, as "cultural imperialism." Globalization to these commentators represents an effort to impose their own, Westernized world view on other peoples, undermining and eventually destroying those cultural systems and resulting in a homogenized, uniform "global culture" that is in reality nothing more than an extension of Western, and predominantly American, culture. Proponents of this view point to the loss of languages, traditional music and literature because of the adoption of modern, "global" cultural perspectives that has occurred among many peoples in the past century.

Furthermore, many argue that there are clearly negative economic effects associated with globalization, both for developed and developing countries. For example, global mobility of labor, especially of highly skilled people, is frequently to the detriment of developing countries, who are the victims of a "brain drain" to the developed world. Doctors, engineers, and other highly skilled personnel, many of whom were trained at the expense of their home country, are able to migrate to the developed world, thereby undermining the potential for development and improved living standards in their countries of origin. In the developed countries, opponents of globalization claim that the wages and bargaining power of workers are weakened and jobs are actually lost, due to the relocation of capital and production facilities to developing countries, where wages are lower.

Those who argue for the positive aspects of globalization point out that at least for some, the process of globalization represents a shift toward equalization of economic opportunities and advancement, as well as providing jobs and economic development to workers in developing countries that otherwise would not be available.

While "outsourcing" results in job losses in the economically developed countries, it rewards skilled workers in underdeveloped economies by creating well-paying jobs, resulting in a global expansion of economic opportunity and ultimately a more equitable distribution of wealth. Regarding the expansion and adoption of cultural values and behaviors, defenders of globalization suggest that these processes actually result in an *increased* appreciation among many for their tradition values, and that the absorption of external cultural attributes need not necessarily result in the decline or loss of existing systems and attitudes. But whether one sees globalization as threatening or beneficial, the process appears to be accelerating and will likely reshape the world of the future.

Global Positioning System

A global positioning system, or GPS, is a mechanism for specifying locations on the **Earth** using a cluster of satellites. Two systems are in use: one developed, managed, and maintained by the United States Department of Defense, and the other under the control of the Russian military called GLONASS. The European Union has had a separate GPS system known as *Galileo* in development for over a decade, but the system is not yet operational. Recent announcements (2008) from EU administrators indicate that the EU expects the system to be ready by 2013. The American system utilizes more than two dozen satellites in medium orbit, which broadcast high-energy signals to receivers on the surface of the planet, or in the lower atmosphere. The U.S. Department of Defense launched satellites throughout the late 1980s and early 1990s, building the capacity of the network toward global coverage. An early form of GPS technology was used by the U.S. military during the first Gulf War in 1990, but the system remained accessible only for military use. In 1993, the system reached full operating capacity, with 24 satellites in orbit. Three years later, the government passed legislation allowing for the civilian use of the system. The result was an explosion of the use of GPS technology because the applications were widespread, and relatively inexpensive GPS receivers were available to the general public in only a few years. By 2000 GPS-related business amounted to over $6 billion, and continued to increase throughout the early 2000s. The advantages of GPS technology are many. Access to the system is free, a GPS receiver will work outdoors in any weather (but typically will not work inside a building), and the system is accessible at all times.

The applications of GPS technology are virtually limitless. Inexpensive GPS receivers can pinpoint the user's location within a few meters, and provide a map showing the location. In addition, the location can be stored, so that the user can return to virtually the exact same place in the future. More expensive GPS receivers can provide the location of the user, or an object being mapped, to within a few centimeters.

GPS is a highly useful tool for outdoor recreation. Those engaged in hiking, fishing, and hunting can find their location if lost, record a favorite spot and the path to reach it, or even map an area using a GPS receiver. Business applications are also numerous. Law enforcement can track the movements of suspected criminals by attaching a GPS unit to a vehicle. Surveyors can use GPS to survey a piece of property much more quickly than when using older techniques. Farmers can use high-precision planting techniques that employ GPS technology, increasing yields and conserving costs of production. Aircraft can be precisely tracked by ground controllers using GPS, and the aircraft itself can use GPS for navigational purposes. There are also applications for forestry, wildlife management, and many other fields. Of course, perhaps the broadest application of GPS technology is to commercial mapping, which can now be done more quickly and more accurately.

There are three components to the GPS system. First, there is the ground control component. This consists of a system of control stations located on the surface of the Earth that regulates the signals being sent from the satellites. There is a master control station, currently located in Colorado, a backup control station used only if the master control station ceases to function, and several monitoring stations located around the world. The master control station is constantly in operation and regularly receives data from the monitoring stations that has been downloaded from the satellites and corrects any errors in the signals. The second component is the network of satellites. The satellites are not in stationary orbit but revolve around the planet at an altitude of about 12,000 miles, and are arranged in space so that at least four are above the horizon at all times, as seen from any point on the Earth's surface. The final component of the GPS system is the user, who accesses the signal from the satellite using a GPS receiver. The receiving unit must acquire signals from four satellites (this is why the satellite system is configured to have four satellites above the horizon at all times), and can accurately locate the position of the user by comparing the time required for each signal to reach the unit. Any errors that may be in the signal must be corrected by the master control station, or the readings provided may be inaccurate. Errors may be generated by disturbances in the atmosphere, which delay the time the signal reaches the surface of the Earth. If the time measurement is inaccurate, the location provided by the receiver will also be inaccurate.

The Global Positioning System (GPS) is an array of satellites that provide a highly accurate means of identifying points on the earth's surface. (NOAA)

One of the most useful characteristics of the GPS is its utility as a mapping tool. GPS has revolutionized **cartography** by enabling data on features to be gathered in the field instantaneously and downloaded into a mapping program or **Geographic Information System (GIS)**, which may then be used for locational analysis, or other types of studies. Data points may be gathered by simply recording the GPS receiver's location and storing these points in the unit. These data can then be transferred to existing spatial information already held in a GIS, which can then display the data points on a map. Many navigation systems available for use in automobiles now use GPS technology, such as TomTom GPS or Garmin. These systems make it nearly impossible to become lost in an automobile in the United States. Similar systems that automatically notify emergency personnel in the event of an accident also use GPS technology. The military applications of GPS have been illustrated in several recent conflicts, dating to the Gulf War of 1990. American cruise missiles, equipped with the GPS coordinates of enemy targets, have been guided directly to those locations by GPS technology. Such accurate delivery of ordnance, achieved without endangering American pilots and directed with pinpoint precision to limit civilian casualties, has obvious military advantages. It is likely that GPS receivers will become more affordable and even more precise as the technology becomes more widely applied and utilized.

Global Warming

"Global warming" are the two words identifying the most controversial scientific topic of the early 21st century. The concept has such vast societal, economic, and governmental implications that it receives major attention in public media outlets. Geographers have become fully engaged with their suite of **climate** change studies and also with their assessments of possible impacts. There are three salient points that a reader must appreciate when learning about global warming:

(1) There is strong scientific consensus that the warming is real, (2) the warming is geographically uneven, and (3) anthropogenic increases of greenhouse gases is a major contributor to the warming. For the first time in human history there is a "very high likelihood" that humans have significantly impacted the greenhouse effect and that all projected greenhouse gas scenarios indicate increases in the warming effect.

The current time is a relatively warm one in recent **Earth** history. The planet has not been as warm for 1,300 years and the last time the planet had extended warmth at the level of the present day, it was 125,000 years ago and sea level was 6–8 m higher. Earth's climate was considerably warmer for most of Earth's history but those eons are separated from us by the Pleistocene Epoch ice age of the last couple of million years.

The mechanism for the current global warming has been the unintentional enhancement of the greenhouse effect by a number of human activities. The concern is that the increase of greenhouse gases has increased the atmospheric equilibrium temperature by causing a greater percentage of the Earth surface and atmosphere's longwave emitted energy to be absorbed and then re-emitted by the atmosphere. Worldwide industrialization and population increase has led to a third more atmospheric carbon dioxide since the middle of the 1800s. Carbon dioxide is a greenhouse gas produced by all combustion processes, including the burning of fossil fuels, so that there are myriad major sources. Other gases such as methane, nitrous oxides, and chlorofluorocarbons (entirely human-made) have also significantly increased; they are trace gases but are far more absorptive of longwave energy on a per mass basis.

Scientific consensus for the global warming effects has increased and been confirmed by virtually all evidence. Perhaps the most publicized warming effect is the worldwide decrease in the mass and geographic extent of ice on the planet. Satellite and on-site surveys have confirmed that the mass of ice is decreasing at increasing rates over virtually the entire planet. Even glaciers that have advancing margins have been shown to exhibit this behavior by the increase of subsurface meltwater lubricating glacial motion.

The current global warming does not seem to have been extreme in the minds of most non-scientists. After all, the world mean temperature has been on the order of three-quarters of a degree Celsius since the mid-1800s. Research does not suggest the greenhouse effect will increase to the point where human life becomes impossible. Yet, the worry extends far beyond temperatures. The vehicle to convey global warming scientific knowledge and the bellwether for ferocious criticism has been the series of *Assessment* reports made by the Intergovernmental Panel on Climate Change (IPCC). The IPCC is an effort of the World Meteorological Organization and the Environment Programme of the United Nations. Over a course of years, the IPCC has produced several assessments of global warming through teams of internationally respected scientists. Their 2007 conclusion leaves little doubt that humans are impacting the atmosphere. Their continued work was awarded the Nobel Prize for Peace.

What does the future hold? The IPCC has used a suite of numerical models to assess global warming over various times and under various scenarios of rate of greenhouse gas release. Using a plausible doubling of atmospheric carbon dioxide, a middle solution from their suite of possibilities is that global temperature would warm about 2°C in the coming century. This is three times the rate of warming during the 20th century. The following are the qualitative likelihoods of future conditions over most land areas as listed in the IPCC's 2007 *Assessment*:

1. Warmer with fewer cold days—virtually certain
2. Warmer with more frequent hot days and nights—virtually certain
3. Increase of warm spells/heat waves—very likely
4. Increased frequency or proportion of heavy precipitation falls—very likely
5. Increased area affected by droughts—likely
6. Increased frequency of intense tropical cyclones—likely
7. Increased incidence of extreme high sea levels (tsunamis excluded)—likely

The projected geography of the warming is of great interest. Changes in temperature would be due to the radiative forcing of increased greenhouse gases, but also shifts in the global wind and pressure systems. To wit, the imbalance of energy between equator and poles energizes large-scale wind systems. If warming decreases the latitudinal imbalance, winds would blow less vigorously in each hemisphere as a whole and displace wind and pressure systems out of their present average positions.

A consistency among numerical climate simulations is the relatively lack of warming ascribed to the tropics. This is not to imply that there would not be tropical effects. For instance, there are some small island countries in the Pacific and Indian oceans that would virtually disappear if the IPCC's middle estimate of upward of a half meter of sea-level rise by 2099 verifies. Upper middle and polar latitudes are projected to experience the highest temperature increases. The IPCC list seems to indicate a plethora of harmful effects, but not all regions or countries stand to be losers. Countries such as Russia and Canada may experience the opening of new agricultural lands as permafrost thaws. Yet, a possible effect during global warming is shifts in wind and precipitation patterns so that other lands become untenable in terms of their traditional crops and practices. Some researchers have raised the specter of regional famines and, even, wars as a result.

Climate change model scenarios are far from certain. Earth's climate system is the product of an interplay of factors, many of them unrelated to the presence of humans. As numerical modeling becomes more sophisticated, the models are better able to incorporate the impacts of the processes by which greenhouse gases are removed from the atmosphere, the potential for increased humidity (from increased) evaporation exacerbating the warming, the impacts of the extent and depths of clouds, and the possible mitigating influence of sulfate aerosols as released by human activities.

Grasslands

Grasslands are a major floralistic association of the **Earth**. They are characterized by grasses (Poaceae) and forbs (non-woody plants) of various heights and densities. On some grasslands, trees are interspersed, especially along watercourses. The climatic cause for widespread grasslands is a greater potential evapotranspiration than precipitation for the year. In such cases, the landscape cannot be dominated by trees because of the moisture stress. Grass species become shorter and sparser as potential evapotranspiration becomes progressively larger than precipitation. Ultimately, the grasslands grade into desert where the potential evapotranspiration is more than twice precipitation. There are also non-climatic reasons by which grasses can dominate the landscape, such as places in which the soils contain high concentrations of minerals such as nickel that are toxic to many tree species.

Geographers usually differentiate grasslands into two **biomes** dominated by grasses. The first is the tropical savanna and the second is the midlatitude grasslands. Their characteristics were given in the biomes article earlier in this *Handbook*. There are major grasslands on all the continents save Antarctica. The regional names are familiar to geographers: Examples include the Great Plains (North America), pampas (South America), steppe (Russia), veldt (Africa), and rangelands (Australia). Each continent has its own "flavor" of grassland floral and faunal species, yet the grassland formations are distinct from those in forests and other floralistic associations.

Within the last few thousand years, probably 40 percent of continental surfaces were grasslands but the percentage is declining as a result of the warming of the planet as it has emerged from the Pleistocene (ice age) and the unintentional destruction of the ecosystem by overgrazing (see **Desertification**). A billion people live in the grasslands of the world, so these areas are quite susceptible to human modifications. Overall, grasslands have declined in their ability to sustain animals and people. For instance, the tallgrass prairie of the eastern Great Plains has all but disappeared. Agriculture is prolific in the more precipitation-rich grasslands with all the major grain crops having originated in grasslands and being variants of grasses. These crops include millet, sorghum, corn, rice, and rye. When considering the human influence, it is apparent that up to a third of the species of grasses in regions such as the Great Plains are not native to the region.

In general, there is lesser diversity of plant and animal species in grasslands compared to adjacent forests. However, this cannot be taken to mean a paucity of life. All sorts of bacteria, fungi, insects, and earthworms are at home in grassland soils. Moreover, plentiful large herbivores are one signature of native grassland conditions. In North America, the bison once populated the Great Plains in untold millions. In today's African savanna, elephants, zebra, and wildebeests are still largely present but in declining numbers. The herbivores may, in fact, help keep grasses dominant in some places as herbivores are attracted to competing plant types such as young trees and bushes.

Grassland dominates in places where moisture is precious. A complete grass cover is very conservative of precipitation because of the slowing of surface runoff to the encouragement of moisture infiltration into the soil. Closely spaced root system mats help hold soil in place. A Great Plains thunderstorm dropping 2.5 cm (1 in) of precipitation may well not generate any significant runoff into streams. The increased infiltration leads to moist soil profiles under grasslands. Complex root systems greatly increase the length of storage and percentage uptake of soil moisture. So too is the leaching of soil nutrients slowed. This means that the mid-latitude grasslands of the world provide some of the most nutrient-rich soils when they are plowed for agriculture. Unfortunately, this is tempered by the fact that the soils dry out and are susceptible to wind erosion once the grass cover has been removed.

Of great importance to grasslands is the presence of fire. In dry seasons and droughts the aboveground portions of grasses brown out, allowing lightning-caused or human-caused fires to spread effectively, sometimes over millions of hectares. Fire represents a sudden change in the mass/energy pathways in grasslands but it does not represent the end of life. Fire releases plentiful nutrients into the top of the soil and the still-living underground portions of grasses are able to poke above the surface to tap this natural fertilizer. Additionally, fire kills trees and bushes helping grasses to dominate.

Native Americans were successful users of North America's grasslands for thousands of years before European contact. Yet, as the United States enlarged to fulfill its "Manifest Destiny" during the 1800s, settlers neglected the Great Plains in search of opportunities on the Pacific Coast. The grasslands were viewed as so desolate compared to the woodlands of the eastern United States that the area was commonly known as the "Great American Desert." With time, settlement accrued in this grassland realm. Three technological innovations played a key role: the invention of the steel plow, the invention of barbwire fence, and the building of railroads. As these innovations became widespread the vast grasslands were subdivided by farms and ranches and the crops and cattle could be shipped by rail to distant markets.

The Great Plains of the United States has a cautionary history that is mirrored in other places around the world. Grasslands are the climatic result of repeated moisture stress punctuated by relatively wet times. Grasslands usually exist between the deserts and the forests; people are tempted to use them as if they had dependable precipitation like the forests. As the Great Plains were fully settled, much grass cover was removed to grow dryland (unirrigated) crops like winter wheat. World War I saw rapid expansion of wheat agriculture as crop prices soared. Ironically, when wheat prices collapsed after the war, farmers opened up even more grassland to make enough money to pay mortgages. When the droughts of the 1930s arrived with the driest series of years in a century and a half in some places, the soil became so dry that much of it blew away, creating the Dust Bowl. The Dust Bowl's droughts had a natural cause, but the social effects were staggering, depopulating entire areas by half or more. The ecological effects were also devastating, ruining grassland areas for centuries to come.

Green Belt

Green belts are **buffer zones** of undeveloped land that are found around urban areas. Sometimes these zones form true belts that completely surround the urban region; in other cases the green areas are not contiguous but are separated into sections, often dividing the city proper from suburbs and satellite towns. The concept of preserving a natural space around urban development first appeared as a component of urban planning in the United Kingdom, just before World War II, as a response to urban sprawl and the decline of farmland and natural areas around the larger cities. Legislation passed after the war enabled city planners to incorporate the concept of undeveloped spaces into the larger urban structure, and green belts became a common feature of many towns and cities in England. The establishment of green belts and other "green spaces" in urban areas in the developed world has become a mainstay in the idea of **sustainable development**. Today it is estimated that about 13 percent of the total area of England is protected in the form of green belts, and some of these are quite large. The green belt that surrounds the greater London metropolitan area, for example, is more than 5,000 sq km in area—one of the largest contiguous urban green belts in the world. Only the green belt associated with Canada's "Golden Horseshoe" conurbation and the Sao Paulo City Green Belt Biosphere Preserve are larger, although the latter is not actually a planned green belt in the classic sense of the term. A proposed "European Greenbelt," a zone linking parks and preserves that would extend from the Baltic to the Mediterranean seas through Eastern Europe, would be the largest such space in the world if it is successfully completed. Promoters of green belts argue that they add quality to the lives of urban dwellers, allowing them to connect with and enjoy the natural environment, while preserving nature, in a limited way, in the urbanized setting. It is claimed that they help to maintain both air and water quality, and of course they provide both outdoor recreational opportunities for city residents and habitat for wildlife.

Some scholars have criticized the green belt approach as economically unsound, and often counterproductive. Some economists suggest that by reserving land for the green belt in urban areas, housing prices are maintained at high levels and the quality of housing actually suffers, because competition in the housing market is stifled. There is some evidence to support this view, especially in England. Moreover, while green belts have preserved a strip of land adjacent to the main urban area, in many cases development has simply skipped over the green belt and exploded on the other side, in the form of suburbs and satellite towns. Critics argue that this shows that the green belt simply forced urban expansion further out from the main urban center, actually increasing transportation costs and pollution levels. Although controversial, the green belt concept in urban planning appears to be widely accepted and applied, especially in Europe, and is likely to be a feature of the urban landscape of the future.

Groundwater

Earth's land surface is less than monolithically solid. The joints, **faults**, and weathered rocks described elsewhere in the book allow rainwater to penetrate the crust, sometimes to considerable depths and become groundwater. Stream channels that intersect subsurface layers containing groundwater can add water to or take water from the groundwater supply.

Far from being an isolated source of stored water, the shallower groundwater supplies are intimately related to the rest of the **hydrologic cycle** in that they are recharged by precipitation. It should be noted that most soil moisture is distinct from groundwater (although there are times and places at which the top of the groundwater supply can be within the soil and so be one in the same). In that groundwater is out of sight and moves much slower than streams, its nature is not appreciated by most people. Yet, its importance looms large in that there is two-and-a-half times more groundwater than in all the streams and lakes on the planet.

Not all groundwater is potable. The contact of underground water with rock leads to the solution of salts into the water. In some cases, the saltiness is much higher than that found in the ocean. In other places with moderately "hard" water containing positively charged ions (cations) of calcium and magnesium, the salts can be dropped out of solution by using water softener, thus making it usable for human purposes.

Groundwater can be found underneath most of Earth's surfaces but not in consistently developable amounts by either depth or geographical region. Most groundwater is within a kilometer of the surface, but some water has been found as deep as 10 km. With depth, however, the pore spaces in rocks and sediments become considerably smaller because of the pressure of overlying materials, and groundwater is essentially trapped in place. The deeper groundwater is connate or "fossil water" that was trapped as the rock layer was laid down. This deep groundwater is usually brought to the surface as an unavoidable consequence of drilling for oil and natural gas. It is almost invariably salty because of the immense amounts of time the water has had to dissolve surrounding materials. Sometimes these brines can be economically tapped and important materials such as iodine extracted.

Groundwater is sometimes conceived as a gigantic underground lake or stream, but this is not close to the truth. Near Earth's surface, there are four zones differentiated with respect to groundwater: the zone of aeration, the zone of saturation, the zone of confined water, and the waterless zone. The depths, amount of water, and flow characteristics have immense variations according to the type and structure of the underground materials.

The zone of aeration is the topmost zone and abuts the surface. It is composed of solid materials with pore spaces occupied by air and water. As the sky precipitates, water infiltrates into the zone of aeration, filling the pore spaces for a while. The pore spaces drain of water via gravity and via evaporation directly to the atmosphere and via transpiration from plants whose roots absorbed water from the zone of aeration. The nature of the zone of aeration changes dramatically after each precipitation event. It has vertical depths that can extend to hundreds of meters or much less than a meter. This is not a zone in which one would normally situate the bottom of a groundwater withdrawal well in that the amount of water present is so highly variable.

The zone of saturation is beneath the zone of aeration, and it is this layer into which wells are drilled to tap the groundwater supply. The zone of saturation has gained its water gravitationally from above and all of its spaces are completely filled by water, which is properly known as groundwater. The top of the saturated zone is the water table. The depth to the water table is highly variable around the planet, and humans have gravitated to the areas of large water supply close to the surface. The depth to the water table varies seasonally and topographically. Summer seasons usually result in the drop of the water table with winter recharge. The depth of the water table generally follows the slope of the land above and is closest to the surface in stream valleys. Humans have made impressive changes in the water table. Around wells, there is usually a cone of depression but as many wells tap the zone of saturation the water table is drawn inexorably lower. In Ft. Worth, Texas, the water table has dropped 125 m in the last century. Such drops require increasing energy use to draw the water to the surface and make use of the water less economically efficient.

In some places, layers of impermeable sediments of rocks surround parts of the saturated zone and so impede water from leaving that zone. If the amount of groundwater so confined is large, this confined zone is an aquifer. Aquifers are composed of materials like sandstone that are conducive to the movement of water. Wells tapping an aquifer are kept supplied by groundwater moving toward the wells. Major aquifers around the world include the Ogallala of the U.S. Great Plains, the Great Artesian Basin of Australia, the Guarani aquifer of South America, and the Nubian aquifer of Africa. In some cases, these aquifers do not have substantial recharge, being the relicts of wetter times at the end of the Pleistocene ice age thousands of years ago.

Beneath the saturated zone is the fourth zone, known as the waterless zone. Usually, this zone begins a few kilometers under the surface and exists because pressure from the overlying materials precludes the existence of pores in which water can be stored.

The flow of groundwater is considerably slower than in surface water. Common rates are between 15 and 125 m per day with some places having rates of centimeters *per year.* Rather than straight flow, groundwater flow is confined to pathways using the tiny openings existing in the surrounding materials. The flow is energy by differential pressures. Unlike surface water, groundwater is able to move up or down depending on the direction of the pressure gradient. Sometimes, there is enough pressure involved that a break in a confining layer (by natural circumstances or by a well) allows the groundwater to escape to the surface where it is known as artesian water.

It has been estimated the world's groundwater use is 600–700 *cubic kilometers* per year and represents the greatest tonnage of any extracted material. Use is quite variable by circumstance. Just over a fifth of the United States water consumption is supplied by groundwater withdrawal. In countries like Saudi Arabia, the ratio exceeds four-fifths. In that many aquifers extend beneath national boundaries, the scene is set for potential international tensions as groundwater becomes scarcer.

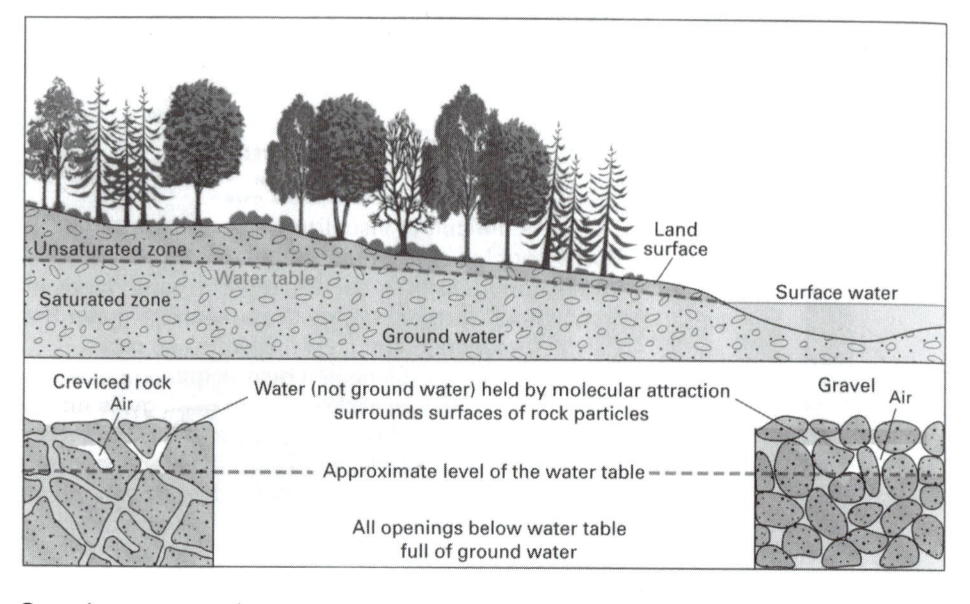

Groundwater is an important source of water for drinking and irrigation in many world regions. Even deserts contain groundwater, although it may be located at considerable depth. (U.S. Geological Survey)

Oases are interesting examples of situations where the water table is close to or at the desert surface (see **Deserts**). Oases have played important roles in human geography in diverse countries such as Egypt, Saudi Arabia, Niger, Peru, Libya, and the United States. In the midst of aridity, these contrasting locales of plentiful water have provided people with shade, water and forage for animals, and irrigation for crops since prehistoric times.

Groundwater depletion is a major problem in places dependent on groundwater for water supplies. It is clear that many groundwater withdrawals represent "overdraft" situations in which use is many times that of replenishment. The world groundwater overdraft is estimated at 200 billion cubic meters per year such that the current rate of use cannot be considered sustainable over the long term.

In places like Long Island and Florida, some places proximal to the oceans have had withdrawal of freshwater allowing lenses of salty groundwater from underneath the ocean bottom to supplant the freshwater and make the well water too salty for use. Land subsidence is an issue in areas with substantial withdrawals of groundwater. The groundwater within fine-grained sedimentary rocks and loose sediments forms part of the crustal mass maintaining the level of the surface. Withdrawal of the mass represented by the water allows the compaction of the water-bearing strata by the mass of the materials overhead. Subsidence has been noticeable and damaging in places such as the Central Valley of California,

Mexico City, Venice, Tokyo, and Bangkok. Tens of meters of subsidence have been documented. Building foundations can fail, underground pipes can burst, and there can be flooding of low-lying areas. The so-called Leaning Tower of Pisa is perhaps the world's most famous example of subsidence caused by the great weight of the tower differentially compacting groundwater out from only a meter or two beneath the surface.

Groundwater pollution is a problem that becomes increasingly salient as more human activities take place over shallow groundwater supplies. There are myriad ways in which groundwater can be contaminated to the point where it is nonusable. These include infiltration from septic tank leaching fills, landfills, animal feedlots, and accidental spills of materials. Additionally, injections of toxic wastes down wells, the spread of agricultural chemicals on crops, and the leakage of underground storage tanks can compromise the groundwater supply.

H

Heartland Theory

A theory of **geopolitics** proposed by J. Halford MacKinder at the beginning of the 20th century. MacKinder taught geography at Oxford University and served in the British parliament for 12 years and was well connected to the policymakers and strategic thinkers of his time. Great Britain was still the world's foremost power when MacKinder put forth his ideas, and it is clear that he viewed British **imperialism** as the most moral and just among the European empires. It is likely that he was deeply influenced by notions of the "Great Game" between Russia and Great Britain that had motivated British policy in south Asia during the second half of the 19th century. MacKinder believed that the heartland theory was in reality a roadmap to global domination in an age when Britain's preeminent position in international affairs was being challenged by emerging powers such as Germany and Russia. His ideas are firmly rooted in **territoriality**, as he held that control of a key geographic space was essential to political power.

MacKinder first publicly presented his theory to the Royal Geographical Society (RGS) in 1904. The title of his paper was "The Geographical Pivot of History," which was published later the same year. In this discussion MacKinder analyzed the distinction between, and relative merits of, "land power" versus "sea power." Moreover, he sought to identify a larger geographical relationship to historical events, postulating a "geographical causation in universal history." He suggested that the Eurasian landmass was the key to global political power, represented by a core "pivot region" that encompassed much of Eastern Europe, Central Asia, and extended across northern Russia. This "geographical pivot" region, according to the theory, had been from ancient times the key to territorial control of greater Eurasia, although during the "Columbian epoch," as MacKinder termed the previous four hundred years, the importance of dominating the pivot region had temporarily been overshadowed by the rise of sea power. MacKinder acknowledged that the dominance of the high seas by the English navy had been vital to the construction of the British Empire, but argued that the development of extensive railway networks in Asia would offset the advantages of controlling the world's sea lanes and **choke points**. The Russian empire had begun construction of the Trans-Siberian Railroad about a decade earlier, and this development very likely convinced MacKinder that the vast expanses of Asia could be connected via rail. His thesis also was influenced by **organic theory**, as he spoke of history and physical geography as being "organically connected," and there are clear undertones of **environmental determinism** in his ideas regarding the relationship of history and physical geography.

MacKinder conceived of world geography as being composed of "natural seats of power," with the pivot region being supreme and representing dominant land power. Arranged in two bands surrounding the pivot area were regional arcs of power, the "inner crescent," or sometimes referred to as the "marginal" crescent, and the "outer crescent," or "insular" crescent. The inner crescent was controlled for the most part by the European colonial powers and swept from Western Europe toward the southeast, including India and the entire eastern coastline of Asia, all the way through eastern Siberia. The inner crescent completely enveloped the pivot region, except on the northern coast, and was controlled by a combination of land and sea power. The outer crescent consisted of Great Britain, Australia, Japan, southern Africa, and North and South America. The countries of this zone, all maritime nations, were completely dependent on sea power to protect their political interests and commerce. Interestingly, MacKinder did not view the United States as a vital player in the dynamic between the inner crescent and the pivot region. He believed that the construction of the Panama Canal would direct American interests toward the Pacific Basin and away from the Eurasian landmass. Great Britain, as part of the outer crescent, was compelled to cultivate political alliances with countries in the inner crescent so as to establish "bridge heads" there that would prevent the expansion of any power from controlling the pivot region. The rise of the British Empire in the previous centuries, noted MacKinder, was "a short rotation of marginal power round the southwestern and western edge of the pivotal area," which only temporarily obscured the pre-eminence of the pivot area.

Although MacKinder did not use the term "heartland" during the initial reading before the RGS, he would later introduce that term in a revision of the theory, using it to replace "pivot region." In 1919, in the aftermath of World War I, he reformulated the theory in a book-length study, *Democratic Ideals and Reality*. In this work MacKinder introduced several new regional terms into his original thesis. The "pivot region" now became the "heartland," with some slight modifications of the original boundaries—the heartland was somewhat larger, as it included western China and more of Siberia. MacKinder now referred to the "World Island," a vast territory that was composed of most of Europe, Asia, and Africa. In the revised theory, the World Island represented a huge repository of resources that could be exploited by an aggressive power bent on world domination. The key to controlling the World Island was control of the heartland's gateway region,

Eastern Europe. Thus, MacKinder articulated the famous encapsulation of his argument:

> Who rules East Europe commands the Heartland; who rules the Heartland commands the World-Island; who rules the World-Island controls the world.

The heartland theory unquestionably had a profound impact on geopolitical thinking in the wake of World War I. Eastern Europe's strategic importance as a **shatterbelt** was now reinforced by the heartland theory, and German scholars like Karl Haushofer and other members of the Geopolitik school certainly considered MacKinder's theory to be a major contribution. Even MacKinder's critics, most notably Nicholas Spykman and his **Rimland theory**, employed a similar **core-periphery** theoretical framework in constructing alternatives to the heartland concept. It is a testimony to the tenacity of MacKinder's ideas that the heartland theory remains the subject of debate among political geographers today.

Heating and Cooling

Change of temperature on hourly, daily, and seasonal time scales is familiar to all of us. In one sense they are quite simple because they are largely driven by **solar energy**. As the sun changes its position in the sky during the day, disappears below the horizon at night, or its path through the sky systematically changes over a course of months, we can readily understand its importance to heating and cooling here on the planetary surface. The full story is more complicated and vital to our understanding of physical geography.

Heat is the energy of molecular motion that is exchanged between molecules. All objects in **Earth's** physical systems contain some heat, including those at very cold **temperatures**. Heat transfers from warmer to colder substances at a rate proportional to the temperature differences between the substances. The calorie is the standard unit of heat and is the amount of energy needed to raise the temperature of 1 gram of pure liquid water 1°C. (This is different than from Calories used in dietetic applications where 1 Calorie = 1,000 calories.) The amount of heat that can be transferred out of an object is related to its total heat content and its emissive properties. The sun is large and is losing its heat content over billions of years. The metal in an automobile may be hot to the touch when exposed to solar energy, but rapidly loses its heat content at night and typically becomes much colder than surrounding, natural materials. In a broad sense, heating and cooling can be understood via energy imbalances. If an object gains more energy than it loses, its temperature increases. If an object loses more energy than it gains, its temperature decreases. If an object remains at a constant temperature, incoming and outgoing energy are approximately equivalent.

Heat can be transferred in three different ways. The first is via conduction. This is a molecule-to-molecule flux of energy and is most efficient in the closely packed molecules of solids and least efficient in the atmosphere. Thus, conduction is important in heating and cooling of the soil while negligible in the atmosphere above shoe top height.

The second mode is via convection. In convection, heat is transferred from place to place along with the flow of a substance. In our environment both the atmosphere and water are capable of convective transfer of energy. Convection is important in bringing energy from Earth's surface up through the troposphere and convection sluggishly causes vertical motion in the denser fluid of the oceans.

The third mechanism of heat exchange is via electromagnetic radiation. All objects above absolute zero (–273°C) radiate electromagnetic energy and our Earth environment is far above absolute zero. The hotter an object, the greater its heat content by volume and the greater its total electromagnetic emissions. The wavelengths of the emissions are controlled by the temperatures. Hot objects radiate at relatively short wavelengths and cool objects radiate at relatively long wavelengths. For instance, the sun's energy has its most plentiful wavelengths in the visible portion of the spectrum while the lesser energy in the Earth system is radiated most plentifully at wavelengths 20 times as long. Electromagnetic radiation does not need any medium through which to propagate. It can be transmitted through the void of space and through media such as water or atmosphere in which the molecules are not closely packed. It occurs at "the speed of light," which is immensely faster than either conduction or convection.

Somewhat more complicated than conduction and convection is the disposition of electromagnetic energy. As electromagnetic energy contacts mass at Earth's surface or in its atmosphere, one of three things can happen: (1) The energy can be transmitted through the mass and not add to its heat content (such as we are able to view sunlight that has passed through the entire atmosphere, (2) the energy can be absorbed by the mass and thus increase the heat content of the mass, or (3) the energy can bounce off the mass in one direction (reflection) or multiple directions (scattering) and not add to the heat content of the mass. It is the absorption of energy by mass that increases its heat and accordant temperature. Different types of objects heat differently because their physical nature makes them transmit, absorb, and reflect (and refract) electromagnetic energy differentially. For instance, snow does not melt as quickly as one might suppose because the large majority of solar radiation is reflected without having a chance to increase the snow's heat content.

The temperatures of the objects around us are based on their internal heat content that is usually closely related to their temperatures. However, changes in receipt of thermal infrared radiation energy are not always definitive in heating and cooling experienced. Another factor involved in heating and cooling is the specific heat of each substance. The specific heat is the amount of heat needed to raise the temperature of a gram of a substance 1°C. Water has a much higher specific heat than does soil or rock so that the identical amount of solar energy falling on water as on land causes much less temperature increase on water. A major factor in understanding the climates of Earth is the large amount of surface water on the planet that heats and cools ever so much more slowly than does rock and soil.

On our water-rich planet, water moderates temperature because of its high specific heat and its ability to change sensible heat into latent heat and retard temperature increases. A landscape with wet soils will warm less quickly during the day than a landscape with dry soils. At high temperatures, the human body employs the mechanism of perspiration to cool itself.

Additionally, rising air cools dramatically and sinking air warms dramatically. This is because of the decrease and increase, respectively, of the average distance between the molecules. This is the adiabatic process and does not require any addition or subtraction of heat from the vertically moving air.

Hinterland

Hinterland is a term that may be used in several ways in geography. It has a more specific meaning to economic geographers, when it is used to designate the region that is economically connected to an economic urban hub, usually a port or city. In this context the "hinterland" in effect refers to the market area, or distribution area, for the economic center. It may also be used in a general way to identify a region generally devoid of urban development that extends beyond the margin of such development but is still influenced by it. In the **core and periphery** concept, the term "hinterland" is often used interchangeably for the periphery, and therefore makes up the outer margin of a functional region. The word is sometimes used in reference to political control as well, to indicate a zone beyond the formal political or legal **boundaries** of a state or other political unit, but nevertheless swayed by its policies.

In a more general sense, the term is equivalent to other English terms such as "backwoods," "outback," "frontier," and "back country." Hinterland used in this way often characterizes an undeveloped region located inland from a strip of coastal settlement.

In economic geography an **entrepôt** is often considered to have a hinterland. The hinterland region is the larger market space through which the goods and services provided by the entrepôt are dispersed. It is also the region that may provide raw materials for export to the entrepôt. The size and function of the hinterland will modify, depending on the rate of growth (or lack thereof) of the entrepôt. Lower-order economic centers may emerge in the hinterland, which function as distribution centers for the larger entrepôt, and often acquire secondary production functions, sometimes developing hinterlands of their own on a smaller scale.

The concept of an economic hinterland is a key part of **Central Place Theory**. Each central place is serviced by a hinterland, or market area. The size of the hinterland is determined by the range of the good or service in question—the hinterland's radius is the equivalent to the range of the good or service in question. The range of a good or service is the distance that consumers will travel to access it. Different types of goods have different ranges. So-called higher-order goods and services, like automobiles, dental care, legal advice, etc., that are expensive and are used infrequently will have a large range, and therefore a much larger hinterland associated with them than lower-order goods. Lower-order goods, like a loaf of bread, gasoline, a car wash, etc., are used frequently and are generally inexpensive. These have a small range, and the hinterland connected to such goods and services is much smaller than with higher-order products. Thus, there is a hierarchy of hinterlands, with smaller hinterlands (lower-order goods and services) clustered within the hinterlands of higher-order goods and services. Central Place Theory suggests that these hinterlands are not circular, but rather take the form of hexagons.

Hoyt Model. *See* Sector Model.

Humidity

Humidity refers to the water vapor content of the atmosphere. Usually, this term does not include the water held as liquids and solids in clouds and precipitation. Instead, water vapor is the gaseous form of water. It is odorless and colorless, but significant amounts of humidity can impair visibility. Water vapor is one of the gases of the atmosphere and freely mixes and moves in the fluid flow we call wind.

To persist in the form of a gas, water must achieve much higher energy content than the liquid or solid states. In the dynamic setting of the free atmosphere water readily changes state to and from its gaseous phase. When vapor changes phase to liquid, this process is called condensation and the water molecules organize themselves into chains because of loss of energy from the vapor. Conversely, when water evaporates and changes its phase from liquid to vapor, this necessitates the addition of substantial amounts of energy to break the chains of liquid molecules. Less well known to most of us is that water vapor can lose enough energy at once to change directly from gas to the regular crystalline arrangement of ice molecules; this process is called deposition and the result can be witnessed as the accumulation of frost on an automobile on a clear winter night. In reverse, ice can directly change to water vapor by gaining enough energy; this process is called sublimation and can be subtly observed by noting the decrease in an ice or snow cover after windy conditions; even when **temperatures** are below 0°C, ice can sublimate into water vapor.

The energy inherent to water vapor that allows water vapor to remain as water vapor is known as latent heat. This name implies that the heat is "hidden" and not directly detectable by thermometer. At sea level, the conversion of ice to water vapor requires about 2,850 joules. This is energy lost from the water's surroundings and converted into hidden, latent form. The latent heat remains with the water vapor until condensation or evaporation takes place. These processes may take place many thousands of kilometers from where the vapor was added to the air. The latent heat effect can be quite dramatic. When water changes phases from vapor to liquid or solid, the release of latent heat can warm the atmosphere. If this phase change is ongoing while air is rising latent heat release retards the adiabatic cooling, making the air rise to much greater heights than it would without this phase change. Latent heat provides the huge amounts of energy needed for thunderstorms and hurricanes.

When the atmosphere is holding all the water vapor it can hold, the condition is known as saturation. This does not mean that there will be **precipitation**, but saturation is a necessary precondition for precipitation. The saturation condition of the atmosphere is a cubic function of air temperature. That is, several times more water vapor is needed to saturate the air at 40°C then at 0°C. This explains why precipitation amounts can be much more out of cold air than out of warm air.

In that water vapor is so essential to the functioning of world weather, several methods are used to describe atmospheric water vapor. The following are measures highlighting various facets of humidity. As one of the gases in the atmospheric mix, water vapor exerts a partial pressure subsumed as part of barometric pressure. In the polar regions, vapor pressure is frequently less than 5 mb at sea level while it averages over 25 mb over tropical oceans. Compared to the average sea level pressure of 1,013.2 mb, this means vapor pressure is, at best, a small fraction of atmospheric pressure.

Dewpoint temperature is another measure of humidity. The dewpoint is the temperature to which air must cool to bring it to saturation (see below). Dewpoint temperature is an indirect measure of the amount of water vapor and latent heat content. If dewpoint temperature and air temperature are the same, the air is saturated. If not, the dewpoint will be lower than the air temperature and the air has some additional capacity with which to store water vapor. The greater the spread between air temperature and dewpoint temperature, the drier is the air; the farther the air is from saturation, the faster evaporation will take place. Dewpoints up to about 15°C are perceived as comfortable by most people, while dewpoints in excess of 20°C cause discomfort because of slowness of evaporation of perspiration. The U.S. National Oceanic and Atmospheric Administration calculates apparent temperature, which quantifies human discomfort based on the combination between air temperature and the relative humidity (see below). The mixing ratio is measured in grams of water vapor per kilogram of air not including the water vapor. This measure is frequently used in scientific work because it does not vary in rising or sinking air. The mixing ratio varies from almost 0 in polar winters to almost 20 over tropical oceans.

Relative humidity is the humidity measure most people hear about. It is the percentage relationship between the actual amount of water vapor in the air and the amount of water vapor possible in the air *at that temperature*. It ranges up to 100 percent (saturation) and can represent varying amounts of moisture because vapor capacity changes with temperatures. Two factors can change relative humidity: amount of moisture in the atmosphere and temperature change. Of these two factors, the atmosphere uses temperature decreases as the most ready way to bring itself to saturation. Relative humidity is affected by the diurnal heating cycle. Under good weather conditions, relative humidity is highest near dawn and lowest in the mid afternoon; this is because relative humidity is inversely related to temperature.

Unlike the lower atmosphere's gaseous mainstays of nitrogen and oxygen, water vapor presents hugely complicated patterns over the planet over time and place. Over 99 percent of humidity is in the atmosphere's first layer, the troposphere. The other layers are very dry and offer little in the way of cloudiness or precipitation. On the average, the water vapor content of the sky is greatest at sea level. Thus, the tops of very tall mountains such as Everest (8,848 m) and Denali (6,194 m) have vapor pressures substantially lower than near the surface. The bottom of the **atmosphere** has the richest supply of water vapor because vapor of the troposphere is from the evaporation, sublimation, and transpiration from land and ocean surfaces. Geographically, vapor is most plentiful over tropical oceans and least plentiful over polar land and ice surfaces. There is a huge daily variability as the **air masses** bring surface air poleward and equatorward over the planet. For instance, winter satellite imagery frequently shows "the Pineapple Connection" from Hawaii to the California coast. This huge "conveyor belt" of vapor brings tropical moisture to become the winter rains and snows. Although exact numbers are not available, the estimate of the total amount of Earth's water vapor is 138 cubic kilometers with an average residence time of 3 to 4 days. This means that there is an unending sun-fired transfer of water into atmospheric vapor and that it is soon precipitated out.

Hurricanes

Hurricane is the name for an intense and dangerous type of cyclonic storm known through most of the tropical regions of **Earth**. In the Atlantic Ocean and in the Pacific Ocean east of the International Dateline the storm is called a hurricane. In the Pacific Ocean west of the International Dateline it is known by the name typhoon although there are local appellations such as baguio (Philippines) and willy-willy (Australia). The storm is known as a tropical cyclone or, simply, cyclone in the Indian Ocean and the western South Pacific Ocean. The name appears to originate from the Spanish spelling of a word "huracan" used by peoples of the Caribbean basin to denote a "god of evil" but is alternatively translated as "big wind."

Hurricanes usually start as tropical waves that are mild disturbances in the trade wind of the tropical latitudes. These tropical waves pass from east to west at speed averaging around 20 km/hr with wavelengths somewhat over 2,000 km. The waves are common, with a hundred or so passing over the North Atlantic during its hurricane season. They are areas of disturbed weather containing rain and thundershowers and pass over a location in 3 to 4 days. A few of these storms develop closed isobars whence they are named tropical depressions. When wind speeds exceed 17 m/s they become tropical storms and when the wind exceeds 34 m/s they become hurricanes.

Hurricanes can exist for days at a time. Some hurricanes have lasted for upward of a month, but hurricane force winds usually inhabit the disturbance for an average of a few days. Hurricanes weaken as they pass over cool water or over land or encounter strong, shearing winds at high altitude; any of these conditions disrupts the latent heat supply on which the hurricane feeds. The strengths of tropical disturbances are rated on the Saffir-Simpson scale. The scale ranges from category 1 through category 5. The category definitions are given in Table 2. The wind definitions are based on sustained winds and not the higher wind gusts.

Table 2. The Saffir-Simpson Scale

Category	Wind speed (m/s)	Wind speed (mph)	Wind speed (knots)
Tropical depression	<17	<39	<34
Tropical storm	17–33	39–73	34–63
1	34–42	74–95	64–82
2	43–49	96–110	83–95
3	50–58	111–130	96–113
4	59–69	131–155	114–135
5	>69	>155	>135

Hurricanes are like many other occurrences in the physical geographic environment: there are many small hurricanes and a few large ones. For instance, in the last century and a half only a third of the landfalling hurricanes in the mainland United States have been major hurricanes (categories 3, 4, and 5). Yet, the vast majority of damage and death are caused by the major hurricanes.

Hurricanes are giant heat engines that feed on heat in tropical oceans. They "live" on latent heat release from ocean water at temperatures in excess of 26°C. Energy is distributed upward in the troposphere thus helping to maintain the Earth's energy balance. Hurricanes are distinguished from **middle latitude cyclones** in that hurricanes are warm-core storms because of latent heat of condensation. They also weaken with altitude above the sea and must develop in environments with weak upper winds.

Hurricanes have well-known structures. They are organized cyclones with centers of relative low pressure around which there is a large pressure gradient forcing air to circulate from higher to lower pressure. The ferocity of the winds is dependent on the pressure gradient so the fastest winds are found in the eyewall that surrounds the placid air of the hurricane's eye. Winds decrease outward from the eyewall and are usually beneath hurricane strength a couple of hundred kilometers out from the center. Winds spiral cyclonically inward toward the eyewall, in a clockwise fashion in the Southern Hemisphere and in a counterclockwise fashion in the Northern Hemisphere.

Apart from the eyewall, the most intense weather is arranged in spiral bands feeding mass and energy toward the center. Typical hurricanes have cloud shields that are about 500 km across (smaller than a middle latitude cyclone) and they track at about 25 km per hour. Hurricanes usually originate from 5° through 20° north and south of the equator. The do not originate over the tropical South Atlantic. They commonly are embedded in winds on the equatorward peripheries of the subtropical highs. Individual storms can take a wide variety of paths that can be quite erratic because of the lack of upper air control associated with hurricanes. Long-lasting storms tend to move to the west and then curve poleward as guided

Barrier Islands

Paralleling many of the world's coastlines are long, narrow stretches of sandbars called barrier islands. These can range from a few hundred meters to many kilometers from a coast and can be well over 100 km long. Average elevations on barrier islands are usually a couple of meters. They are present on the shallow portions of continental shelves and most result from plentiful sand deposited by breaking waves. Coastward from barrier islands there are areas of shallow, protected saltwater known as lagoons. Protected from the greatest motions of the oceans, lagoons produce fertile and biodiverse ecosystems and are key to the life cycles in some oceanic fish. Prominent barrier islands in the United States include Padre Island of Texas and the Outer Banks of North Carolina. Elsewhere, Colombia, the Netherlands, and India are among countries with extensive barrier island systems. It is apparent that many barrier islands have had a complicated history as sea level fell dramatically during the Pleistocene and then rose during recent times. Barrier islands tend to be only a few thousand years old and are capable of migration and disappearance as sea levels rise and/or sand supplies lessen. Large storms, particularly hurricanes, are capable of breaching barrier islands to make inlets to the lagoons. Humans have built resorts and cities (e.g., Miami Beach) on barrier islands, thus presenting an inherent locational danger.

by the position of the subtropical high. Hurricanes have a definite seasonality. Even though it can be said that it is "always summer" in the tropics, different portions of the year have different flow regimes. The months just after the summer solstice of the respective hemispheres are the times of hurricanes. For instance, the North Atlantic hurricane season is defined as June 1 through November 30. In this region, hurricanes are most common at the beginning of September. The temporal occurrence of hurricanes is closely related to the extent of the warm sea surface temperatures needed to cause and maintain them.

Around the world, hurricanes are named through the auspices of the World Meteorological Organization, which keeps six-year rotating alphabetical lists of upcoming names for each ocean basin. There is an alternation between men's and women's names appropriate to countries surrounding the ocean basin; the letters Q, U, X, Y, and Z are not used. Storms notable for damage or loss of life can be permanently retired and other names substituted. Physical damage from hurricanes can result from one of several causes. The most utterly destructive in terms of property and life is the storm surge. The storm surge is a large lens of water pushing onshore and does not refer to the wind-generated waves of a hurricane. Sea level is a vertical function of atmosphere pressure. Near the center of the storm, water is piled up by the combination of dramatically low surface pressure and high winds. Storm surges have been known to top 8 m with individual events being governed by the structure of the individual storm and the shallowness of the near-coastal ocean bottom. The storm surge gives no quarter and it is in the storm surge that most lives are lost. The horrific cyclone of 1970 killed more than 300,000 people in the low-lying country of Bangladesh. Winds, of course, are also a major cause of hurricane damage. Winds are strong in a minimal hurricane but can become incredibly ferocious under Saffir-Simpson scale Category 5 conditions. The power of the wind is a cube of its velocity so that small increases in sustained wind velocities make for huge differences in destructive power. A third type of damage is flooding caused by copious rains. Hurricanes frequently produce rainfall totals in excess of 500 mm per day. These rates cause free-flowing streams to quickly rise and exceed flood levels. Finally, the large-area cyclonic rotation in hurricanes frequently makes some of its thunderstorms tornadic.

In the United States, there has been an average of about one-and-a-half hurricane strikes per year. Inflation-adjusted damage has been rising and this is because of the long-term increase in the population and value of structures and possessions. Hurricane Katrina of 2006 was the most damaging of hurricanes with estimated losses placed at $125 billion. It came ashore as a strong Category 3 but with a large storm surge along a heavily populated coast.

The U.S. loss of life has been modest since the unnamed Galveston storm of 1900 that killed over 6,000 people. Even the catastrophe of Hurricane Katrina was responsible for less than 2,000 deaths. Over the years, knowledge of hurricanes has grown and populations receive notice of impending storms days in advance. Although exact landfall locations cannot be reliably forecast more than a day in advance, there is enough foreknowledge to evacuate people inland if there exists an infrastructure with which to quickly remove large numbers of people. Large evacuations are feasible in developed countries like the United States but problematic in poor populations such as that of Bangladesh.

Hydrologic Cycle

Water is continuously involved in transport and exchanges between ocean, land, and atmosphere: this is the hydrologic cycle having no beginning or end. **Earth** is covered by a world **ocean** energized by **solar energy** and causing the cycling of water out of and back into the oceanic reservoir. Water's ready ability to gain and lose energy and change the spatial arrangement between its molecules makes this possible. As water is able to circulate around the planet it is easy to appreciate the importance of the hydrologic cycle that provides **precipitation** to continental interiors. Out of sight to the casual observer is water's role in transporting latent heat thus having a major impact on the global energy balance and providing energy needed to create atmospheric disturbances such as thunderstorms and hurricanes. Of tremendous importance is water vapor's major role in causing the greenhouse effect.

Homer's *Iliad*, written in the eighth century BCE, provided the first recorded notion of the connected nature of water over the planet. It remained for modern science to measure the various components with instruments. Although all of non-ocean water is about 3 percent of the world's supply, the hydrologic cycle moves huge quantities of water over time. Annually, the cycle has been estimated to have over 400,000 *cubic kilometers* of water entering the atmosphere. The hydrologic cycle is not to be viewed as a smooth transference of water but works in "fits and starts." One week the Great Plains of North America might receive flooding rains from the Gulf of Mexico and the next several weeks can be very dry with the steering patterns of the upper troposphere blocking the flow of low-level moisture. So, too, the hydrologic cycle has geographic variations of its components. For instance, the evaporation may be prodigious from a tropical ocean surface, whereas cold **temperatures** and ice in polar waters make the hydrologic cycle much more sluggish.

At any slice of time, 97 percent of Earth's water is in the ocean. The ocean is large compared to the total surface of the planet and also quite deep. The deep ocean water is away from the direct influences of the atmosphere. It is in the ocean for thousands of years at a time because of its very slow circulation; in spatially limited upwelling areas these bottom waters come to the top. Near the surface the waters are mixed by wind and heated by solar radiation so these waters have a shorter average residence time.

Solar energy and energy in the wind evaporate liquid water from the surface and add it to the air as water vapor. Additionally, there is a transformation called sublimation that changes ice directly to water vapor even at air temperatures well below freezing. Worldwide, this latter process generates less water vapor than evaporation because of less total energy locally available in colder places and the fact that most of Earth's ocean is unfrozen. The vapor added to the atmosphere leaves behind the impurities of liquid water and ice so that eventual precipitation will be fresh water.

Water vapor is circulated as part of the wind and sometimes is condensed and deposited onto hygroscopic nuclei to form **clouds**. Cloud formation can occur very far from the oceanic moisture source. It is thought that water vapor has an average atmospheric residence time of about nine days. The vast majority of clouds are non-precipitating, but in selected places with rising air, the hydrologic cycle is furthered through precipitation processes.

As precipitation reaches Earth's surface, about 77 percent of it falls back into the ocean. The other 23 percent makes life on the land possible. Precipitation

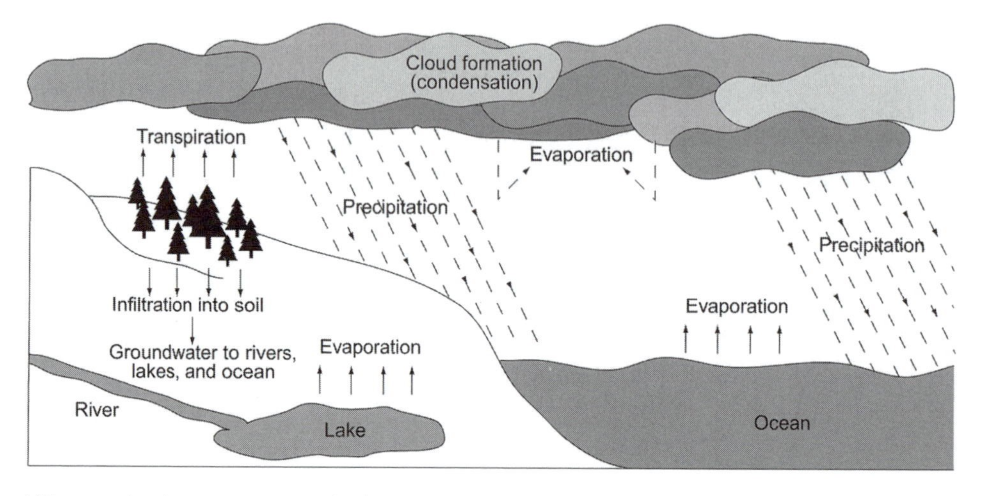

Water molecules are continuously changing physical state and location, moving between solid, liquid, and gas phases. This hydrologic cycle affects climate, ecosystems, and geomorphology. (ABC-CLIO)

falling on the land can be in solid or liquid form. Solid precipitation—snow, sleet, ice pellets, and hail—remain in storage on the surface for various lengths of time. Some solid precipitation will immediately melt while some falls at high altitude or high latitude and becomes part of a year-round snow cover. In some cases the snow falls, is compressed by snowfall after snowfall over a course of many years, and metamorphoses into glacial ice. Glaciers represent storage in the hydrologic cycle. Water may be stored in large glaciers for over 400,000 years. Eventually, glacier ice returns water directly to the ocean by melting in contact with the ocean, melting into a stream, or evaporating and sublimating into the air. Ironically, glaciers and snow account for 78 percent of the world's freshwater supply, making much of it not readily available for the benefit of life.

In places where there are seasonally warm temperatures, there is summer melt from the ice and snow and much of this passes along streams. The American Southwest is arid at low elevations but experiences significant snowfall at high elevations. The winter snowpack provides summer stream flow so human concerns such as hydroelectric generation and irrigation are closely tied to the nature of the snowpack.

Liquid and solid precipitation falling on the ground and vegetation might follow several paths. The simplest is that rain will run via gravity over land and into streams. Most streams are organized into drainage basins connected to the ocean and this is the simplest completion of the hydrologic cycle. Yet, streams comprise only about .0001 of a percent of planetary water. The water in a stream might not reach the ocean directly. There is loss into the groundwater supply, evaporation along streams, and some streams flow into natural and artificial ponds and lakes that can impound water for many years.

Precipitation falling as rain, or falling as snow and then melting, infiltrates **soil**. The amount involved is widely variable even over small areas because soils are so heterogeneous. Some of the infiltrated moisture is held against gravity for days and weeks in the soil pores and so becomes soil moisture. Soil moisture is one key to life on Earth. Plants rooted in soil usually do not tap precipitation directly from the sky. They have evolved to use soil moisture, which is a steady supply as long as precipitation reliably fills the soil pores. Plants bring in the liquid water through the roots and let it go in vapor form, usually from small openings in the undersides of leaves; this path back to the atmosphere is known as transpiration. Of course, some water evaporates from the soil itself. Evapotranspiration is the combined name given to water vapor from all sources.

Sometimes, infiltrated liquid water passes entirely through the soil profile and enters the joints, cracks, fissures, and other openings present in "solid rock." In this case, it is known as **groundwater** as opposed to soil moisture. Given certain rock types, especially sandstones, large quantities of groundwater can be stored. About .5 percent of Earth's water occurs as groundwater. The residence time can be millions of years and it is difficult for groundwater to find as convenient a path to the sea. In selected places, groundwater can emerge at the surface in the form of springs and that water can enter streams. More generally, the rocks containing groundwater intersect with the levels of stream channels and provide steady baseflow to streams. It is generally estimated that about half of Earth's shallow groundwater returns to streams in this way.

I

Imperialism

A set of policies, processes, and strategies that contribute to the construction of an empire. Imperialism has been pursued by some states since the Egyptian and Sumerian empires, dating to as early as 3500 BCE. Usually imperial states subjugate weaker states around them, or establish colonies that they then dominate politically and economically in a relationship known as *colonialism*. Numerous empires have been formed across the span of history, including the Assyrian, Roman, Mongolian, Ottoman, British, and Soviet. States pursue expansion into empires to secure additional **natural resources**, establish markets through mercantilism, spread a religious ideology, or acquire advantages in **geopolitics**. The drive to absorb other territory and peoples into an empire is often accompanied by **ethnocentrism** on the part of imperial powers, who subjugate others in an attempt to "civilize" them. Some historians identify a "New Imperialism," often called the "Age of Imperialism," a period roughly corresponding to the Industrial Revolution from about 1825 to the conclusion of World War I. During this era, the United Kingdom, France, Belgium, Germany, Japan, and the United States all seized additional territories overseas. The European powers came to control virtually the whole of Africa, during the so-called Scramble for Africa—only Liberia and Ethiopia retained independence by the end of the 19th century. Japan built its empire along the east coast of Asia, taking parts of China and Korea, while the United States inherited a sizable portion of Spain's colonial holdings, after defeating that country in the Spanish-American War, as well as a handful of additional territories, including Hawaii.

Imperialism invariably involved the subjugation of less powerful groups by those who are numerically superior, possess greater force of arms, or have more advanced technology. The economic relationship between **core and periphery** was exploitative. The margins of the empire, frequently represented by overseas colonies in the case of European empires, supplied the economically advanced "home country" with a supply of raw materials, extracted with either slave labor or a work force that was frequently underpaid and brutalized. The degree of exploitation and brutalization varied greatly, depending on the nature of the governing power. In the cases of Great Britain and France, an effort to establish local institutions of education and governance based on those in the home country was

The Berlin Conference

In 1884, Otto von Bismarck, the leader of Germany, organized an international meeting of European colonial powers, with the aim of dividing Africa into realms of control and exploitation. Germany was in the process of developing an overseas empire, and von Bismarck's objective was to secure as large a share of the resources and land of Africa as possible for his country. Great Britain and France had already taken control of large swaths of the continent, and Germany, along with Belgium, Portugal, and other European powers, was determined to acquire control of a significant portion of the territory that remained unclaimed. The United States was invited to the conference, but the American administration did not send a representative. The agreement that resulted from the conference assigned virtually the entire continent, with only the exceptions of Liberia and Ethiopia, to some European power, as well as regulating trade and traffic on some African rivers, and establishing other standards between the colonizing powers. One of the most pernicious sections of the agreement allowed the Belgian king, Leopold III, to acquire much of the Congo Basin as his personal holding. The brutalization of native Africans under Leopold's rule would serve as an infamous example of the excesses of **imperialism**.

pursued, and both countries had officially abolished slavery in their colonies prior to 1850. This is not to suggest that the treatment of indigenous peoples living in these empires was nondiscriminatory and humane, but compared to policies and practices elsewhere, the imperialism of both appear almost enlightened. One of the most horrific examples of brutalization transpired under Leopold II, King of Belgium, in the so-called Congo Free State from 1885 to 1908. The entire colony of Congo Free State was Leopold's personal estate, and much of the local population was enslaved to produce rubber. Those who failed to meet the production quotas of Leopold's administrators were frequently shot or had their hands cut off, and some sources estimate that half of the population was eliminated as a result of the cruel administration of the colony.

The economic imbalance between the imperial center and the remainder of the empire was only one aspect of the relationship. A much greater influence was the imposition and adoption of languages, religion and other traits of the European powers, via the process of **cultural diffusion**. Seizing territory and riches was the focal point of imperialistic policy in many instances, but accompanying this was a paternalistic notion of a "civilizing mission," the assumption of the "white man's burden" designed to bring "primitive" natives into the sphere of European social and cultural standards. In all instances of European imperialism a concentrated effort was made to convert the newly incorporated populations to Christianity. This was not unique among empires; Islam spread from the Arabian Peninsula after AD 700 in a similar fashion, but the European effort was more sustained and took place across a larger geographic area. This process completely remade the cultural **landscape** of much of the colonial world, as Christianity, shaped by **religious syncretism**, displaced the existing religious worldviews. In addition, the use of European languages spread to all corners of the globe, in some cases virtually completely replacing the indigenous tongue for a majority of the population, as in Latin America, or becoming the language of the educated elite in Africa and southern Asia. Other cultural elements, from sports and cuisine to modes of dress, also diffused from the European center throughout the various empires. Imperialism in this manner represents an early variation of coercive **globalization**.

The United States acquired overseas territories in the late 19th century, primarily as a result of the Spanish-American War. In 1895 a revolt against the Spanish administration broke out in Cuba, an island territory that the United States had attempted to purchase from the Spanish government several times in the mid-19th century. The mysterious explosion of the battleship *Maine* in Havana harbor in February 1898 set the stage for war with Spain, and after less than a year the United States had acquired the Philippines, Puerto Rico, and Guam as colonial territories, with Cuba as a protectorate. Cuba was granted independence in 1902, but the remaining islands all stayed under U.S. administration until after World War II, when the Philippines became an independent country. Puerto Rico and Guam are still part of U.S. sovereign territory, although Puerto Rico has held plebiscites several times over the issue of independence.

One of the most evident geographical results of the Age of Imperialism was the formation of a multitude of **nation-states**, especially in Africa, the Middle East, and Southeast Asia. The imperial powers that controlled these regions established administrative **boundaries** that rarely took into account the political and ethnic relationships that had existed before the advent of colonial authority. Rather, the boundaries separating these territories were frequently based on natural features such as rivers, mountain ranges, or some other aspect of the landscape; or in some cases followed established lines of latitude and longitude, resulting in geometrical patterns that had no relationship to the cultural histories of the groups living in the **region**. For example, it was commonplace for imperial borders to divide ethnic groups between various colonial powers, resulting in the political fragmentation of such peoples. A case in point are the Yoruba, most of whom were incorporated into Nigeria, a British colony, but sizable communities of Yoruba were included within the colonies of Togoland (Togo), originally a German possession, later transferred to France; and Benin, a part of the French colonial empire in West Africa. Today, the Yoruba people remain divided among these three countries, because the borders adopted by the European imperial governments were retained after the imperial powers withdrew and granted their colonies independence. These political boundaries, superimposed on the cultural geography of Africa, have been blamed for much of the centrifugal tendency and **balkanization** that have characterized African politics since the 1950s because they often clustered groups that historically were adversaries within the confines of a single state, or conversely, diluted the influence of many groups by dividing them among several countries.

Although the overt political control of imperial powers has faded over the past half century, some scholars argue that this has been replaced by more subtle forms of imperialism, including cultural and economic imperialism. Some consider globalization to be simply a more indirect means for the developed world, representing former colonial powers, to control the global economy and secure markets for their products by controlling the flow and supply of capital and technology. **World systems theory** and dependency theory both are anchored in the position that the exploitive economic relationship between the industrialized, imperialistic countries and their colonial holdings was not ameliorated by political independence of the colonies, but rather was simply replaced by a more insidious system of control and manipulation that perpetuates the dominance of the imperial states. For these commentators, the Age of Imperialism has not come to an end, despite what they consider to be a superficial independence for most of the colonial territories. Another school of thinkers holds that cultural imperialism, the ubiquitous application and promotion of external cultural values that erodes the foundation of traditional culture, is the actual legacy of political imperialism. The proponents of this view assert that Western, and especially American, culture and value systems have usurped the traditional cultures of much of the world, and that this process results in both a corruption of traditional value systems, and the elimination of non-Western cultural systems, or at the very least a hybridization of culture. The process is reflected in everything from the loss of indigenous languages to the adoption of Western modes of dress, and is typically viewed as leading to a loss of diversity in the human family and resulting in a homogenized **cultural identity**. The agents of cultural imperialism are primarily the media, and in recent decades the Internet, which now (according to the supporters of this theory) has the ability to project Western culture to every corner of the planet.

Infant Mortality Rate

The Infant Mortality Rate (IMR) is defined as the number of infants who perish during the first year of life per one thousand live births. The IMR is typically computed for a given country or state and reported on an annual basis. Because it is a rate and not a percentage, it is reported as an integer. For a child to be included in the IMR calculation, the infant must exit the birth canal or womb (in the case of a cesarean delivery) with a beating heart or other visible signs of life. Stillbirths, miscarriages, and abortions are not included in the IMR. The cause of death is not specific—a child may perish from any cause and still be included in the calculation. There is some controversy over how various countries report the IMR. Some countries like the United States report all infants who show signs of life on exiting the womb or birth canal and subsequently die as infant deaths, but other countries employ different standards, including minimum infant weights or lengths. The Infant Mortality Rate in countries using the latter standards will appear to be artificially low when compared to those from countries where such reporting techniques are not utilized. In addition, the IMR reported for many developing countries may be lower than the actual figure because a significant number of infant deaths may go unreported in rural areas, due to lack of contact with medical providers or others responsible for reporting IMR data.

The IMR is often used as a measure of quality of life in a country, and may be linked to continued viability of a political state. Indeed, the IMR is sometimes used as a surrogate standard for economic and social development, and therefore has much wider application than simply as a component of demographic dynamics. Japan and several Scandinavian countries currently report the lowest IMR figures among countries having a population of at least 1 million, and for 2008 both Japan and Sweden are estimated to have an IMR below 3. On the other end of the scale, many countries in Africa report IMR numbers exceeding 100. In fact, of the 30 countries with the highest estimated IMR computations for 2008, 28 were African states. Infant deaths in these and other developing countries are often the result of preventable causes, like dehydration due to diarrhea and dysentery. The introduction and distribution of relatively simple and inexpensive treatments for these diseases have resulted in significant advances in limiting infant mortality. Encouragingly, in recent decades the IMR has declined considerably in both developed and underdeveloped countries. In the United States, for example, the IMR fell from a figure of almost 10 in 1990 to slightly above 6 in 2008, nearly a 50 percent drop in less than two decades. Likewise, the global IMR has fallen dramatically, diminishing by more than half between 1960 and 2008. And, although some countries continue to report alarmingly high infant mortality in comparison to others, it should be noted that the rate of infant death now is at historic lows in all portions of the world.

Irredentism

The desire of a state to "recover" territory previously lost to another state, typically on the basis of uniting similar ethnic groups or on the grounds of prior historical claim. Irredentism is a policy rooted in **territoriality** and the **organic theory** of the state, and had a profound effect on the **geopolitics** of the 19th and 20th centuries. Irredentism is a common aspect of the foreign policies of countries occupying a **shatterbelt**, because **boundaries** often shift in such locations, leaving ethnic groups divided among two or more **nation-states**. A synonym is revanchism, a term derived from the French word *revanche*, meaning "revenge." Irredentism comes from the Italian term *irredenta*, which referred to lands populated by ethnic Italians that were not incorporated into the Italian state during the *Risorgimento*, or Italian unification, during the 19th century. A number of such territories existed, including Trieste, Nice, and Corsica. Several of the irredenta lay within the Austro-Hungarian Empire, and the Italian government entered World War I on the side of the Allied Powers with the expectation that a defeat of the Austrians would open the way for the absorption of these places into Italy. Over the course of the 20th century, several of the disputed territories in fact were included in the boundaries of Italy. The collapse of empires usually results in the expression of irredentist motives on the part of the remnant states, since the existing cultural **landscape** of ethnicity rarely correlates directly with the new political borders marking off the new states.

The consequences of irredentist foreign policies have frequently been severe, leading to larger conflicts. Both the world wars of the 20th century were at least partially caused by irredentist motivations. Gavrilo Princip, a member of a radical Serbian irredentist group, killed Archduke Franz Ferdinand, the heir to the throne of the Austro-Hungarian Empire, on June 28, 1914, in the city of Sarajevo. The assassination set into motion a series of events that quickly erupted into World War I, and ultimately the death of approximately 16 million people. Ironically, the creation of Yugoslavia as a result of the war partially fulfilled the ambition of Princip and his fellow irredentist conspirators. In the 1930s, Germany under Adolf Hitler began pursuing an aggressive irredentist policy under the guise of *lebensraum*, or "living room," for the German state. Hitler's remilitarization of the Rhineland in 1936, his incorporation of Austria into German control in 1938 (known as the *Anschluss* in German), and the invasion and absorption of the Sudetenland region of Czechoslovakia in 1938 all were motivated by irredentism, and the desire to create a "Greater Germany" from regions holding a sizable German population. All of these regions were occupied by German troops with little resistance from other major European powers. Some historians argue that Hitler's successful pursuit of an aggressive irredentist policy toward these territories emboldened the German regime to eventually attack Poland and initiate World War II, and also bought valuable time for the strengthening of the German war machine before the outbreak of hostilities. The death toll from Germany's irredentism was truly catastrophic, estimated at 60 million worldwide.

Today there are at least 25 active irredentist claims around the world, and a number of serious conflicts stem directly from these issues. The war between Azerbaijan and Armenia over the territory of Nagorno Karabakh has killed perhaps as many as 10,000 people and resulted in hundreds of thousands fleeing their residences. Nagorno Karabakh lies within the borders of Azerbaijan, but is populated mostly by ethnic Armenians. The Armenians in the region, with the support of the Armenian government, began agitating for unification with the Armenian Soviet Socialist Republic in 1988, three years before the collapse of Soviet authority. Ethnic violence between Armenians and Azerbaijanis increased, and after the disintegration of the USSR, outright civil war erupted in Azerbaijan between the Armenians living in Nagorno Karabakh and the Azerbaijani government. As of early 2009, the status of the territory was not settled. Other potentially violent irredentist claims involve the Peoples Republic of China's claim to Taiwan, the co-claims of India and Pakistan to Kashmir, and Argentina's claim to the Falkland Islands. In the case of the latter two examples, wars have been fought in recent decades, and complete resolution of the disputes does not appear to be imminent. Irredentism has been a volatile element of foreign relations for many centuries and remains problematic even in regions that indicate a strong trend toward **supranationalism**, like Europe. The costs of such a foreign policy, as the history of the 20th century has proven, may be quite high.

K

Karst

Karst is an unusual word used to describe an interesting set of landforms and land-forming processes. The word is the Germanic version of the Slovenian *Kras*, which refers to a particular limestone plateau in Slovenia and Italy that was named for being barren. It was there that the classic scientific descriptions of the **land-scape** were made and, so, the Kras landforms were used as archetypes for features in many other places in the world. Several other languages, including Latin and Chinese, have their own terminologies, but the scientific community uses *karst* and other words to describe the landforms of the Kras Plateau or to any similarly formed surfaces.

Karstic landscapes are widely spread and found in areas underlain by particular rock types, structures, and goodly amounts of precipitation under tropical and middle latitude **temperature** conditions. Additionally, some of today's areas are the products of paleoclimates no longer in existence at those locations. Prominent areas of karst are found in Morocco, the United Kingdom, France, Australia, China, Slovenia, Italy, the United States (the Ozarks, Appalachians, and Florida), Russia (Urals and eastern Siberia), Brazil, Mexico, Costa Rica, South Africa, Namibia, Thailand, Vietnam, and Papua New Guinea. It must be recognized, though, that karst regions represent a minority of the surface areas of these places. Karst landscapes can be quite rugged or very gentle depending on the subsurface rock, amount of precipitation, and the length of time over which the landscape has evolved.

Most landscapes result from the erosion and deposition caused by a complicated interplay of factors energized by climate. The absolutely distinctive identifying feature of karst landscapes is that they are places where underground solution of bedrock has been dominant in shaping the surface. Water is sometimes called the "universal solvent" because it is able to dissolve all natural materials given enough time. Whereas we think of rock as being permanent, it is hardly so in the context of **Earth** history. The solution rates of various rocks vary by millions of times and it is some of the fastest rates that are associated with karst.

Water is much more effective in dissolving rock when the water contains impurities. Water infiltrating the rock incorporates atmospheric carbon dioxide and it is this weak carbonic acid that is a potent dissolver of some types of rocks. Also, contact of groundwater with rock introduces other impurities which hasten solution. Solution rates are fastest where groundwater is not stagnant and passes over interior rock surfaces without becoming saturated and slowing solution. Rocks containing calcium are particularly vulnerable so that the plentiful limestone, dolomite, halite, and gypsum exposed near Earth's surface are candidates for forming karst. Aluminum-, iron-, and silicon-based rocks are much more resilient to erosion. Bedding, jointing, and faulting structures can be vital because they provide avenues of least resistance through which the water can infiltrate vertically and move laterally. Vast amounts of rock can be removed via solution so that over time the near-surface bedrock takes on a "swiss-cheese" configuration with all materials eventually dissolved.

An irony of active karst landscapes is the lack of surface streams in rainy climates. The "plumbing" is dominant, and underground drainage is the path taken by rainfall. Dry streams, which flow only during extreme rain events, are common, and some streams are actually swallowed into the ground. Such flows have been traced many kilometers to outlets in springs and streams. The water is not channeled into an individual underground stream but flows slowly through myriad interconnected tunnels. This underground water is dissolving materials to create all manner of cavities, some of which are expressed as landforms on the surface.

There are many types of distinct landforms in karst topography. The most common of these is a sinkhole (doline). Sinkholes are underground cavities that have had their roofs collapse and are open to direct entry of water. They are steep-sided with inward slope angles of 20–30 percent and range from a few meters to many kilometers across. Prodigious numbers of sinkholes exist in some karst landscapes such that surface activities are severely limited. Sinkholes can be either dry or beneath the level of the water table and therefore filled with water. A uvala is a series of interconnected sinkholes forming a steep-sided valley with no outlets.

As the landscape "rots" from within, the sinkholes and uvalas enlarge and leave behind some steep-sided erosional remnants. The most common of these are haystack hills (magotes), and they are quite visible in parts of western Kentucky, Cuba, and Puerto Rico with heights averaging up to 25 m with diameters of up to 200 m. Tower karst is, perhaps, the most dramatic of the karst forms. Haystack hills are nubs compared to the grand spires of tower karst. These are the erosional remains of much higher landscapes and represent landscapes in late stages of topographic development. Through the process of solution over eons, the rock has been left in steep-sided cones and these are undercut and further steepened into towers. Some of these towers can exceed 300 m in relief and are riddled with cavities. Perhaps the most phantasmagorical landscape on the planet is the karst towers along the River Li in southern China. The juxtaposition of the lowlands

These eroded pillars of limestone, located near Guilin, China, are a classic example of karst topography. (Steve Allen/Dreamstime.com)

along the river with the bare rock walls of the towers is dramatically echoed in several other places such as Malaysia and Thailand.

Caverns are underground expressions of karst. These are large openings and tunnels formed along bedding planes and can extend many kilometers. Careful mapping points to the interconnectedness of some caverns in a grid network because of prominent jointing in the rock. Caverns are formed, at first, by solution and then sometimes enlarged by the presence of a stream. In some caverns, water dripping from small openings in the walls and ceilings undergoes dissolution of its minerals because the carbon dioxide comes out of solution as the water enters the cavern. The dripping water leaves behind mineral deposits, most usually of calcium, that decorate the cave ceilings, walls, and floors with fanciful, delicate forms known as speleothems; these include stalactites hanging from ceilings and stalagmites growing from the floors.

Human interaction with karst has been long and tenuous. Because of the underground drainage, running water at the surface and soil moisture are limited even in **precipitation**-rich areas. Agricultural activities avoid the sinkholes but the landscape can be used. An interesting example can be found in western Ireland. The Burren (Boireann), a karst plateau of 250 sq km, is used for grazing in the winter season. The relative lack of winter grasses in the surrounding valleys forced pastoralists to move their herds to use the plentiful winter grasses that grow in the joints of the limestone; these grasses are active because the microclimatic heating in the joints is greater in the winter than in the lowland surroundings.

Urban areas are at some risk because large sinkholes sometimes open under the weight of human surface modification. Parts of Florida offer many examples of this phenomenon. Even after a building or road has been in existence for many years, the mass of the construction can collapse the roofs of cavities, especially in times of drought when the water table is low. Florida's houses, roads, and car dealerships have been swallowed from time to time.

The relatively free flow of surface water into the groundwater makes groundwater quite susceptible to contamination from the surface. This pollution can spread many times more quickly than in other bedrock structures and becomes a potential hazard as water is tapped from wells and springs. Because of such difficulties, karst areas are frequently thinly settled compared to their surroundings.

L

Landscape

A landscape to a geographer consists of an area or small **region** and the features that appear there. These features may be tangible and physically present and observable; or the landscape may be conceptual, representing a structure of cultural or social characteristics, such as a "linguistic landscape" or an "ethnic landscape." The term in English may be derived from a Dutch noun, *landschap*, which indicated a specific parcel of land; or the German word *Landschaft*, which has the literal meaning of "land shape." Various landscapes may play a pivotal role in the establishment and perpetuation of national identity in a **nation-state**. The notion of "landscape" as a conceptual tool for approaching and analyzing a **location** was established in academic geography in the United States by Carl Sauer and the Berkeley School in the early 20th century. Sauer's seminal article "The Morphology of Landscape" appeared in 1925 and initiated an entirely new theoretical framework that challenged the prevailing paradigm of **environmental determinism**. Sauer and his students pioneered the study of **cultural ecology,** which emphasizes the impact of culture on the physical environment, rather than the reverse, which had been the basis for the deterministic ideas that had dominated geographic thought before the late 1920s. Sauer held that the cultural history of a place could be "read" from an examination of the landscape, and that the relationships between the features on the landscape revealed how the place had evolved through a series of cultural influences over time. Most of the work of the Berkeley School geographers was grounded in **particularism**, and addressed landscapes as unique phenomena that were not subject to "laws" that dictated their nature or circumstances.

The notion of landscape is engrained in the research and ideas of many cultural geographers. Two of the foremost thinkers who have promoted the study of geography via the examination of landscapes are Wilbur Zelinsky and Donald Meinig. Zelinsky earned his doctorate working under the guidance of Sauer, and landscapes of various scales and dimensions inform all of his research and publications, including his magnum opus, *The Cultural Geography of the United States*. Zelinsky has studied and written widely on the landscapes associated with **folk culture** and **toponymy** in the United States, as well as applying the concept to

George Perkins Marsh (1801–1882)

A Vermonter and a lawyer, Marsh served in the diplomatic service of the United States in the middle decades of the 1800s. He was a remarkable scholar who was widely traveled and could read over 20 languages. His travel and studies led him to think about human destruction of **landscapes**. He had widely read contemporary European geographers and was struck with ideas that connected humans and nature. He crystallized his thoughts in *Man and Nature, or Physical Geography as Modified by Human Action* (1864) and *The Earth as Modified by Human Action* (1874). He argued that human interaction with the landscape had both conscious *and* unintentional consequences and that some actions of humans were clearly deleterious to landscape. Though well received, Marsh wrote in a time when America was completing its "Manifest Destiny" of human domination of the supposedly infinite resources from Atlantic to Pacific. His thoughts did not resonate in political circles to be translated into public policy. He was an intellectual father of the conservation movement and his works were highly influential on the intellectual growth of the discipline of geography in the United States. Marsh's work is sometimes evoked as seminal in the man-land tradition of the geographic discipline.

spatial manifestations of popular culture. Donald Meinig was a central figure in the emergence of new perspectives in cultural geography in the late 1970s, which was partially a reaction to the **quantitative revolution** that had redirected geographical research a decade earlier. One of his most significant contributions has been to integrate intellectual approaches from the humanities, especially literature, into the spatial perspective of geographical research.

The work of Meinig, Yi Fu Tuan, and other cultural geographers has given rise to humanistic geography, a subdiscipline that is centered on the landscape perspective. *The Interpretation of Ordinary Landscapes*, an edited collection of landscape research Meinig produced in 1979, has had a lasting influence on subsequent directions in cultural geography.

The past two decades in fact have witnessed the emergence of innovative ways of critically analyzing landscapes. How a landscape is interpreted is an important consideration in the so-called new cultural geography that is theoretically grounded in the philosophy of the French scholar Michael Foucault and the writings of other post-modern and post-structuralist thinkers. Some who are engaged in the "new" school of landscape interpretation regard a landscape as a kind of textual feature, which may be "read" in a number of different ways, depending on the perspective and background of the observer. A given landscape may be interpreted quite differently based on the gender, race, social class, age, sexual preference, or even political affiliation of those viewing it, and therefore the significance and meaning of the landscape varies accordingly. Other key issues for cultural geographers regarding landscapes concern how they are constructed, why they are constructed, and how they affect the behavior and attitudes of those who live within them.

Landscape ecology is a field closely related to geography that attempts a holistic analysis of the interaction between human activity and the natural environment. Studies in landscape ecology often are vital to understanding the methods and dynamics of achieving **sustainable development**. Geographers, ecologists, and regional planners use landscape ecology to construct a broad image of the changes brought to an ecosystem over time by human settlement and economic development. Landscape ecologists are concerned with the **scale** at which changes in the landscape take place. For instance, the landscape under study might be no larger than several fields, or it could be as large as a county, state, or province. Various areas within the landscape that are spatially similar are termed *patches*, and the number and variety of these determine the degree of *heterogeneity* a specific landscape indicates. Heterogeneity is an indicator of the spatial diversity of the landscape, which landscape ecologists consider vital to maintaining its stability and avoiding the degradation of local ecosystems.

Organizing space through the spatial structure of the "landscape" has applications and importance beyond academic geography. The recognition and preservation of "cultural landscapes" is a major component of the United Nations Educational, Scientific and Cultural Organization's (UNESCO) World Heritage Program. The World Heritage Program designates regions that fit a specified group of characteristics as "cultural landscapes," which do not fit either strictly "cultural" or "natural" criteria. Cultural landscapes in this context are a blend of "natural" features and "cultural" features (which is precisely how Carl Sauer conceived them) and show the interaction of both over a considerable stretch of time. One of the first locations awarded this status by the World Heritage Program was the national park that surrounds Uluru (Ayers Rock) in Australia. This is perhaps the quintessential cultural landscape because it represents a combination of spectacular natural features with the **sacred space** of Australia's native people. This landscape has been utilized and modified by humans for thousands of years, and today continues to be changed by human activity through the visitation of tourists. The single constant that all landscapes share is the element of change brought about by both natural and human processes.

Latitude and Longitude

Understanding of the location of physical and cultural features of **Earth** is based on knowledge of the geographic grid, otherwise known as latitude and longitude. This knowledge is vital. For instance, the changing angle of the sun in the sky is responsible for seasons and is dictated by latitude. Time, with all its implications for life, is kept by the relative longitudes of locations. The world system of latitude and longitude can be used to precisely specify absolute location on the planet and, so, is essential for mapping and navigation.

Latitude and longitude is a spherical grid, and Earth's surface is not spherical because of its oblate shape and topographic features. However, the nonspherical properties are ignored in the use of the grid. Latitude and longitude are based on the classical Greek system of angular measurement. A circle is divided into 360°. Staring clockwise from point A, one-quarter of the way around represents 90° (B), one half is 180° (C), three-quarters is 270° (D), and a return to the start at point A is 360°. Degrees are divided into minutes and seconds. One degree includes 60 minutes and one minute is composed of 60 seconds. The notation for degrees, minutes, and seconds are, respectively °, ′, and ″. An example of a location on Earth's surface is 37°07′13″N, 97°04′20″W. Such a notation will direct one to within a couple of meters from any point on the planet.

The system of latitude and longitude is universally accepted. Latitude has been certain since the time of the classical Greeks. They appreciated the roundness of the Earth and that there was an equator. The equator is key to the numbering of latitude because it provides a physically based "0" latitude. The equator is in the center of the rhythmic migration of the solar declination between the Tropic of Cancer and Tropic of Capricorn. There is 90° of angular arc between the equator and the north end of the axis of Earth's rotation and between the equator and the south end of the Earth's axis of rotation. So, latitudes are numbered from 0° to 90° in the Northern Hemisphere and the Southern Hemisphere, designated by their positions relative to the equator. Any individual line of latitude represents a circle around the globe oriented west and east. The equator has the largest circumference possible and other lines of latitude have progressively shorter circumferences with increasing angular distance from the equator.

Lines of latitude are always parallel with each other and, therefore, are also known as parallels. Any number of parallels can be designated using degrees, minutes, and seconds. Neglecting Earth's oblate shape, one degree of latitude is anywhere equivalent to approximately 111 km. Important lines of latitude usually shown on maps or globes are the Tropic of Cancer, Tropic of Capricorn, Arctic Circle, Antarctic Circle, and the equator (see **Seasons**). The tropic lines delimit the poleward boundary of the tropics, 23.5° north and south of the equator. The Arctic and Antarctic circles denote the equatorward limits of polar latitudes and are the limits for the regions in which the sun can stay above the horizon for more than 24 hours in the summer and be below the horizon for 24 hours or more in the winter.

Lines of longitude are also known as meridians. As is the case with lines of latitude, any number of lines of longitude can be drawn on a map or globe. Yet, there are substantial differences between meridians and parallels. Meridians cross parallels at right angles and as such, are north-south lines. Importantly, meridians converge to the North and South poles. That is, the Earth distance associated with one degree of longitude approximates 111 km at the equator and becomes progressively less by latitude. At 45° the distance lessens to 79 km and at 90° the distance is 0 because of the convergence of meridians.

Time is reckoned from the daily appearance of the sun overhead of the local meridian. A day, apart from a few minutes' variation caused by the elliptical Earth orbit around the sun, is 24 hours long. This is because the sun appears to pass westward over 15° of longitude per hour. It is not a coincidence that 24 hours multiplied times 15° per hour yields a circle of 360°. Indeed, the configuration of clocks with hands for hours, minutes, and seconds is an analog to Earth/sun relations.

Where is 0° longitude? Unlike latitude, there is no physical argument for any meridian to garner this label. Originally, most cities reckoned their time using their local meridians. As human society became more time aware and time dependent the need for global timekeeping became vital. By the 19th century, matters of increasingly rapid communication and navigation were confused by the plethora of central meridians. A passenger traveling the transcontinental railroads of North America was subject to a new time at every major stop! In 1884, an international convention adopted a universal time scheme with time zones. The prime meridian of the world was set as the longitude of the Royal Observatory at Greenwich, United Kingdom. This was a result of the United Kingdom's long-standing astronomical observations at this location and the fact that the International Dateline, at which point calendar days change any time it is crossed, was relegated to the population-sparse Pacific Ocean.

Longitude is numbered as the angular difference east and west of the prime meridian. Eastward, longitude is numbered from 0° to 180°, and this is the Eastern Hemisphere, which contains Asia. Westward, the numbering is also from 0° to 180°, and this is the Western Hemisphere, which contains the Americas. The International Dateline nominally follows 180° longitude but deviates around islands and countries so as to not be a nuisance to population centers.

Accurate measurement of latitude and longitude are essential to location and navigation. Historically, latitude was measured by use of a sextant, a device that determines the angle of the sun above the horizon. Using this device when the sun is overhead at the local meridian allows a ready determination of latitude from knowledge of time of year and solar declination. This device was limited to use on days without overcast and was difficult to employ when a ship's deck was vigorously pitching. Alternatively, the angles of stars above the horizon in the night sky could be checked against astronomical tables. Longitude was impossible to measure with precision until the invention of the first portable chronometer in 1722. The chronometer is a clock that keeps solar-based time for a specified meridian (now universally the prime meridian). At local noon when the sun is directly over the meridian, the chronometer is checked. The hours and minutes of time difference between local noon and the chronometer reading is converted into longitude at the rate of 15° of longitude per hour. Sextants and chronometers can provide accurate measurements down to seconds of the geographic grid. There has been a revolution in latitude/longitude determination. This was enabled by the 1993 launch of the last of a constellation of 24 geopositioning satellites. Ground units, hand held and inexpensive enough to be owned by virtually anyone, can be used to determine latitude, longitude, and altitude by geometrically determining the angles between the receiver and—usually four or more—satellites.

Law of the Sea

The Law of the Sea refers to a collective body of jurisprudence that has developed largely over the past three centuries. The fundamental question involved how much **territoriality** a state could claim over the sea, as an extension of its shoreline. Some European countries began claiming a "territorial sea" adjacent to their coastlines in the late 16th century, but no internationally recognized standard existed, and such zones were arbitrarily established. As European states became more involved in maritime commerce and exploration, the indeterminate character of the territorial sea led to frequent wars, especially between England and the Netherlands. The two countries fought four wars between 1652 and 1784, and one of the main causes of this extended string of conflicts was the struggle over control of shipping lanes, maritime resources, and territorial waters. By the beginning of the 19th century, a number of countries were claiming a stretch of water adjacent to their coastlines of three to four nautical miles, although some demanded a somewhat wider swath of water. No wars erupted from these conflicting claims, although disputes sometimes occurred. Most of these disagreements were resolved by bilateral or multilateral treaties, but it still remained the purview of any sovereign maritime state to declare and enforce the extent of its territorial waters—no universally accepted and recognized code governed such claims. The formation of the League of Nations in the wake of World War I led to greater efforts to establish global standards governing the seas, although few treaties were actually formalized, and the generally accepted limit of a three- to four-mile territorial sea was maintained. However, the relative stability brought about by this arrangement was shaken by the so-called Truman Proclamation of 1945, in which the United States asserted sovereign control over the resources of the continental shelf.

The Truman Proclamation set in motion a cycle of claims by various countries regarding control of their coastlines and adjacent waters. Some states rushed to declare a territorial sea far greater in extent than any previously recognized, in some cases reaching 200 nautical miles from the shoreline. Others extended control to 12 miles and some retained the old limit of 3 miles. The confusion and friction resulting from such conflicting claims spurred the newly established United Nations to attempt a universally recognized codification of the Law of the Sea. This was especially important because it had become evident that ocean-going commerce and natural resources lying below the water's surface were becoming increasingly vital to the economy of many countries. Accordingly, the United Nations convened an international conference in 1956 with the goal of establishing and formalizing a body of law that would apply globally to all maritime states. The United Nations Convention on the Law of the Sea, or UNCLOS I as it became known, was more successful at creating a code for governing the territorial sea than previous efforts. Four treaties addressing issues of the territorial sea, fishing rights, exploitation of the continental shelf, and conduct on the high sea were ratified, and all went into effect between 1962 and 1966. The larger question of the acceptable extent of a territorial sea remained unresolved, however. Most countries by the mid-1960s were claiming a zone of 12 nautical miles, but some countries in Latin America and elsewhere continued to claim sovereignty over a region 200 miles from their coastline.

A second convention was held in 1960, UNCLOS II, in an effort to address some of the issues that remained unresolved from the previous meeting. This conference lasted only six weeks, however, and accomplished little except to reinforce the status quo. Issues of legal jurisdiction and use and sovereignty concerning the territorial sea remained arbitrary and unclear, and in 1973 UNCLOS III was convened in an effort to finally resolve the most pressing of these matters. Although it required negotiations lasting almost a decade, in 1982 the conference proposed a broad treaty that offered international standards for claims and use of the seas and oceans. A number of important components of the Law of the Sea were resolved in the treaty. A limit to the sovereign territorial sea was set at 12 nautical miles, but a "contiguous zone" of an additional 12 miles was allowed, providing limited legal control over the environment, the right to collect taxes and customs duties, etc. Rules governing the use of **straits, passages and canals,** particularly in so-called archipelagic waters, were promulgated, as well as a new concept, the

Exclusive Economic Zone

The **Law of the Sea** promulgated in 1982 allows each state with a border on an ocean or sea to claim an Exclusive Economic Zone. This region may extend outward from the coastal baseline for up to 200 miles, and grants the controlling country exclusive rights to the economic resources contained within the EEZ. This includes fish and other food sources, as well as mineral resources located on and under the seabed. Countries in possession of an EEZ are expected to follow international conventions and law governing the preservation of marine resources within the EEZ. Island nations of course are permitted to claim an EEZ, and much of the world's oceans now fall within the jurisdiction of either a mainland or island EEZ. In regions where maritime states are located in close proximity, such as straits and passages, then the line delineating jurisdiction must be bilaterally determined. In the case of some supranational organizations, the EEZ is shared by all member states, as it is in the European Union. Although no country has officially declared war over a conflict involving EEZ claims, a number of conflicts have erupted since the concept was implemented, including a serious dispute over the Spratly Islands in the South China Sea.

Exclusive Economic Zone, or EEZ. In general, the EEZ represents a zone extending 200 nautical miles from a coastal baseline and grants exclusive economic control to all mineral resources on and under the seabed of the continental shelf. The treaty also enshrined the concept of "innocent passage," allowing the ships of countries to transit the territorial waters of other states as long as they do so without hostile intent, or breaking laws governing pollution, smuggling, or engaging in other prohibited activities.

The treaty produced by UNCLOS III was ratified in the early 1990s, but not all states recognize its provisions. A few countries, many in the so-called "territorialist group," continue to claim territorial seas far in excess of the 12-mile limit, although the number of states doing so has declined since the convention treaty went into effect in 1994. Some countries have signed the treaty but have not officially ratified its provisions. Interestingly, the United States Senate has to date (2009) failed to formally ratify the treaty, although the U.S. government is abiding by most of the treaty's requirements. U.S. objections have to do mostly with some of the laws governing the exploitation of resources within the EEZs, and it is likely that this issue will be resolved to the satisfaction of the U.S. government in the near future. The UNCLOS III treaty even recognizes limited maritime rights for landlocked states, including the right to innocent passage. Although some issues regarding the Law of the Sea remain ill-defined, UNCLOS III has succeeded in crafting a broad legal framework that should result in fewer conflicting claims over territorial waters and the resources they may hold.

Lingua Franca

A lingua franca is a language that is used as a common medium of communication between two or more unrelated groups. Such a language may be a remnant of **imperialism**, when the colonial power imposed a common language of administration and education on diverse linguistic groups living within the **boundaries** of a colony. For example, English is a lingua franca in Nigeria and India, both former British colonies holding dozens of distinct linguistic groups within their borders. In both countries, English is an "official" language and is widely taught in the public educational system. The use of English as a common language in both Nigeria and India allows people from many diverse ethnic backgrounds to share information, express points of view, and seek mutually acceptable solutions to problems. Colonial imposition is not the only means whereby a lingua franca may develop. A lingua franca may emerge from economic or social necessity. A good example is a language of various pidgin tongues, widely used between groups that have trade relationships, but not long-term contact with one another. Unlike the example of English cited above, pidgins that operate as a lingua franca are not spoken as first languages by any group.

For many post-colonial **nation-states,** the lingua franca in use is a key element in overcoming **centrifugal forces,** and crafting a unitary **cultural identity**. In the case of Indonesia, the official language of Bahasa Indonesia has proven to play a vital role in uniting the linguistically diverse population of over 250 million. The Indonesian government claims that almost 100 percent of the country's population is literate in Bahasa Indonesia, a lingua franca officially developed when Indonesia became independent in 1949. Many scholars credit the emergence of this common tongue, in a nation where several hundred languages and dialects are spoken, with crafting a national identity where little had previously existed. A functional, common language appears to be essential in maintaining the geographic integrity of a nation-state, and in promoting regional economic development. States in which the use of local languages is maintained and the adoption of the official lingua franca is resisted (Yugoslavia, the Soviet Union) often fall victim to centrifugal forces and collapse.

A lingua franca may also help to establish and maintain a regional relationship among a group of countries—French in West Africa and Russian in Central Asia are examples. Some have attempted to develop a global lingua franca, in the hope that an increased ability to communicate would lead to greater cooperation among countries. In the 19th century a lingua franca was created with the intent to promote world peace and integration. This was Esperanto, a language spoken today by several million people, although it has failed to achieve global standing over the last century. On the other hand, the influence of **globalization** in the 20th century has made English into the most widely used lingua franca in history, a trend accelerated by expansion of the Internet and the dissemination of English-language music, film, and other entertainment media.

Linguistic Geography

One of the "pillars" of human culture is a complex language. Human language and dialects are spatially distributed over the **Earth's** surface, and linguistic geography is the study of this distribution. Geographers who focus on the locational dynamics of language may examine many aspects of linguistic diversity. Where are the **boundaries** of language families, how do languages migrate across space, and what is the relationship between language and **cultural identity** are only a few of the relevant questions that linguistic geography addresses. Linguistic **landscapes** are constantly changing on multiple **scales**, a fact that forces linguistic geographers to approach their subject in a variety of methodological ways, and from a range of perspectives. Linguistic geography may involve the study of a local dialect, a national language, the relationship among languages in a given place, the impact of English as a global **lingua franca**, or many other characteristics of spoken and written communication.

Languages and dialects typically dominate a given space, or *language* ***region***. The location of such regions is frequently changing and is not always determined simply by political boundaries. In addition, variations in dialect and usage may often be associated with a specific space. There are almost 7,000 languages spoken in the world today, but approximately 500 of these tongues are classified as "near extinct," meaning that there are only a few individuals who continue to speak and/or write the language. Hundreds of languages have died out over the course of history, but in many cases remnants of such tongues continue to "survive" in living languages. For example, Latin is often labeled a "dead language," but in fact dozens of phrases in English are either directly incorporated from Latin or are modified forms of the original word. In specialized fields such as law, Latin phrases such as *pro bono, quid pro quo*, and *habeas corpus* are part of the everyday vocabulary. Languages move through space by the process of **cultural diffusion,** and the historical **migration** of ethnic groups may sometimes be traced through linguistic connections. It is believed that successive waves of Celtic, Germanic, and later Slavic peoples all migrated into Europe from western or central Asia, because of the linguistic linkages these groups share with other speakers of Indo-European languages. The migration pattern of peoples like the Zulu in southern Africa may be identified in a similar fashion, through similarities with other groups using

Treaty of Tordesillas (1494)

Two years after the initial voyage of Columbus to the New World, Portugal and Spain were vying for possession of lands outside of Europe in both hemispheres. Although both were Roman Catholic countries and neighbors, tensions arose over the conflicting and confusing claims on territory made by explorers from both powers. In order to circumvent warfare between the two Christian states, Pope Alexander VI decreed that all lands west of a meridian located approximately 1,300 miles west of the Cape Verde islands would belong to Spain, while lands to the east would lie within the realm of Portugal. The exact location of the line as it cut through the land mass of South America was ill-defined, and Portugal was able to expand the western border of its claim in what became the colony of Brazil. Through mutual agreement between the courts at Lisbon and Madrid, the line was later shifted so Portugal could lay claim to India. The treaty resulted in Spain receiving virtually no holdings in Africa, although about half of South America and all of Central America belonged to the Spanish. A simple line on a map determined why an entire continent has a linguistic geography divided between Spanish and Portuguese.

Bantu tongues. Thus linguistic geography articulates both contemporary language patterns and those of the past in many instances.

The geographic study of languages provides understanding of the dynamics of cultural and political interaction between different ethnic groups. In such an interaction, the language of the politically dominant group frequently becomes the language of the less-influential group, with a corresponding loss of usage of their native tongue. This is a process called *assimilation*, and has occurred many times in history. A classic example is the loss of Native American languages after the advent of European settlement, roughly in the years between 1600 and 1900. Although estimates of the number of languages spoken in North America in 1600 vary, there is no question that dozens of indigenous languages and dialects disappeared from the North American continent over the next three centuries. It is generally accepted that there were at least 300 languages in use at the beginning of European settlement, but by the end of the 20th century only 175 were still spoken, and a large number of those were in danger of becoming extinct. Indigenous peoples in the region abandoned their native speech for the dominant language of English, resulting in a loss of their cultural history. Not only did the languages die, but the songs, stories, and other elements of native culture also largely disappeared. Assimilation is by no means unique to the indigenous languages of North America, or even to the period of **imperialism**.

The language geographer Charles Withers has offered two theories concerning how and why linguistic assimilation happens. The first theory is what Withers termed the *social morale model*, in which the subordinate group progressively

These Indian children are celebrating Mother Language Day, commemorated by the United Nations to celebrate linguistic diversity across the globe. (AP/Wide World Photos)

declines over successive generations to use the native tongue, due to a perceived lower social standing of that language. Knowledge of the language may even be taken as a sign of backwardness or of being uneducated and unsophisticated, and the language is abandoned by its speakers. In other words, the tongue becomes socially stigmatized, and its use shameful or embarrassing. An example of the social morale model would be the fate of Manx, a Gaelic language unique to the Isle of Man in the British Isles. Manx was spoken by thousands of people there until the 19th century, when its use dramatically declined and was replaced by English. A second method of assimilation suggested by Withers is the *clearance model*, which argues that changes in the economic structure of society lead to disruption in the use of a language by causing the native speakers to migrate to areas offering more economic advantages, where they may be linguistically assimilated. Or alternatively a process of *changeover* may occur, in which speakers of the dominant language settle in large numbers among indigenous speakers, diluting the use of the native tongue and eventually resulting in assimilation. Assimilation is not the only process that may transpire when language regions overlap. Languages may maintain their integrity, but may acquire linguistic elements from surrounding or overlapping speakers, resulting in a *sprachbund*, or linguistic convergence area. When this happens, the languages in question incorporate not only vocabulary from neighboring language regions, but over time may actually change some characteristics of syntax, grammatical rules or structure, or other aspects of usage. This may happen even in cases where the languages are not closely related. For example, Romanian, a Romance language, has for centuries been surrounded by Slavic language regions and over time has incorporated many elements from these tongues. Although still considered a member of the Romance subfamily, Romanian has unique characteristics that differ significantly from other Romance languages.

Rarely are the characteristics of a language uniform across the region where it is used. Internal spatial variations in a linguistic region are called *dialects*, and language geographers study these variations in an effort to distinguish and understand the geographic qualities of dialect use. Dialects may be marked by changes in pronunciation, a nonstandard use of vocabulary or grammar, or some other set of characteristics unique to the location under study. In the United States, for example, use of the term "fixin' " as a verb equivalent in meaning to "preparing" ("I'm fixin' to do that.") is unique to a dialectical region that conforms closely to the Old South and a few states west of the Mississippi River, such as Texas and eastern Oklahoma. Speakers of American English in other parts of the country do not use this construction, and its use marks off the boundaries of a dialect region. Guy Bailey, a linguist, and Tom Wikle, a cultural geographer, have published several recent studies of dialect geography in the United States. Linguistic geography may also include the examination of **toponymy**, the study of the origin and distribution of place names. Toponyms, or place names, can reveal much about a cultural landscape, and the people who created it.

Location

Location identifies a place in space. The notion of "location" is a key concept to geographers, because it provides a context for situating data or phenomena relative to each other in space and is the basis for analyzing such data, because recognizing specific location allows spatial patterns to be ascertained. Location is crucial to the nature of a place and often determines at least some of the characteristics that make it noteworthy, distinct, or perhaps even unique. **Cartography**, the art and science of making the **map**, uses location as the basis for displaying all spatially distributed features and without the concept, space would become a meaningless void, with the discipline of geography losing its *raison d'etre*. In general terms, location can be approached as having either an *absolute* quality or a *relative* quality. Absolute location refers to a mathematically defined position using some system of reference coordinates, typically based on a grid covering the **Earth's** surface—a commonly used grid would be that of **latitude and longitude,** but others such as the Universal Transverse Mercator coordinate system might be used to identify the absolute location of a place as well. Absolute location provides a precise reference for a place or feature, but it tells us nothing about the specific qualities of that place. Relative location, on the other hand, indicates how a place is related to other places around it, how it interacts with those places, and brings to light other factors of location that absolute location ignores or cannot identify.

The idea of location lies at the heart of every branch of geography. Urban geography uses the locational aspects of "site" and "situation" to analyze patterns of urban development. This approach is a more sophisticated extension of "absolute" versus "relative" location in regard to the location of urban places. A central line of inquiry for urban geographers is the question of why cities develop in certain places, and what factors help to determine whether a city grows and possibly emerges as a **primate city,** or fails to increase in size and function, or perhaps even declines and vanishes from the **landscape** completely. The site of a city or town is a key aspect of its location, because it includes the physical characteristics of the place, and the manner in which the city is organized and functions, including its morphology (layout) and internal spatial dynamics. When considering aspects of site, the emphasis is on the immediate vicinity, meaning the *local* geographical considerations. The situation of the city, on the other hand, refers to the locational context of the urban place in relation to the surrounding **region**. The situation of the urban settlement considers its connectivity to the broader environment, including the available **natural resources,** the city's economic and political integration with other cities in the region, its relationship to the regional transport system, and other factors. The situation may change quite rapidly, of course. A city that has a major highway built adjacent to it, or is declared an administrative capital, will likely reap significant benefits from its new situation.

Economic geographers are concerned with how location influences the activities of development and commerce in a place, and the concept is central to much of the theoretical underpinnings of the subdiscipline. "Location theory" in economic geography has focused on the relevance of location in both agricultural production and distribution, as well as how locational factors dictate the distribution, shape and function of markets, and patterns of consumption. A well-known example of the former is the **von Thunen model,** developed in the early 1800s in Germany to explain the locational distribution of various zones or rings of agricultural production that formed around the locus of an urban market. Von Thunen was the first scholar to observe the organizational structure and the spatial dynamics of agricultural output in relation to transport costs, demand, and other market forces. The location and functionality of market centers has also attracted a good deal of attention on the part of economic geographers. Two pioneering theorists in this area were Walter Christaller, who is credited with developing **Central Place Theory,** and August Losch, who, working two decades after the initial articulation of Christaller's theory, offered a modified version that placed more emphasis on the behavioral patterns of consumers, but nevertheless maintained most of the assumptions Christaller had originally employed in regard to location. Both Christaller's original theory and Losch's modifications continue to serve as the foundation for the **locational analysis** of retail activity and have made important contributions to the fields of urban geography, regional science, and marketing analysis.

Several theoretical concepts in social and cultural geography stem directly from the notion of location. One of these is the idea of *locale*. A locale is somewhat similar to a "place" in the minds of social and cultural geographers, but whereas place to some scholars signifies just the physical attributes of a location (this is not the approach of all geographers, by any means), the term "locale" offers a sense, according to the British scholar Anthony Giddens, of "a setting for interaction," in which both the physical attributes of the place and its social functions combine to present a platform for communication between those encountering it. A related, but not identical, idea is that of "sense of place," which can carry several connotations for students of geography. As a subject of academic study, "sense of place" emerged in the 1990s in the geographical literature, rooted in studies in behavioral geography and spatial perception. Geographers have become increasingly aware of the mental impact a particular location may have on visitors and residents, and how this influence shapes the image and emotional connectivity that humans hold toward specific places. The concept has particular relevance to studies in social geography and the geography of tourism. Of special significance is the process of how people construct a sense of place, why we do this, and ultimately how this shapes the character of the place itself. Sense of place may be influenced by gender, age, social class, or historical experience, but the basis of the perspective lies in the concept of location.

Locational Analysis

A theory and methodology of spatial relationships, their flows and interrelationships that served as the core of the **quantitative revolution**. The intellectual approach of locational analysis did not arise with the new momentum of the revolution in the late 1950s, but could be found in much earlier efforts to investigate patterns manifest in space that appeared rational and explicable. The **von Thunen model** of agricultural land use is an example from more than a century earlier, and the work of theorists working in both urban and economic geography in the first decades of the 20th century may be taken as precursors of the later focus as well. Those scholars who promoted locational analysis as a framework for geographical study employed an array of mathematical and statistical techniques, including multivariate analysis, probability models, and other aspects of what was later termed "geomatics."

Academic departments of geography in which locational analysis became prominent as a philosophy of spatial inquiry pioneered the use of early computers as a means of evaluating the patterns of spatially arranged data. This cutting edge technology solidified the claim for many university administrators and federal agencies that locational analysis was a more scientifically based kind of geography, and at least for a time garnered the young scholars who were the foremost practitioners of the methodologies support in the form of grants and contracts, as well as acceptance in many of the leading scholarly journals. Locational analysis became the primary set of methodologies in economic and urban geography and was widely used in the study of **population** dynamics, **migration**, and related topics.

The rise of locational analysis in American geography in the 1960s was initially triggered by the adoption of mathematical models and approaches in several American academic departments in the 1950s. At the University of Washington Professor William Louis Garrison began training students in the use of quantitative techniques in the 1950s, and by the early 1960s a host of newly minted specialists in locational analysis had received doctorates in geography from the university. Additional advocates came from the University of Wisconsin, Ohio State University, and several other institutions. Several of these dispersed to geography departments across the United States, and in many cases, began taking on their own students. By the late 1960s research rooted in locational analysis was rapidly displacing the descriptive, observational narratives that had been the rule for much of the literature in geography for several decades. One of Garrison's students, Brian Berry, became one of the foremost proponents of incorporating locational analysis into geographical research and earned the distinction of being the most cited geographic scholar in English language publications for many years. The 1970s witnessed a backlash against what many human geographers, especially those trained in cultural geography, saw as the overly deterministic nature of locational analysis, and a dehumanization of **landscapes** and their meaning, leading to the emergence of so-called humanistic geography. Nevertheless, locational analysis remains a methodological approach that many geographers embrace when investigating spatial phenomena, and led to the development of **Geographic Information Systems (GIS)**.

M

Malthusian Theory

A theory of **population** growth and its consequences, first articulated by Thomas Malthus in 1798. Malthus is generally credited with being the first scholar to undertake a scientific examination of the dynamics of population growth. Trained as a clergyman, Malthus eventually took a position in academe, after the publication of his groundbreaking work, *Essay on the Principle of Population*. Malthus's work set in motion a vigorous debate over how and why human population increases, and what the consequences of such increases may be. Between 1798 and his death in 1834, Malthus published five revised editions of the original book in response to his critics. He was familiar with the theories of utilitarianism, especially the writings of William Godwin, and also those of the French philosopher Jean Jacques Rousseau, the British economist Adam Smith, and others who conceived of a boundless capacity of the human species to better its condition under favorable circumstances. Malthus directly challenged this optimistic perspective, arguing that humanity was in reality condemned to a cycle of recurrent overpopulation and subsequent famine, regardless of the economic or social system in place. In other words, Malthus holds that human populations inevitably exceed the **carrying capacity** of the land they occupy. The debate Malthus initiated has continued, to one degree or another, for the past two centuries and his theory has been both reviled and revived by a host of social commentators.

Malthus argued that the rate of growth for the human population would invariably outpace the ability to produce food. This is due primarily to the sexual drive of humans, leading to high rates of procreation, and to the subsequent principle, according to Malthus, that a population will increase at a *geometric* rate of growth, while the ability to increase the food supply can only grow at an *arithmetic* rate. That is, while population will increase by units of 1, 2, 4, 8, 16 . . ., agricultural production can be enlarged only at a rate of 1, 2, 3, 4, 5. . . . Thus, at certain points in the history of a population, represented by the early stages of the cycle, the rate of food production is equivalent to the rate of population increase, and there is enough food to feed everyone. But over time the two rates diverge, and the gap between the increase in the food supply and the increase in population widens, resulting in ever-increasing food shortages. At some point, this results in widespread famine and a large-scale loss of population due to starvation. This event is often referred to as a *Malthusian catastrophe*, an episode that then triggers a return to equilibrium and the beginning of the cycle. The food supply available after the catastrophe is sufficient for the reduced population, but the instinct to reproduce will soon, other factors not intervening, result in a population growth rate that will exceed the ability to feed that population. Malthus suggests that this dynamic has always been in place in history, but that the relationship between food production and humanity's drive to overproduce resulting in "gigantic, inevitable famine" is often masked or delayed by other events that reduce the population growth rate, at least temporarily.

In the earlier editions of his treatise, Malthus labeled events that slowed or interfered with the "principle" he described as "positive checks." These factors included war and disease, which obviously reduced the population, as well as more general limiting factors like poverty, hazardous working conditions, and infanticide. He did not believe that the perils of overpopulation could be alleviated by implementing legislation that assisted the poor. On the contrary, Malthus was opposed to reforms that were directed at improving the lot of the masses, because he held that such efforts would only increase the capacity of humanity to reproduce, making the path to the Malthusian catastrophe that much shorter. He was not indifferent to the condition of the poor in England but believed that private charity was more effective in improving their lives. Moreover, he did not advocate contraception in general, although he recognized that the practice would also serve as a "check" to some degree on population growth. As his theory evolved through the years, Malthus eventually offered the view that expansion of the population might also be limited by couples delaying marriage, some individuals not marrying at all, and limiting sexual activity. He was not optimistic that humanity would adopt these suggestions, however, as such voluntary checks had seldom been applied historically and had little record of extended success. Even in regions where the resource base appeared virtually limitless in the early 19th century, such as North America, Malthus propounded that the population would eventually reach a level of growth that would exceed the ability of the land to feed everyone, without the frequent mitigating effect of the "positive checks."

The influence of the Malthusian argument has penetrated many fields of thinking and decision-making. In biological science, it is clear that the theory of natural selection and evolution developed by Charles Darwin was shaped by the ideas of Malthus. Darwin himself characterized his theories as representing the Malthusian model at work in nature, rather than in human society, and other theoreticians in evolution were also admirers of Malthus. In economics, the influential British economist David Ricardo was a contemporary of Malthus and corresponded with him. Ricardo was especially interested in how the precepts of Malthusian theory applied to the labor supply and subsequent influences on economic growth.

A century after Malthus lived, another British economist, John Maynard Keynes, based much of the theory of *Keynesian economics*, an approach widely used in the United States and other developed countries in the 1930s, on Malthusian concepts. The writings of Malthus also had a ready impact on the politics of his own day, especially regarding the so-called Poor Laws of England, which were designed to address the poverty generated by the dramatic shifts in employment and economic development brought about by the **agricultural revolution**. In addition, the *Essay on the Principle of Population* was a major stimulus for the implementation of a regular 10-year census in the United Kingdom, starting in 1801 and continuing through the present day.

Malthus had many critics in his own time, and his theories remain controversial today. The periodic Malthusian catastrophes predicted by the theory have failed to materialize in the economically developed world for the last two centuries. For example, no country in Western Europe or North America has experienced massive population losses due to famine, except in isolated cases such as the Irish Potato Famine of 1847–52, since the *Essay on the Principle of Population* was first published in 1798. Malthus, writing before the mechanization of agriculture and accompanying advancements in food production brought about by the Industrial Revolution, failed to anticipate the massive increases in farm output that would result from technological innovation. Indeed, in 1800 global population is estimated to have been slightly less than 1 billion people, and by 2008 had reached almost 7 billion. As modern critics of Malthus point out, according to Malthusian theory, such an increase would surely trigger a Malthusian catastrophe of global proportions. In fact, although food shortages occasionally strike many developing regions, these "positive checks" have had little impact on overall rates of population growth. Furthermore, Malthus failed to account for some of the major geographical and social trends of the 20th century, especially the **urbanization** of populations in North America, Europe, and parts of Asia; and the social advancement of women in education and employment.

In spite of the persistent criticism of Malthusian theory in the 19th and early 20th centuries, in the 1960s the theory was revived in the form of neo-Malthusianism. This school of theorists suggested that although the evidence of the cyclical principle that Malthus had articulated in the early 1800s had not materialized, in fact, Malthus was accurate in his approach. According to Paul Ehrlich and other leading proponents of the neo-Malthusian perspective, Malthus had erred only in his timing: the Industrial Revolution had only postponed the cycle, and humanity was once again approaching the point where population growth would outstrip food production. The benchmark work of the neo-Malthusians was Ehrlich's 1968 book, *The Population Bomb*, which had an immediate effect on social policy and legislation in the developing world, especially the United States. Ehrlich predicted massive food shortages that would result in the deaths of "hundreds of millions" of people in the ensuing decades. Many of these deaths would not be confined to the developing world, but would occur in North America and Europe as well. He forecast a dramatic decline in life expectancy in the United States, as well as an overall drop in the total American population. Other catastrophic predictions were a severe water crisis in the western United States and pervasive shortages of certain resources, particularly metallic ores. None of Ehrlich's predicted calamities came to pass, and in the developed world, one of the greatest health challenges to emerge in recent decades is widespread obesity, along with an overabundance of food consumption.

Map

The most important analytical tool of the geographer. A map illustrates some characteristic or set of characteristics of a specified space; frequently, that space is a **region** or some other delineated area. Maps are the most direct means of displaying and analyzing geographic information, although a map is only a representation of spatial reality, and all maps contain some element of distortion or inaccuracy. This is primarily due to the fact that a map is typically a two-dimensional representation of a three-dimensional surface, and therefore it is impossible to reproduce the latter without introducing some level of error. Moreover, a printed map shows spatial relationships for only a specific point in history, and therefore the accuracy of the information illustrated may be degraded as time passes. The past 20 years have seen the advent of maps that are stored as computer files, and such maps may be interactive and readily updated, but nevertheless these maps, just like their "hard copy" counterparts, must be periodically revised, because the features associated with any place on the **Earth's** surface never remain static. Of course geographers are not the only users of maps—tourists, travelers, and anyone attempting to find an unfamiliar place has probably employed a map. Virtually everyone carries with them a series of **mental maps** that enable them to visualize spatial information that they frequently require. The art of making maps is called **cartography**, and has been practiced since the prehistoric era. In 2009, a team of scholars revealed the discovery in a cave in northern Spain of a map etched on a piece of stone that was approximately 14,000 years old.

A map contains three components that affect its utility, accuracy, and the degree of detail it may illustrate. These are the **scale**, legend, and the **map projection**. The scale of a map indicates the spatial ratio between distance represented on the map and the actual distance. Such a ratio is necessary, because in almost all cases a map reduces the mapped area to a more compact format. As a general rule, the larger the area mapped, the smaller the scale ratio. The map legend provides information on the spatial pattern of the data or phenomena that are displayed on the map. Typically, some system of symbols is used to represent features on the map. Such features might be tangible (major buildings in a town) or may not exist physically on the landscape at all (dominant political views in a state; incidence of a disease by ethnic group, etc.). If individual data points are presented on the map, then discrete symbols are used. If the data are continuous across the area shown on the map (such as political attitudes or infection rates), then colors or patterns are used to indicate the regions where the data are present, or the colors or patterns may show the degree, level, or intensity of the data. The legend may also feature a symbol indicating the cardinal direction of north, typically as an arrow pointing northward. The convention is to place north at the top of the map, and almost all maps are constructed using this rule, but this was not always the standard. In the Middle Ages, Europeans often placed east, or the direction of the Orient, at the top of their maps, and this practice is echoed today in the expression to "orient" the map—but by placing north at the top!

The projection of a map is a key element that influences how the map appears, and for what purpose the map should be employed. The two-dimensional character of a flat map, representing the three-dimensional surface of the Earth, invariably introduces some distortion. No map can simultaneously indicate true direction, distance, areal relationships, and angles between two discrete points—at least one of these properties will be inaccurately represented on the map. Various projections may be selected that correctly indicate one, or more than one, of these properties but not all of them. Moreover, projections may be centered on any point on the globe's surface, and the point chosen will dramatically affect the perspective offered by the map. For example, a projection centered on the South Pole in Antarctica, representing a polar projection, will in addition to that landmass show only the southern tips of South America and Africa, and most of Australia. In contrast, a projection centered on a point on the equator will show all of the Earth's major landmasses, but only a portion of each polar region.

Two commonly encountered types of maps are general maps and thematic maps. General maps are constructed with the intent to convey geographical information indicating the **location** and relationship of multiple spatial features on the **landscape** as a reference. Any road map is an example of a general map, as are topographic sheets (in the United States, produced by the U.S. Geological Service), maps showing diverse physical phenomena such as lakes, rivers, mountains, etc., and maps portraying political states and their boundaries. Maps showing basic features of an area, such as outlines of country **boundaries** or major physical characteristics, are often used as base maps for mapping other information. Thematic maps differ from general maps in that they typically are designed to illustrate the spatial characteristics of a single characteristic or those of a closely related set of phenomena. The purpose of the thematic map is to provide a spatial representation of the distribution of such features and the relationship(s) between them. A map showing the average per capita income of each state in the United States for a given year would be an example of a thematic map. Such a map would indicate the state boundaries like a general map, but the addition of the specific attribute of per capita income, unique to each state, creates a specific theme for the map. The possibilities for thematic mapping are infinite, as a limitless number of spatial phenomena and data exist that might be mapped, and new data are constantly being generated.

The process of creating a map is complex and requires great skill and care. Frequently the cartographer must decide on a proper scheme of generalization and symbolization of the data that are to be displayed. Information must be classified into categories, because mapping all the variation in the data points would result in a map that would be overly complex and confusing. There are numerous techniques for dividing the data into appropriate classes, and the way this is done often depends on the nature of the data itself, as well as what facets of the information the cartographer wishes to emphasize. One of the most common approaches when working with statistical data is to produce a *choropleth* map. On this type of map the data is presented as continuous classes that highlight the magnitude or intensity of the data within the defined boundaries on the map. For example, a map showing the average income of residents in Florida by county would likely have five or six classes ranging from the lowest average county to the highest. Such a map will clearly show where the wealthiest residents live in the state, and where incomes are relatively low. The counties within the classes can be indicated using a distinctive pattern for each class, or by using different colors.

For thousands of years maps were drawn by hand using the same techniques employed by artists producing sketches and pictures, and indeed many maps produced this way could be considered works of art in their own right. This meant that map production, especially of maps showing a great deal of data or detail, was a time-consuming process that required not only geographic knowledge but considerable artistic skill. The revolution in technology that transpired at the conclusion of the 20th century dramatically changed the way maps are generated. Computer cartography, the ability to construct maps using computer software packages and sophisticated graphics, has revolutionized map production. Maps can now be created by anyone using mapping software, and may be updated instantaneously and frequently without the need to redraw the entire map. Maps are the main output generated by **Geographic Information Systems (GIS),** and the development of GIS has meant that the analytical utility of maps has increased exponentially. Using the multiple layers of data contained in GIS, a researcher can quickly display and compare any number of spatial attributes, producing an array of thematic maps. Technology has also increased the degree of accuracy of maps—sophisticated Global Positioning Satellites (GPS) units are capable of locating features to within centimeters of their actual position, resulting in maps that are the most precise ever made. Maps are vital to our lives, as they help us find our way to new places, fight the spread of diseases, produce more food, control damage to the environment, and have thousands of other crucial applications. Most importantly, they provide a means of visualizing and understanding the world around us.

Map Projections

The basic problem in displaying the real world, which appears in three dimensions, onto the two-dimensional surface of a **map** is addressed by the map projection. Because there is no way to accomplish this transformation without distorting at least one of the basic elements indicated on the map (angles, areas, distance, and direction), a means of projecting the information contained on the map must be selected. Although no map can show all of these basic relationships in their true dimensions, many projections are able to display at least one of them accurately and truly. For this reason, it is quite important that when using a map for a specific purpose, attention is paid to the projection of the map. Geographers and others working with maps must be aware of the limitations of the projection they employ to display their data, and how it may affect the spatial relationships they wish to analyze. In theory, a limitless number of map projections might be constructed, and with the development of computer **cartography**, many options are readily available. Today, with a click of the computer mouse the projection of a map and the information it contains may be changed instantaneously, an option made available to geographers only in recent years. A number of projections have limited application and are used only for highly specialized work. However, over the past several centuries, some projections have become "standard" in the sense that they are used more than others for certain purposes such as comparing relationships between areas, navigating on the high seas or during transoceanic flights, or other types of uses.

Most map projections feature a graticule, a grid of lines running in both a north-south direction and in an east-west direction. The graticule may be based on lines of **latitude and longitude**, or can be based on some other coordinate system. The purpose of the graticule is to provide a series of reference points and lines that may be useful to those working with the map, but it must be remembered that the graticule typically contains the same problems with distortion and inaccuracy that the projection itself does. This is connected to the general problem of **scale** on map projections. Due to the inaccuracies and distortions of casting a three-dimensional surface onto a two-dimensional map, true scale cannot be maintained from a given point in all directions. Some projections are constructed so as to present scale accurately between two points *along a given line*, but scale will still be inaccurate in other directions from the point of reference. Another aspect of a projection affecting its appearance, accuracy and utility, is how the projection is centered; that is, what point or latitude represents the main line of reference in the center of the projection? Frequently the equator is chosen for centering a projection; this is called an equatorial projection, but for certain reasons, one of the poles may be chosen as the centering point, resulting in a polar projection. Any other point or line may be used as well, and changing the centering point will dramatically alter the appearance of the map.

In mathematical terms, to create a projection of the globe, a "developmental surface" must be achieved, meaning a surface that may be flattened out into two dimensions. This may be accomplished in a large number of ways, but most projections use certain geometric shapes that are aligned with the axis of the **Earth**, which are then "opened" into a two-dimension representation. The most common of these shapes are a cylinder, a cone, and a plane—the latter is also sometimes referred to as an azimuthal projection. Each of these has certain advantages, depending on the nature and purpose of the map being constructed. Generally, the graticule of each of these follows certain properties. On a cylindrical projection, the lines of the graticule will remain parallel, with each pair of lines meeting at a right angle. On such a projection, the poles are impossible to represent. A conic projection will often display the graticule lines as converging toward one pole and diverging in the opposite direction. The azimuthal projection shows a graticule that radiates outward from a point or set of points, representing where the plane intersects that portion of the globe the projection represents. Azimuthal projections are not particularly well suited to showing large areas but are quite useful in calculating direction from a known **location**; therefore, they have wide application in navigation and scientific studies.

One of the most famous map projections is also one of the oldest. The Mercator projection was introduced in 1569 by Gerardus Mercator, a Belgian cartographer who was interested in designing a chart that would be accurate for determining true compass bearings for oceanic navigation. Mercator's projection is no doubt the most widely used projection in history, as it was adopted not only by mariners but over the centuries became popular with other map users as well. The projection is a cylindrical projection centered on the equator, and thus the graticule lines, both for latitude and longitude, appear as parallels running at right angles. Maps constructed using this projection contain very little distortion in regard to direction, and the Mercator projection or some derivative remains the basis for many, if not most, nautical charts used by oceangoing traffic even today. On the other hand, the Mercator projection tends to distort the relative size of **regions** especially toward the poles, while the latter of course are not represented on the projection. In reality, the Mercator projection shows areas that lie in equatorial regions close to their true proportions but, when employed for a world map, leads to misrepresentations of the actual geographic relationships. The classic example is the comparison of the relative sizes of Greenland and South America on the Mercator projection. Greenland, located at a northerly latitude, is shown to have about the same area as South America, when in reality South America is more than eight times larger. The Mercator projection then is a poor choice for teaching regional geography in most cases.

A better choice for most geography teachers and students is the equal area projection. The question of accurate representation of regions and countries on maps is not merely academic. In the 1970s, the German activist and academic Arno Peters claimed that continued widespread use of the Mercator projection in textbooks and academic publications was in fact an indication of cultural imperialism, and in essence a subtle attempt to undermine the **cultural identity** of peoples in the economically developing world. Peters claimed to have developed a new projection that corrected the misuse of maps based on the Mercator projection, called the "Peters projection," which was actually nearly identical to a much earlier projection developed by James Gall, a Scottish cartographer. Although Peters's claims about the accuracy of his map were exaggerated, his main point concerning the appropriate use of map projections, and the implications of such use, was well taken and generated much discussion in the cartographic community and among geographers in general. The Peters projection, ironically, is not widely used today, but several well-known equal area projections of this type are frequently encountered. Albers's equal area map is a conic projection centered on two standard lines of latitude located in the middle latitudes. The Albers's projection is especially useful for regions like the United States and Europe, where the east-west distance is greater than the north-south extent of the region. Other common equal area projections include Mollweide's equal area and Lambert's equal area projections, both of which are frequently used in atlases.

A map that is to be used for plotting the course of a moving object should be based on a conformal projection. Conformal projections, because of their mathematical properties, preserve the true angular relationship between points on the map. This type of projection may be used to accurately calculate direction and distance, properties that are crucial to moving from one location to another as represented on the map. The aforementioned Mercator projection is a prime example of a conformal projection, and others would include topographical sheets and aeronautical charts. Because the angular relationships shown on the map are the same as those on the globe, a navigator may draw a bearing, or a line of movement, between two locations to plot a course. Even a tiny amount of distortion in a conformal projection could result in a journey ending hundreds of miles off course, if the distance to be traveled is great. Technological advancements like **GPS** systems have made navigation much easier in recent years, but selecting the proper projection remains an important part of a navigator's skills.

Medical Geography

Medical geography is the study of the spatial aspects of disease and human health. Like almost all phenomena associated with human activity, the incidence of disease exhibits a spatial pattern. By analyzing such patterns, a causal relationship between disease and behavior, or disease and environmental factors, may be discerned. Such recognition may then be used in combating the spread of the malady, and limiting the infection rate if the disease is contagious, or lowering the incidence of the sickness if the environment or human behavior is responsible for the disease. The individual often credited with founding this subdiscipline of the field is Dr. John Snow, an English physician who treated patients suffering from cholera during a severe outbreak in London in 1854. Snow acquired data on known victims of the epidemic and used these figures to plot the number of deaths from the disease on **maps** of the city. He thereby determined that many of the victims lived quite close to a water pump used as a water source, located on a busy thoroughfare called Broad Street. At the time, little was known about how cholera was transferred from person to person, and Snow theorized that the source of the infection was the communal water pump. He called for authorities to shut it down, and once it was closed, the number of deaths and rate of contagion dropped abruptly. Snow had stopped a serious killer in its tracks by applying geographical techniques to the study of disease.

In the era of **globalization**, diseases may spread quite rapidly following international transportation routes, and modern medical geographers focus their efforts on tracking the spatial distribution of diseases, determining their source **regions**, and formulating strategies to block and contain new outbreaks. The prime example of such research is the tracking and monitoring of the global AIDS epidemic. By using the techniques of medical geography, investigators have been able to geographically pinpoint the source region of the HIV virus as lying in East Africa, providing valuable clues to when and how the virus first entered the human **population**. It is now believed that the HIV virus crossed over from primates, especially chimpanzees, into the human population at some point in the 19th or early 20th century in that part of Africa. Medical geographers have also been able to sometimes track the spatial movement of the disease within specific countries, leading to strategies on how to slow or stop the spread of the virus. For example, in Tanzania in the early 1990s it was discovered that the incidence of HIV infection dropped significantly as distance from a major roadway increased. This spatial distribution led health care officials to theorize that long-haul truck drivers, frequent customers of the country's sex workers, were primary agents of spreading the disease along transportation routes, especially roads. Truck drivers therefore have become an important group targeted by government educational programs, designed to combat the spread of AIDS by promoting the use of condoms and limiting high-risk sexual behavior.

Megalopolis

A term from urban geography that identifies a large, contiguous urban region. The word was first used by the French geographer Jean Gottman in describing the strip of **urbanization** running along the eastern seaboard of the United States from Boston, Massachusetts, to Washington, D.C. Another term for this concentrated zone of urban development is BosWash, an amalgamation of the names of the terminal cities. The label of "megalopolis" is now applied to any large urban conurbation anywhere in the world, and many such locations may be identified. Typically, a megalopolis is characterized by a large **population**, dense and sophisticated transportation systems, a concentration of industrial and service economic functions, and numerous urban problems, such as higher levels of pollution, greater congestion, and loss of green space. A 2005 study by Robert Lang and Dawn Dhavale identified 10 "megapolitan areas" in the United States. To qualify for this label, an urban corridor had to include a combined population of at least 10 million by the year 2040, occupy a similar physical geography, represent an identifiable cultural region with a common sense of identity, and be "linked by major transportation infrastructure," among other criteria. Lang and Dhavale identified 10 such urban regions within the United States, and highlighted the economic, demographic, and political power such mega-cities represent. According to their study, approximately two-thirds of the U.S. population lives within the **boundaries** of a megapolitan area, and 80 percent of the members of the U.S. House of Representatives have some portion of their congressional district lying within such a region.

The **toponymy** of the 10 areas delineated by the research of Lang and Dhavale reveals both the **location** and identity of the megalopolis. The "I-35 Corridor," for example, is an emerging strip of urban development that extends from San Antonio, Texas to Kansas City, closely following the interstate highway that links these cities and a string of others lying in between. The "Valley of the Sun" is a megalopolis centered on Phoenix, Arizona, while "Piedmont" is the urbanized region emerging along the southeastern flank of the Appalachian Mountains. But the development of such "super cities" is not limited to North America, of course. In Japan the *Taiheyo Belt* is an enormous urban concentration of more than 80 million people, running from Tokyo on the island of Honshu to the northern end of the island of Kyushu, connected by high-speed trains. The Taiheyo Belt represents one of the most dense concentrations of industrial and financial development in the world, all focused in a strip of territory the size of southern California between San Francisco and Los Angeles. Indeed, megalopoli appear to be emerging in many corners of the planet, and it seems likely that such enormous urban clusters will dominate the urban landscape of the future. This will present new challenges in the form of transportation infrastructure, integration and governance, environmental degradation and quality of life, and very likely others that have yet to appear. Changes in the **cultural landscape,** as well as in **cultural identity,** also seem inevitable as the megalopolis evolves into the defining feature of urban life.

Mental Maps

Spatial information about the world, carried in one's mind. Studies in behavioral geography have shown that all humans utilize mental maps, although to varying degrees and at different levels of ability. The formation of a mental map is generally a subconscious process and begins at an early age. This "brain **cartography**" provides a means of organizing our activities, recognizing our "home" terrain, and negotiating through places that are new and unfamiliar. Without the ability to form mental maps, humans would lack the capacity to venture forth to locate food (done today at a local supermarket, but only a few thousand years ago in a field or forest), find a mate, and other vital behaviors—in other words, the formation of mental maps is crucial to human survival. The average person forms a mental map through experience—as one explores an unfamiliar landscape, various spatial "cues" or markers are noted and recorded, providing points of reference that later can be recalled when encountered again. The process works exactly the same way, whether one is exploring a highly organized and complex urban environment or a pristine wilderness. Street signs in a city provide the same type of mental benchmark as an unusual looking tree, a spring, or a pile of rocks does in a forest—both allow an individual to organize and traverse a new spatial environment. As that environment is repeatedly encountered, new features are added to the "map," and the level of detail increases.

Scholars of behavioral geography are especially interested in mental maps, because such maps form the core of spatial perception and reveal a great deal about how people view the world around them, as well as the world at large. The mental maps one carries may play a significant role in the formation of **cultural identity,** as well as how other regions are perceived. Studies have shown that **nation-states** emphasize the development of a more detailed domestic mental map among students in their public school systems than mental maps they form of adjacent countries, even in students who live adjacent to international **boundaries**. Views of **regions** that are not precisely defined provide interesting insight into spatial perception. For example, the boundaries of the "Deep South' in the United States vary considerably, depending on where the question is asked. The mental maps of the "South" that the average resident of Alabama holds will differ markedly from that held by a resident of Nebraska. Interestingly, many people form mental maps not only about places they have encountered, but also about places they *have not* yet been to. These sorts of maps may be formed from media reports or other sources of information that are not geographical, yet in many cases people form a spatial concept from them. Some research has found, for example, that a significant number of people link spatial extent to the frequency of hearing information about a place, so that the greater the times a country is mentioned in the media, for example, the larger it is perceived to be.

Middle Latitude Cyclones

A middle latitude cyclone (MLC) is a large disturbance that lasts for several days, and is the most significant weather-maker of the middle latitudes. This storm is also known as an extratropical cyclone or a wave cyclone. The term "cyclone" denotes a central low pressure with closed isobars. Middle latitude cyclones, **hurricanes**, tornadoes, and dust devils are cyclones of various causes, sizes, and strengths. In particular, a middle latitude cyclone is the largest of all cyclones and is a complicated admixture of the polar front jet stream, **air masses,** and **fronts**. The MLC is a seminal concept in understanding the middle latitude weather of our planet and was first proposed by Norwegian scientists shortly after World War I.

The origin of an MLC, called cyclogenesis, is directly related to energy imbalances between equator and poles. As such, the strongest MLCs are usually reserved for the winter season when the imbalance is greatest. Near the polar front jet stream, cold, dense air is to the polar side and warm, less dense air is to the tropical side. These air masses do not mix unless they are energized by a disturbance in the polar front jet stream. As such a disturbance passes overhead an air aloft diverges (pulls apart horizontally). This creates a surface low pressure zone into which air converges and rises to make cloudiness and **precipitation**. However, this is not all: the convergence of air into the surface low makes the polar and tropical air masses push on each other thus creating active fronts. The strength of an MLC cyclone is largely dependent on the upper air flow. If the flow aloft strengthens, divergence is favorable for evacuating air from the surface cyclone and strengthening the storm. If the upper divergence decreases, the surface circulation is damped. The paths of MLCs are readily forecasted because they are steered in the direction of the air in the polar front jet stream.

A MLC experiences several stages as it crosses the United States. At first, tropical and polar air masses encounter each other with neither being the aggressor. An upper disturbance in the polar front jet stream causes a center of low pressure (the cyclone), a cold front, a warm front, and an incipient wave. With the passage of a few hours the MLC forms a large wave and there is a circulation established (a counterclockwise spiral toward the center in the Northern Hemisphere). Within the wave is, typically, a maritime tropical air mass while pushing the cold front along is a continental polar or Arctic air mass. It is at this mature wave stage that the cyclone is strongest and the most severe weather is expected. With the passage of more time, an occluded front starts to form. The occluded front occurs because cold fronts travel faster than warm fronts. As cold air overtakes the maritime tropical air within the wave, the warm air is forced aloft ending the air mass contrast along that part of the front. Occlusion progresses southward from the cyclone's center in the Northern Hemisphere. As the cyclone occludes its strength abates and all that remains is residual cloudiness and light precipitation. The cyclone has "died" but, before this, much polar and tropical air has mixed and the boundary between the polar and tropical air is established in another position.

The air masses, frontal types, and the cyclone center make for a wide variety of weather. The cold front can create tornadic thunderstorms while there is heavy snow to the north of the cyclone's center. As an MLC passes there are significant weather changes. Indeed, the recognition of existence of MLCs allows some very good weather forecasting. Suppose we are interested in the weather forecast for Dayton, Ohio. As a MLC passes, Dayton is in the maritime tropical air ahead of the strong cold front. As the MLC approaches, the weather will become cloudy from upper parts of frontal clouds blown along in the upper air. The front will pass with rain and Dayton will experience a "dry slot" composed of cold air wrapped around the cyclone's center. As the cyclone's center passes Dayton, the city will experience snow and Arctic temperatures. However, the forecast would be quite different further to the south, where there would be a round of severe thunderstorms followed by the passage of the cold front.

MLC's travel beneath the flow of the middle and upper troposphere. In general, their centers move along at about 40 km per hour and can traverse thousands of kilometers during their life cycles. At any one **location**, the disturbance takes about two days to pass. Because the associated clouds and **precipitation** spread hundreds of kilometers in all directions from the center, huge areas are affected at one time. Considering their motion, there are some large MLCs that affect virtually all of the continental United States.

Middle latitude cyclones occur in families, and it is a common wintertime occurrence to have a cyclone crossing North America while another three cyclones are over the Pacific Ocean in various stages of development. As these storms swing onshore there will be dramatic changes of weather every few days.

There is a definite seasonality to MLCs. In that they are energized by the winds aloft, they are strongest and most plentiful in the winter. In winter, the polar front jet stream brings occasional storm centers to the subtropics. In summer, MLCs are fewer, weaker, and have an average position of about 50° latitude. Additionally, the Earth's surface geography plays a role in defining average storm tracks. For instance, winter MLCs frequently form just downwind (east) of the Rocky Mountains because the air has been forced to bend to the right to compensate for the presence of the mountains; this effectively becomes a disturbance aloft that produces a MLC at the surface.

MLCs can sometimes achieve the strength of hurricanes. In 1921, the Pacific Northwest experienced the "Great Blowdown," a MLC with winds exceeding 160 kph toppling seven billion board feet of valuable old growth timber and killing hundreds of farm animals and elk. As winds become strong, MLCs are able to move significant quantities of dust and snow off of the surface. Strong disturbances passing through without much precipitation are sometimes labeled "dust storms." A blizzard, defined in the United States as winds of greater than 46 kph with blowing snow, is a serious social concern with loss of life and property likely. The largest blizzards have been responsible for massive delays in air and ground traffic and the deaths of grazing livestock. Urbanized areas are not immune to these large storms. In 1888, a blizzard dumped over 60 cm of drifting snow in the New York area. Over 400 people lost their lives and commerce was at a standstill for over a week. The "Storm of the Century" was a cyclone affecting Central America to Ontario in March 1993. Its immense intensity and scope cost economies several billion dollars while ruining properties with tornadoes and snowdrifts over 10 m.

Migration

The movement of any multicellular animate life from one **location** to another. Various species of insects, birds, and mammals all migrate, but the discussion here will focus on human migration. **Population** geographers and demographers study the migration of people because this movement transpires across space and directly influences the dynamics of population growth in many parts of the world. *Homo sapiens sapiens*, or modern humans, have migrated throughout their history, at times in quite large groups. Anthropologists who support the "out of Africa hypothesis" (sometimes called the "candelabra theory") believe that modern humans populated the **Earth** over many centuries by migrating from a central location in eastern Africa, starting about 50,000 years ago. If accurate, this would represent the largest and most important migration in human history, but there are many additional episodes of massive migration in the human experience. Migration may be temporary or permanent, as some people migrate to another place for a few months or years, but return to their point of origin after some period of time. Migration within a developing country's borders often takes place between the **core and periphery,** as people move to urban areas seeking economic opportunity, but often must settle in **squatter settlements**. Migrants frequently cross international **boundaries**, either legally or illegally, and controlling the level of migration, regulating who is allowed to migrate, determining the status of illegal migrants and other aspects of migration policy often become important and controversial political issues in many countries. In the United States, policy decisions aimed at controlling illegal migration have recently received a great deal of attention at the federal level as well as in some states located near the border with Mexico.

People relocate from one place to another for a great range of reasons. In many historical instances, thousands or even millions of people were forced to migrate against their will. Some well-known cases of such forced migration include the so-called "Babylonian captivity" of the Jewish people in the sixth century BCE, and the enslavement and transport of 8 to 10 million Africans from their home continent to the New World from the 16th to the 19th centuries. Many others move to flee religious or political oppression. Today the majority of those relocating are economic migrants, who seek better wages and living conditions, either within the confines of the **nation-state** they live in or abroad. Some economic migrants represent seasonal laborers, who frequently are employed to harvest certain labor-intensive agricultural products, as is the case with sugar cane harvests in some Caribbean countries. Generalizations about migratory motivation are difficult to support, given the various types and rationales for migration and the many different historical circumstances in which movement has occurred. The geographer Everett Lee developed the **push-pull concept** in attempting to identify

Turner Thesis

Sometimes called the Frontier Thesis, the Turner Thesis is named after Frederick Jackson Turner (1861–1932), a professor of history who first proposed his theory in 1893 at a professional meeting of historians. Turner's concept of American identity focuses on the role of the frontier in sculpting a unique American character. Turner argued that the interaction between the untamed wilderness and "civilization" along the frontier's margin had conferred a special set of qualities on the American identity, in that the struggle to master the wilderness beyond the frontier provided confidence, strength, and rugged self-reliance. As the line of the frontier advanced westward over the course of the century and a half of American history, these characteristics intensified. American **cultural identity**, although initially derived from the European continent, deepened the qualities that differentiated it from its origins, as the boundary shifted to the west. Turner was strongly influenced by **environmental determinism**, and some of his critics suggest that his ideas were designed to justify American **imperialism** in the Pacific Basin and elsewhere. Although his theory has been largely abandoned by historians, it represented one of the first efforts to articulate the foundations of an American identity, a topic that continues to generate debate today.

factors that affect the tendency of people to migrate. Simply put, Lee holds that potential migrants will consider two sets of criteria when making a decision about migration. The "push" factors are negative characteristics of the place of origin of migrants and might include high unemployment rates, poor living conditions, a limited number of potential spouses, and numerous other conditions. The "pull" factors represent the positive qualities of the potential destination, such as a higher standard of living, being closer to family members who have already migrated, greater political or religious freedom, etc. These considerations may work either individually or in concert to motivate a person or group to migrate. The incentives to migrate must overcome the **friction of distance** between two locations, which represents the costs of moving, both in terms of actual financial investment as well as the time required to migrate. In general, the "friction" associated with migration increases as a function of distance, but many other factors besides just distance may increase the friction; the necessity of crossing an international border, for example, even if located nearby, functions to inhibit migration.

Migration has played a large role in both human tragedy and progress. Migration to the New World, both voluntary and forced, completely changed the **landscape** of that region between the 16th and 20th centuries, leading to development of vast stretches of wilderness, but often with catastrophic consequences for indigenous peoples. The United States is frequently cited as a "nation of immigrants," a statement that is quite true in that except for Native Americans, who comprise about two percent of the population, all citizens or their ancestors immigrated to the United States in the past four centuries, with the majority arriving in the past 150 years. This trend continues, as in 2009 more than 38 million American citizens, or well over 10 percent of the total, were foreign born. Of course, other countries also are "nations of immigrants" as well, most notably Australia, Canada, and some Latin American countries like Brazil and Argentina. Rural to urban migration in Great Britain as a result of the **agricultural revolutions** in the early 19th century provided labor that fueled the industrial revolution, a process that was repeated to some extent in every other economically advanced country, and that continues to occur to a significant degree in the developing world. This migration provided the labor resources to build the modern global economy, and labor migratory flows continue to be vital to the economic progress of many nations.

On the other hand forced migration, or population shifts brought about by war or decolonization, resulted in millions of deaths in the 20th century alone. At the conclusion of World War II, millions of ethnic Germans living in the **shatterbelt** of Eastern Europe were forced to migrate back to Germany, even though their ancestors had lived in the region for generations. The partition of British India in 1947 set in motion one of the largest migrations in modern times with some 12 million people shifting between India and Pakistan, along with an estimated one million deaths. Human beings migrate because of both hope and fear; and indeed, the mass movement of people is a recurrent theme in the history of humanity, bringing advancement and development along with calamity and cruelty.

Monsoon

Somewhat less than half of **Earth's** population lives in monsoon zones and is highly dependent on monsoon **precipitation** for agricultural production. Religious festivals are timed to the onset of the wet and dry monsoons. Failure of monsoon rains or flooding from monsoon rains raise the specter of famine.

Monsoon is a commonly used but misunderstood term. One might hear an acquaintance talk about "today's monsoon" out of a Great Plains thunderstorm, but the term was not originally used to convey the sense of prodigious wetness. The word is derived from the Arabic *mausam* meaning "season" and referring to the seasonal reversal of wind directions in the arid Arabian Sea and western Indian Ocean. Sailors were able to reliably time their trading trips by this expected feature of the atmosphere. Pronounced seasonal wind reversals are present in much of the world, but the term monsoon now most properly refers to a climatic regime of reversal of seasonal winds bringing alternating seasons of wetness and dryness.

The monsoon regions of the planet encompass wide swaths of the tropics and subtropics and are, classically, experienced in Asia and parts of central Africa. From December through February the primary wind direction is northeast to southwest, while June through September brings the reversal of this flow. The former period is quite dry while the latter is prodigiously wet.

Monsoons have their roots in the tropical and subtropical parts of the global wind and pressure belts. The shifting declination of the vertical sun causes parts of the global circulation to expand, contract, and change latitudes. Monsoons are features in the tropical and subtropical portions of the planetary circulation. Differential heating of land versus water is the simplest explanation of the wind reversals. Continents heat more dramatically in the summer than do oceans and these **temperature** gradients set up wind gradients transporting air from the oceans to the continents. In the winter, the thermal gradients reverse and cause the surface air to reverse flow. Continental air is dry while oceanic air is moist and associated with great amounts of precipitation. The monsoon is most prominent around Asia because of the immense size of the Asian land mass and the complexities of its coasts.

The air flow high in the troposphere is known as a partial cause of monsoons. The subtropical jet stream is most active in the winter season because of the great temperature difference between equator and poles. With its mean winter position to the south of the Himalaya Mountains and Tibetan Plateau this strong upper flow complements the cooling of the Asian continent and keeps the oceanic air out of the continent. In the summer season, the subtropical jet weakens and jumps to the north of these southern Asia highland areas and the oceanic air is allowed to incur into the continent.

In Asian monsoon terminology, the Intertropical Convergence Zone is known as the monsoon trough. This broad zone of shallow low pressure is able to make exceptionally large forays about the equator (especially to the north) because of the great size of the Asian continent. The surface mechanics of the wet (summer) monsoon are that the monsoon trough arrives and behind it is a deep, moist maritime **tropical** or equatorial air. These air masses are unstable and host many slow moving low pressure disturbances containing many convectional storms with high rainfall rates. Monsoon lows and monsoon depressions are common with the former moving over an individual **location** about once every 5 days while the latter can be expected every 15 days. Both have mean diameters that can approach 3,000 km and can produce copious precipitation. The monsoon depression is the more organized and has winds that can reach as high as 18 m per second. The summer monsoon "bursts" as the monsoonal trough passes a location. There is another burst of precipitation along the retreating monsoon trough, and this is closely followed by the low sun season of dryness.

There is an expected geography of the timing onset of the monsoon. The classic Indian subcontinent situation is that southern India experiences the wind shift heralding the onset of the wet monsoon in early May and the rains progress northwest as far as Pakistan by the middle of July. At that time, the wind shift backpedals to the southeast. Thus, the wet monsoon lasts longer with higher rainfall totals in the south and east than in the northwest. In the rest of monsoon Asia the topographic/sea configurations are very complicated, and there are many nuances to the monsoonal flows in these realms.

The South Asian monsoon is notoriously variable. The timing of onset and finish vary by several weeks and amount of precipitation shows large interannual variance. El Niño conditions, occurring on the eastern side of the Pacific Ocean, have long been shown to be closely tied to the occurrence of weak Asian monsoons. There is keen interest in predicting monsoonal rains for upcoming seasons, but these efforts have been only moderately successful because of limited scientific understanding of monsoon causes. Increasingly good mathematical modeling holds the promise of reasonable monsoon forecasts in the near future.

Monsoon climate is one of the important regional atmospheric characterizations of the planet. In a classic monsoon climate there are three seasons: the relatively mild season of the winter monsoon; the increasingly stifling heat and humidity leading to the onset of the wet monsoon sometime near the summer solstice; and the warm, prodigiously wet summer monsoon rains. The monsoon climates are located in parts of southern Asia, tropical Africa, northern Australia, and Central America. Freezing temperatures are quite uncommon with average monthly temperatures ranging from the low- to mid-20s C. Precipitation is expected to be hugely variable by season. The cool season months commonly average less than 20 mm of precipitation, while the wettest months of the high sun season can average in excess of 300 mm. Monsoon climates tend not to be as wet as tropical rainforest climates for the year as a whole but tend to be the wettest places on the planet in the high sun season.

Cherrapunji is a city in the Khasi Hills of Assam in northeastern India. Cherrapunji sits at 1,290 m above sea level on the first significant topography encountered by the summer monsoonal flow as it leaves the Bay of Bengal. With this mix of monsoonal rains and orographic effect, Cherrapunji has received some truly remarkable rainfall totals. In 1861, the city recorded almost 23 m of precipitation with a stunning 9.3 m falling in July of that year.

In the United States, the Southwest has a minor monsoon. In this case the term refers to the rainfall that peaks in the high-sun season. This effect is not related to the classic Asian monsoon by physical connection. The precipitation is enabled by the formation of a summertime thermal low over the Southwest drawing in maritime tropical air from the Pacific Ocean and Gulf of Mexico. These moist air streams represent a seasonal shift in the winds. Yet, these precipitation amounts are not on the same magnitude as those of the Asian monsoon. For instance, summer is the wettest time in Phoenix, Arizona, but the average precipitation is only about 25 mm in the wettest month. The southwestern United States is still a desert at low altitudes!

N

Nation-State

A unit of political organization of space, enclosed by legally established **boundaries**, and representing, at least in theory, a territory linked to the **cultural identity** of an ethnic group in the form of national identity or nationalism. **Territoriality** provides the link between a defined space and identity in most instances. Key to the notion of the nation-state is the concept of the *nation*, which in this sense is not a synonym for "country," but rather identifies a group of people who share common attributes of language, religion, historical development, or a combination of these. Nations may exist without states—the Kurds, Sikhs, and Basque are all examples of nations who do not possess their own state, but rather live within the confines of one or more nation-states based on some other national identity. However, there is considerable debate among scholars on exactly what constitutes a nation-state, when the first nation-states arose, and how a nation-state comes into being. Benedict Anderson, for example, suggests that nations and therefore nation-states are in essence "imagined communities" (the title of his famous book on the subject) brought about by the advent of "print capitalism," and did not really exist in their modern form until the end of the 1700s. Ernest Gellner argued that nation-states were the product of the industrial age and emerged first in Europe in the 19th century. Other theories abound, but what seems clear is that the collective identity of some ethnic groups, bound to a given territory, had initiated the basic concept of the nation-state by the early modern era, if not before.

What distinguishes a nation-state from an empire is the character of the polity; in an imperial system, many diverse peoples are absorbed and controlled who hold little in common with the culture of the ruling classes. In the Roman Empire, for example, only a small elite group of Egyptians, Jews, Celts, and others spoke Latin, the imperial language, and all of these peoples for the most part retained their own religious identity—they were never made into Romans, nor was there any serious effort to make them so. In the nation-state, institutions are created that cultivate and promote the concept of a shared national identity, a sense of kinship that binds the citizenry together, using themes of shared culture, historical antecedents, and common destiny. The institutions that create this kinship are public schools, the military, various types of media, and the government itself in many cases. Symbols of the nation-state take on an elevated importance and respect—these include the national flag, anthem, heroic figures (both real and mythological), and even elements of the popular culture ("as American as baseball and apple pie. . . ."). The institutions of the nation state therefore serve in an integrative capacity, attempting to craft a unitary identity among all members. If taken to extremes, this process may result in national chauvinism, **ethnocentrism,** and racism, as was the case in Germany under the National Socialist (Nazi) government in the 1930s. Some scholars attempt to draw a clear distinction between national identity, seen as a positive phenomenon, and nationalism, which they view as extremist and discriminatory.

Today there are slightly more than 200 independent states in the world, and while scholars might not consider all of them to qualify as "nation-states" under a strict academic definition, the fact is that most of the governmental officials of these states would hold that they represent a unitary national identity, and thus are indeed "nation-states." This global system of nation-states greatly influences the dynamics of **geopolitics**, as states tend to join together into alliances, either formal or informal, when they believe doing so is in their national interest. This "Westphalian system" of nations may be traced to the Peace of Westphalia, a collection of treaty agreements promulgated in 1648 that ended both the Thirty Years' War and the Eighty Years' War in Europe. The Peace of Westphalia was instrumental in defining what came to be known as the "balance of power" concept, a means of maintaining order among the diverse set of sovereign states in Europe and regulating relations between the emerging nation-states there.

In the past two centuries the collapse of **imperialism** has invariably resulted in the establishment of multiple nation-states. Spain lost most of its colonial holdings in the New World in the first decades of the 19th century, and although the territory stretching from Mexico to Uruguay shared a common language and religion, eventually 16 Spanish-speaking, Catholic nation-states appeared on the **map** of Central and South America. The decline of the Ottoman Empire at the conclusion of World War I did not result in the appearance of nation-states in all of the **regions** that the Sultan administered, as much of this territory was simply absorbed into the British and French empires, but ironically it resulted in the emergence of a strong Turkish national identity and the founding of the Republic of Turkey, certainly an example of a modern nation-state. The aftermath of World War II stimulated the disintegration of all of the European empires, although the British and the French were quicker to part with their colonies than the Dutch and Portuguese. The British began jettisoning their empire as early as 1947 with the independence of their colony of India, and the following quarter century witnessed an explosion of nation-states across Africa and Asia. In 1960 alone 14 nation-states were established in former French West Africa, and by the end of the decade, virtually the entire continent was divided by the borders of more than 50 nation-states. Perhaps the most remarkable facet of these states has been the resiliency of their borders, which have changed very little in three decades. Despite the recent rise of **supranationalism** in Europe, those who have declared that the nation-state is "dead" appear to be premature. Indeed, the world's newest nation-state, Kosovo, is located in Europe, and the advocates of a "Euro" identity that supersedes the national identities of European nation-states appear to have met with very little success. The nation-state continues to be a viable means of dividing political space.

Natural Hazards

A type of event, triggered by forces in nature, that has the potential to cause significant loss of life, damage to property, and other natural hazards. To be considered a "natural hazard" the event cannot be caused by human activity or negligence, i.e., a fire caused by a burning cigarette in an apartment building has the potential to cause multiple deaths, but it is not a natural hazard because the fire is the result of human carelessness and is preventable. On the other hand, a similar fire caused by a lightning strike would be considered the result of a natural hazard. Natural hazards are generally not preventable, and in some cases cannot even be predicted with any level of accuracy, a characteristic that makes preparation for and response to this class of hazards particularly difficult. If a natural hazard results in an event that causes significant loss of life and/or property damage, the event is termed a *natural disaster*. In the United States, relief from the damage resulting from natural hazards is addressed initially by local and state agencies, but if the consequences of the event are severe enough to warrant additional assistance, the governor of a state may request federal disaster relief by declaring a specific **location** or group of locations (usually cities or counties) "disaster areas" and requesting assistance from the Federal Emergency Management Agency (FEMA). The financial losses and cost in lives of natural hazards can be enormous, with a single significant event resulting in billions of dollars of damage and thousands of deaths. Geographers have become increasingly interested in recent years in the study of natural hazards, especially in the **locational analysis** of the patterns associated with the occurrence of hazards, and in the spatial aspects of response to such events.

Hazards may be classified according to the forces that trigger the event. These forces may be geological, climatological, and hydrological. Diseases may also be considered to be natural hazards and are considered as a separate category. Extremely rare events, which may indeed be the cause of massive death and destruction, are not typically considered by scholars and policymakers studying natural hazards, simply because such events happen so infrequently that they are not grouped with other, more common events. An example would be a large meteor strike of the **Earth**—certainly a potentially "hazardous" occurrence, but so rare that in general it is not included in studies of natural hazards. Hazards that have geological origins are often termed "geohazards" to differentiate them from other types of hazards, although some commentators regard any hazard located in the biosphere to be a geohazard. The major hazards that are geological in origin are **earthquakes**, landslides, volcanic eruptions, and avalanches; and although other events may occur, they typically are limited in scope and cause death and destruction only on a local **scale**. Climatological hazards include **hurricanes** (called typhoons in the Pacific and Indian Ocean basins), tornados, blizzards, heat waves, extended periods of drought, and geomagnetic disturbances, which have extraterrestrial origin, as they are caused by solar activity. The primary hydrological hazards are represented by floods, and tsunamis, although other localized hazards exist.

Enormous destruction may result if a natural hazard causes a natural disaster. In terms of single disastrous events, earthquakes and floods historically have caused the greatest loss of life. The most catastrophic loss of human life in history was the result of extended flooding along the Huang He River in northern China in 1931, which killed an estimated 1 million people. Combined with the results from other flooding in 1931, the total number of deaths from natural disaster that year in China approaches 2.5 to 4 million. The greatest loss of life due to an earthquake also occurred in China. In 1556 a massive tremor struck northern China in a region where thousands of people lived in caves carved from the local loess soil. The earthquake is believed to have killed in excess of 800,000 people, with most of the dead buried alive by the collapse of their dwellings. But these examples, while stunning in the magnitude of loss of life, pale in comparison to the death toll of infectious diseases, especially virulent outbreaks of maladies with a high mortality rate. The Black Death, an epidemic of bubonic plague in the 14th century, killed almost a third of the population of Europe in only two years, or an estimated total of approximately 100 million people. Just in the 20th century, influenza pandemics, smallpox, malaria, and AIDS have resulted in well over 1 billion deaths. Famines also account for a large number of fatalities. While earthquakes, volcanic eruptions, and hurricanes receive a great deal of media attention when they occur, the most devastating natural hazards clearly are disease and famine.

Natural hazards are a major concern for those in the field of emergency management and response. Although natural hazards are found everywhere on the planet, they take a disproportionate toll in the developing world, where a larger population may be potentially affected by natural hazards, and fewer precautions are available to limit loss of life. For example, hundreds of thousands of people in Bangladesh live in the delta and floodplain of the Ganges River, only a few feet above sea level. This precarious location makes them quite vulnerable to periodic typhoons that sweep inland from the Bay of Bengal, resulting in tragic loss of life from extreme winds and massive flooding. In 1970 the Bhola typhoon killed approximately half a million people, and destroyed hundreds of villages and towns in the **region**. Only two decades later, in 1991, a similar storm took more than 100,000 lives, again with catastrophic damage. A poor country like Bangladesh lacks the resources and infrastructure to mount an evacuation of the population when a typhoon threatens the coastal areas, and moreover, does not have the ability to respond quickly and adequately to the needs of hundreds of thousands of victims in the aftermath of such a disaster. Some view the matter of ameliorating the impact of natural hazards as a component of **sustainable development,** by raising awareness among especially vulnerable populations of potential of hazards and helping to develop the means to reduce their impact.

Natural Resources

Natural resources are those materials that are present in, on, or above **Earth** have economic value and are not the product of human endeavor. They constitute the raw materials that are extracted by activities in the primary sector of the economy or they may represent substances that are vital to human existence or quality of life, such as water, forests, or other types of ecosystems. Natural resources may be found in abundance and in multiple **locations**, or they may be quite rare and obtainable in only a few places. Furthermore, the value and utility of natural resources arise and change with modifications and advancements in technology. Today, there is no global demand for whale oil, and in fact it is quite difficult (and often illegal) to obtain in most parts of the world. But in the early 19th century the oil obtained from whales was a common fuel source for lamps and also had applications in the processing of wool. Today this once vital natural resource has been all but completely replaced by other substances, and for the most part has lost both its value and utility. On the other hand, before the industrial age, rubber, both natural and synthesized, had virtually no utility and little value. Since the advent of the automobile the demand for rubber has exponentially increased, making it a valuable natural resource indeed. More recently, new sources of energy like sunlight and wind would not have been considered "natural resources" in the traditional sense of the term, but now may be classified as such since they are now used in the production of energy.

Access to natural resources is vital to economic development. In the modern global economy, the primary sources of energy remain hydrocarbon fuels, primarily petroleum, coal, and natural gas. In 2007 these so-called fossil fuels accounted for about 86 percent of global energy usage, a startling figure that highlights the essential function such resources serve in maintaining economic standards and driving growth. Countries that lack deposits of these fuel resources must acquire them if they are to remain competitive, and the **geopolitics** of energy is the subject of study for many scholars. In a similar fashion, strategic metals, many of which are found in only a handful of locations, are vital to modern industrial economies, and are therefore quite valuable and sought after. The country of South Africa, for example, controls 80 percent of the world's production of platinum, a highly valuable metallic ore that has many industrial applications, as well as being used for jewelry.

There are a number of ways of classifying natural resources. In broad terms, resources may be separated into renewable and non-renewable resources. Renewable resources are those that are not fundamentally destroyed in their consumption, although they may be degraded or change form to a certain extent; or they are materials that may be reproduced by nature in a relatively short period, even if they are consumed or destroyed by use. An example of the first type of renewable resource is water, which may be drunk by humans or animals, used to irrigate farmland, piped around machinery to cool it, or employed for many other purposes, yet is not destroyed or consumed in the process. Wood is an example of the second kind of renewable resource. It may be destroyed through use (burning wood for heat, or to make charcoal, for example), but if managed properly, it is renewable because new tree seedlings may be planted and new wood produced indefinitely.

Non-renewable resources are materials that are fundamentally altered or destroyed when they are used and cannot be reproduced readily by nature. All fossil fuels are examples of non-renewable resources. When coal is burned to heat a home, generate electricity, or used as the raw material to make more complex and sophisticated chemicals, it is broken down at the molecular level and cannot be recovered or reproduced. Because it requires millions of years and tremendous pressures found below the Earth's surface to make coal, it is considered non-renewable. Natural gas is also classified as a non-renewable resource, because the large, commercially valuable deposits currently exploited were the result of natural processes, and not from human activity. On the other hand, natural gas (methane) may be produced in small quantities for personal or local consumption via compost piles or from other organic sources, and therefore natural gas may one day be considered a renewable resource. Metallic ores are also non-renewable, although they may be recycled and reused once they have been consumed.

A second way to classify natural resources is to group them as either biotic or abiotic. Biotic resources are those that consist of living matter or are derived from organic material that once was living. Agricultural products, wood, commercial fisheries, and other similar products are all biotic resources, as are hydrocarbon fuels, because they are derived from living material. Abiotic resources are obtained from locations outside the biosphere and are not alive, such as metallic ores that are mined below the Earth's surface, salt, and other minerals obtained from sea water, etc.

The management of natural resources is a vital process in the modern world. Before the age of industrialization and rapid growth of the human **population**, resources could be and often were exploited as though they were essentially limitless. The enormous demand for resources of all types by modern economies has meant that most resources must be carefully monitored and maintained, a key approach in the concept of **sustainable development**. Moreover, poor management of one resource may lead to degradation of others. A simple illustration of this is the clear-cutting of forest, rather than using a more sustainable approach like selective harvesting of trees. Clear-cutting not only eliminates habitat for wildlife, but also typically results in heavy erosion of the soil, thus degrading another resource and damaging the entire ecosystem further. Soil erosion then in turn may lead to a decline in the quality of local water resources. Proper management of natural resources attempts to approach the issue in an integrated fashion, because poor management of one resource frequently has negative consequences for others. In the developing world, resource management is sometimes tied to the concept of **carrying capacity**. Unless resources are utilized carefully and conservatively carrying capacity may be exceeded, for example, when land is overgrazed or when water for irrigation is wasted or poorly allocated. Exceeding the carrying capacity of an area is not simply a matter of poor resource management; it may well be a matter of life and death for some local people, because the result may be a decline in food production.

Natural resources are thus essential to economic growth and innovation but must be conserved and utilized wisely. Because resources have economic value, as they become scarce their value increases, driving consumers to use them more carefully and efficiently. On the other hand, the value of some resources may not be initially apparent, as may be the case of the tropical rainforests, which serve as a vast repository of potentially vital and useful genetic material. Market forces alone may not be sufficient to protect such fragile and irreplaceable ecological resources, and other means must be found to preserve them.

Neo-Malthusianism. See Malthusian Theory.

O

Ocean Currents

Earth is a water planet and the world ocean is always in large-**scale** motion, slowly rearranging its waters over large extents of **latitude and longitude**. Whereas tides and waves are vertical motions, ocean currents are horizontal. Currents can be directly forced by surface wind or occur more slowly and deeply because of gradients of **temperature**, pressure, and salinity. Humans have long been knowledgeable about currents because they are large and readily associated with fishing conditions and navigation. More recently, science has connected changes in currents with global changes in weather and long-term changes in **climate**.

There is a known geography to the major currents of the world. Because of the wind emanating outward from the subtropical highs of the global wind and pressure belts, kinetic energy from the fluid that is the **atmosphere** is imparted to the much denser fluid of the ocean. In that the subtropical highs are semi-permanent features of the atmosphere, this means that air blows from preferred directions over large stretches of oceans for days and weeks at a time. The ocean water starts to move in response though not as fast or in exactly the same direction because of the increased force of friction and the correspondingly lesser Coriolis effect.

The main geographic regularity that can be observed is the presence of oceanic gyres, immense circulations around ocean basins. These gyres have a clockwise flow of water in the Northern Hemisphere and a counterclockwise flow of water in the Southern Hemisphere. This means that there is considerable water being transported from tropical to polar latitudes and vice versa. The main subtropical gyres are joined by Equatorial Countercurrents in the Pacific and Indian oceans. The strong trade winds on the equatorward side of the gyres cause the westward motion of the water and the piling up of water level in the eastern parts of these ocean basins. The higher water in the east literally runs downhill into the linear area of light and variable Intertropical Convergence Zone winds, and it is this eastward-moving water that is the Equatorial Countercurrent.

There is the tendency for poleward transport of water near east coasts of continents and the equatorward transport of water along west coasts. When water temperatures are studied it is apparent that east coast currents are "warm" and west coast currents are "cold" compared to the average of the ocean temperatures of each latitude. These temperature adjectives do not have absolute meaning. For instance, the Gulf Stream leaves the vicinity of Florida with surface temperatures in excess of 27°C in the summer and the temperature decreases with the trip northward. By the time the remnants reach Ireland, a dip of the human hand would confirm that it is cold to the touch at 13°C. Yet, the current is still considered "warm" because it is warm relative to the rest of the water at that latitude. There are many notable ocean currents in the world. In general, ocean currents are set up wherever wind is strong and prolonged from one direction, seasonally or yearly. The largest flow of water on the planet is the Antarctic Circumpolar Current. It girdles the globe in the region of the Southern Hemisphere's westerly winds and does not experience significant interference from land masses. So, too, there are many turns and splits of currents because of the particular shapes of the various ocean basins. The currents have undoubtedly changed with the drift of continents caused by **plate tectonics**. Central America did not exist as an isthmus until about three million years ago. Thirty million years ago the region was a string of volcanic islands between widely spaced North and South America and there was a west to east ocean current where today there can be none; this change in current configuration doubtless had some important climatic impacts.

As currents are examined it can be seen that there are several fluid characteristics in common with rivers. Ocean current meanders are progressively moving features that take months to form, progress, and dissipate. They are readily seen on **satellite images**. Additionally, great eddies swirl at the edge of the currents mixing current water with the ocean waters surrounding them. With great contrasts of water characteristics at the edge of ocean currents, these moving features are often the site of large biologic productivity.

The Gulf Stream is a warm current supplying water to eastern North America. It was discovered in the early 1500s and thereafter used as a boost to sailing ships returning to Europe. The water has tropical origins in the Atlantic and the Gulf of Mexico and is tightly focused as it passes through the straits between Florida and Cuba. It has an average speed of about 3 to 4 kph. It has been estimated that the Gulf Stream near Newfoundland transports 150,000,000 cubic meters per second, making it much larger than the combined flow of all the river discharges into the Atlantic Ocean. As it travels across the North Atlantic it becomes known as the North Atlantic Drift and comes against the European continent to split north and south. The northern extension, the Norwegian Current, keeps the Barents Sea ice free year round, which played a pivotal role in keeping the Soviet Union port of Murmansk supplied with arms during World War II. Although there is scientific controversy regarding the nature of the climatic influences, there is no doubt that these waters of tropical origin play an important role in the mildness of climate in Western Europe.

The Peru Current, also known as the Humboldt Current after the geographer Humboldt, is a huge flow of cold water off of the western coast of South America

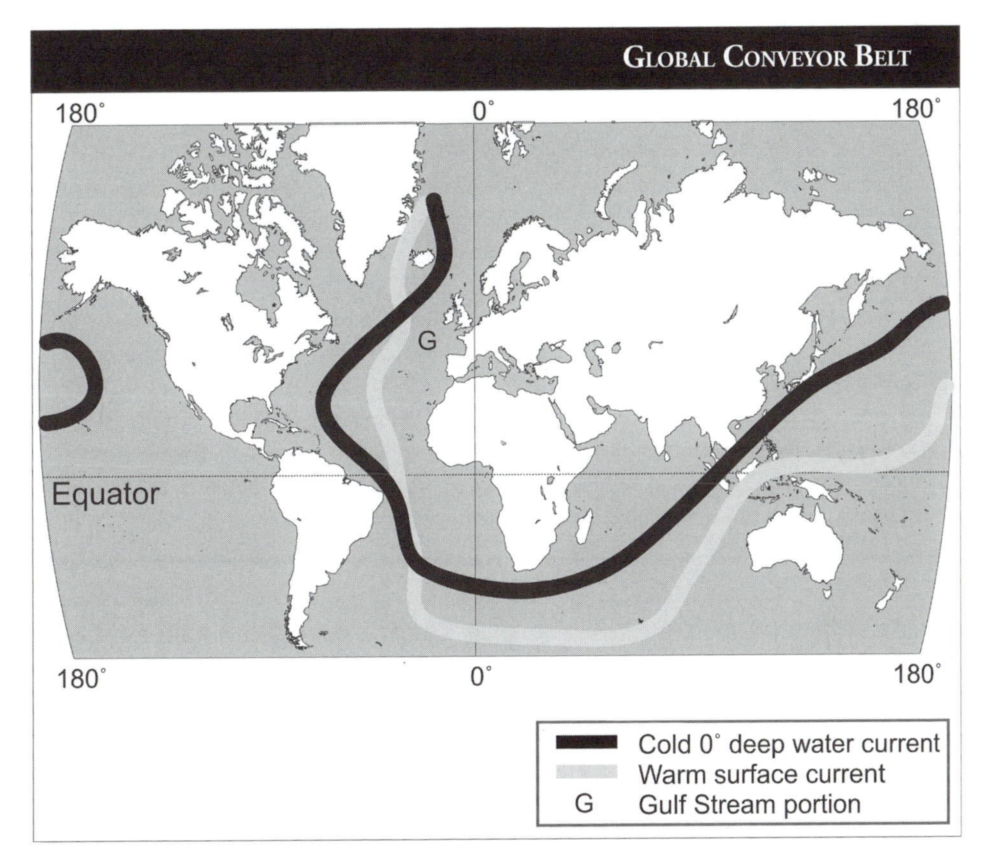

The Thermohaline Circulation, also called the Global Conveyor Belt, is a global current that transfers energy and material between the world's major oceans. (ABC-CLIO)

and especially pronounced in the latitudes of Peru and northern Chile. Its origins are at the latitudes of southern Chile (43°S) proceeding into tropical waters where it turns westward at about 5°S to feed the South Equatorial Current. The current can be as much as 1,000 km wide and a couple of hundred meters deep flowing at an average rate of about .4 km/hr. Compared to the remainder of the Pacific Ocean at the latitudes of Peru, the Peru Current represents a swath of water cooler by 8°C. This pronounced coolness stems not only from the arrival of water from the higher latitudes but from coastal and continental shelf configurations encouraging upwelling of deep, cold waters. The coolness helps to stabilize the lower atmosphere and is a partial cause of the extreme aridity of the Atacama Desert in the narrow coastal strip west of the Andes Mountains. More importantly, the upwelling waters transport nutrients and life upward making them available to an unusually diverse chain of life near the surface. The result is that the Peru Current is associated with about one-fifth of the fish tonnage caught in the world's

El Niño

An occasional cessation of the flow of the Humboldt Current and upwelling off of Peru. The phenomenon was first documented by the Spanish in the 1600s, though South American fisherman knew about it long before that. The name means "the child" for the Christ child in that the onshore effects of El Niño are visible starting in December. El Niño leads to abnormal flooding rains and the local collapse of the oceanic food chain. Appearing irregularly every 3 to 6 years, El Niño usually lasts 9 to 12 months. It has some Northern Hemisphere effects such as an association with wet winters in the southern United States and low numbers of hurricanes in the North Atlantic. El Niño is now known to be part of the gigantic Southern Oscillation, which is a pressure reversal over the Southern Pacific Ocean affecting weather conditions in Australia and impacting the South Asian **monsoon**. El Niño sometimes alternates with La Niña which is associated with very dry conditions in western South America. El Niño is now monitored by satellite so that onset can be predicted months in advance. The ultimate cause(s) of El Niño is unknown and under investigation.

oceans. At times, the Peru Current weakens causing rains in the Atacama Desert and catastrophic die-offs of ocean life; this phenomenon is known as El Niño.

A thermohaline circulation connecting the ocean currents of the topwaters with the currents' deep waters has become better known in the last couple of decades. Whereas the surface ocean currents are driven by winds, deep currents are driven by differences in density. There are two ways in which the density differences are maintained: the first is by salinity differences and the second is by temperature differences. Mainly in the North Atlantic and near the Antarctic periphery cold surface water sinks into the deep waters. The ocean bottom geography is key as the dense flow seeks the lowest ocean bottoms. The deep currents are prevented from flowing to certain locations because of the relative high ocean bottom, like mid-oceanic ridges blocking the way. The rate of travel of these waters is a mere fraction of surface currents. At these tortoise-like rates, water sinking in the North Atlantic may take on the order of 1,000 years to become surface water again in the North Pacific and the Indian Ocean. The thermohaline is sometimes likened to a conveyor belt. Long-term variations in the thermohaline circulation are speculated to play a role in climate change.

Oceans

Approximately 71 percent of **Earth's** surface is covered by a world ocean. The water is salty, deep, and some of it is in transit. In that human life is largely confined to continents and islands, and navigation was strictly on the surface until the 20th century, the ocean has been a realm of considerable mystery. In the last several decades, the exploration of the sea bottom, sensing by satellites, and research into connections between sea surfaces and weather over the land half the globe away has started to correct the deficiencies in our knowledge. Still, some writers have lamented we know less about the world ocean than the cosmos.

As the most common substance on Earth's surface, ocean water provides an immense thermal stability for the planet. Water is the natural substance with the highest specific heat. This means that water heats and cools much more slowly than atmosphere or land. The great mass of the ocean represents a tremendous capacity for heat storage so the presence of the ocean greatly moderates solar heating of the **atmosphere** and slows atmospheric cooling. Geographically, ocean water heats and cools according to patterns of insolation and warm and cold **ocean currents**. This is evidenced in the observation that seasonal **temperature** differences over land are much more extreme than over the ocean. In a scenario of **global warming** the ocean is a "sink" dissolving and sequestering much of the carbon dioxide released into the atmosphere; it is unclear as to the capacity of the ocean in this respect.

The ocean is the most important source of water feeding into the atmospheric portion of the **hydrologic cycle** with some water waiting in the oceanic storage for thousands and thousands of years. The evaporation from tropical oceans is in net terms transported from the tropics toward the poles helping to balance the ongoing global energy imbalance. The latent heat captured by evaporation can be transported thousands of kilometers and transformed into atmospheric storminess. Because most global **precipitation** originates from oceanic evaporation and because the ocean is so large, approximately 77 percent of the world's precipitation falls over oceans.

Geographers and others have subdivided the world ocean into several large parts commonly called "oceans." Water exchanges between the oceans to the extent that the composition of seawater is quite similar over the planet, but some passages are constricted thus decreasing free circulation by ocean currents. A case in point is the Bering Strait, which is shallow and has a width of only 80 km thus impeding circulation between the Arctic Ocean and the Pacific Ocean.

By far the largest ocean is the Pacific Ocean comprising a surface of 167,000,000 sq km, which approximates one-third the surface of the planet and contains 46 percent of the world ocean's water. It extends from the Arctic to the Antarctic, roughly between 90° west longitude and 120° east longitude. The Atlantic Ocean is the second largest and, like the Pacific, extends from the Arctic to Antarctica. However, the Atlantic's longitudinal extent is only about half of the Pacific's. The Indian Ocean is a bit smaller than the Atlantic and extends from the

Tides

The sweep of tides in and out from seacoasts is obvious to anyone who lives near the sea. This motion is caused by gravity associated with the moon and, secondarily, the sun. The moon's attraction causes the oceans to bulge on the side toward the moon and the side away from the moon. Earth's rotation means the bulges progress around the Earth. Earth's rotation and the moon's revolution are not synchronized, causing two high tides and two ebb tides every 25 hours at each location. Additionally, the highest of high tides is called a spring tide and occurs twice a month when Earth, moon, and sun are aligned. The height of tides varies over the world's oceans. In the deep ocean, tides tend to be about 1 m (3 ft) while along coastlines the local depth of the sea bottom and the shape of the coastline play major roles. Along shallow coastlines, tides sweep in and out over horizontal distances of many kilometers. World record tides of 15 m (50 ft) are found in the Bay of Fundy in eastern Canada, but other locations such as the British Isles, southeastern South America, and east central Africa have tides well in excess of the world average of 2 m (7 ft). In human history, tides have had considerable importance. Fisheries and boot moorage are intimately tied to the depth of water in the coastal zone.

Tropic of Cancer to Antarctica between 30° east longitude and 120° east longitude. The smallest ocean delineated is the Arctic and it is only 8 percent of the size of the Pacific; this polar ocean presents an interesting counterpoint to the more harsh regimes associated with the southernmost Pacific Ocean's boundary with Antarctica.

The various large undulations in the coasts of the continents have caused parts of the oceans to be identified as "seas." For instance, the largest of these is the South China Sea, which is a tropical body of water bounded by southern China, the Philippines, and Vietnam and is part of the Pacific Ocean. There are numerous other seas on the planet. In a couple of Asiatic cases—the Caspian and Aral Seas—the bodies of water are not connected to water and are actually saltwater lakes. Bays are indentations in coastlines and Hudson Bay is the largest of these, representing an Arctic Ocean incursion into the land mass of North America. Gulfs are another major subdivision of oceans. The Gulf of Mexico near southern North America is a notable example. There is no precise geographic definition dividing the sizes and shapes of seas, bays, and gulfs.

The oceans are circumscribed by the land masses and have changed their shapes and sizes over long amounts of Earth history because of **plate tectonics**. For instance, a quarter billion years ago the Atlantic Ocean did not exist and is still in the inexorable process of widening through seafloor spreading. In other places, ocean floors are subducting causing oceanic area to become less. Over much shorter times the extent, volume and area can undergo appreciable changes. Since deglaciation heralding the end of the last glacial episode 18,000 years ago, the sea level has risen 120 meters and the ocean has covered much of the continental shelves of the continents.

Ocean water is salty and these impurities reflect the nature of the continental crust and ocean bottom. Water is capable of dissolving any rock (though at widely varying rates), and precipitation flow to streams carries dissolved rock components to the ocean basins. In addition, the contact of the ocean bottom waters with the oceanic crust is responsible for further solution of rocks. Not surprisingly, ocean water contains most all known minerals in solution and some of these can be reclaimed for use. Several minerals are economically extracted. Yet, there are limitations to the use of the mineral content of seawater. A major consideration is that most minerals are present in trace amounts. It is technologically possible to extract gold, yet there is only a milligram of gold in every ton of seawater. The cost to concentrate the gold into usable amounts is far from being economically possible.

Although the various **climates** and **locations** associated with the ocean mean that there are some geographic differences in the waters, ocean water is impressively similar from place to place because the world ocean is subject to slow mixing. By weight, seawater is approximately 96.5 percent water and 3.5 percent dissolved minerals that are called "salts." The most common salt is sodium chloride or table salt. Other salts contain potassium, sulfur, magnesium, and calcium.

The salinity varies geographically. Places that are sunny, warm, and arid have high rates of evaporation leaving an increased concentration of salts at the surface. In the narrow Red Sea, which has limited mixing with the rest of the ocean, the concentration of salts is about 4 percent. Rainy places tend to be a bit less salty than average. In places where there is little evaporation the water tends to be a bit less salty; a prime example is polar ocean water. So, too, there are local variations in saltiness resulting from the debouching of freshwater streams to the ocean. The Amazon River was discovered by the Spanish who found potable water outside sight of the coastline.

The ocean is vertically zoned. The top waters of the ocean are subject to the vagaries of climate and vary in temperature and circulation as forced by the atmosphere above. Deeper than the first hundred meters exists the thermocline, a precipitous shift to the colder and more uniform characteristics of the deeper waters. In this same few hundred meters of transition, density and salinity greatly increase. Underneath, the waters are impressive for their relative homogeneity of characteristics.

In general, except for limited areas of upwelling of immense importance to ocean energy balance and fisheries, the ocean is thermally stable. That is, the deep waters are cold and near freezing while the top waters are far above freezing over most of the planet. This provides a lack of impetus for the ocean to circulate via convection against the force of gravity. Interestingly, the troposphere—the lowest portion of Earth's fluid atmosphere—is completely opposite, maintaining well-mixed major gases because of convection resulting from the heating of the atmosphere from Earth's surface.

Organic Theory

In general, the term "organic" implies a relation to a living organism. In social science theory, this usually means a philosophical attempt to conceptualize some process as analogous to the life cycle of a living creature. It is often associated with the philosophy of Social Darwinism, a theory popularized in the 19th century that applied the precepts proposed by Charles Darwin regarding "natural selection" and "evolution" to human social, cultural, and political systems. But in fact, organic theories date to a much earlier era, and suggestions of the application of "organic" qualities to human institutions and behavior may be found as far back in history as the writings of Aristotle and Plato, and continue well into the Middle Ages in the work of Niccolo Machiavelli and Ibn Khaldun. Organic theory has been frequently directed at the origin and nature of the state as a means of organizing political space and regulating activity. The concept reappeared in the philosophy of Friedrich Hegel at the beginning of the 19th century, who clearly had a significant influence on a wide group of scholars and writers. Hegel proclaimed the state as an "ideal" means of organizing political authority and made clear that he regarded the state as a living thing, when he stated directly that the "state is an organism." To Hegel, the state could be examined and understood only in this context. In the discipline of geography, this concept was adopted in the positions of many early theoreticians in political geography in the late 19th and early 20th centuries.

Friedrich Ratzel, a prominent thinker in political geography in the 19th century, produced a detailed organic theory of the state and international relations. Ratzel developed the Hegelian concept of the state as an organism to a much greater degree, declaring that states experience a kind of "life cycle" with stages representing "youth, maturity and old age." His claim that states, like living organisms, require *lebensraum* ("living space") would have enormous ramifications in the 20th century, when the notion was incorporated into the **geopolitics** of Adolf Hitler and the National Socialist Party (Nazi) in Germany. Ratzel identified what he termed "laws" that governed the dynamics of state growth and expansion, including the idea that states must grow by absorbing smaller, weaker neighbors (a view seemingly derived from Darwin's observations on natural selection), and that "youthful" states are

Ibn Khaldun (1332–1406)

A Muslim social scientist, Khaldun was a broad thinker and writer who had a significant influence beyond the **boundaries** of the Islamic realm. His seminal work was a multi-part history of the world, in which he expounded several theories concerning the dynamics of human civilization, the impact of environment on human society, and the foundation of **cultural identity**. He profoundly shaped the thinking of many geographers who followed him, as one may find the antecedents of the **organic theory** of the state in his writings, as well as early notions of **environmental determinism**, especially in his comparison of urban residents and nomadic peoples. He was heavily involved in the political intrigues of his time, both in Muslim Spain and North Africa, and served in numerous posts for a multitude of rulers. Even as an elderly man, he engaged in a series of famous and intense negotiations with the Central Asian despot Amir Timur (Tamerlane). His political experience provided a vantage point from which he made many keen observations concerning relationships and the human condition, and many regard him as one of the first scholars to apply a strong sociological perspective to the study and analysis of history and geography.

characterized by **population** growth and a stable **cultural identity**. Ratzel's great admirer, the Swedish scholar Rudolf Kjellen, carried the metaphor even further in his study entitled *The State as an Organism*, published during the First World War. Kjellen believed that the state was sustained by "organs" just as a living body would be, and that over time more powerful states would simply absorb weaker ones, leading eventually to a global political geography consisting of only a few large powerful states, thereby applying the concept of "survival of the fittest" to international relations. Political geographers today mostly reject the application of organic theory to the functions of the state and relations between countries as an oversimplification of much more sophisticated processes.

P

Particularism

A broadly applied term with several meanings in the social sciences and humanities. Particularism in geography usually is used in reference to cultural geography and **cultural identity**, and is borrowed from cultural anthropology. At the beginning of the 20th century, American anthropology was in the throes of a great philosophical debate between those scholars who favored the established view of "universalism," and those who argued for a new position based on "particularism." The most influential proponent of particularism was Franz Boas, who opposed the notion of Lewis Henry Morgan and others that cultures develop in a stepwise manner following a universal pattern, an approach known as *cultural evolution*, which in turn was derived from the philosophy of Social Darwinism that had been prevalent in the second half of the 19th century. Boas, who had training as a geographer and who had conducted extensive fieldwork on Native American peoples, held that cultures are influenced by the individual circumstances presented by their environment and history, and that such development did not follow any universal law. Particularism did not completely reject the Darwinian concept of evolution of cultures and civilizations, just the notion that all cultures were destined to follow the same pattern of evolution. Boas, who had been born and educated in Germany, was influenced in his thinking by some influential geographers, including Fredrich Ratzel, who had applied **organic theory** to the development of human institutions. His opposition would soon transfer to other disciplines that had also adopted variations of universalist thinking.

One of Boaz's most influential students was Alfred Kroeber, who had established the Department of Anthropology at Berkeley University in the early 20th century. Kroeber had been on the faculty for almost two decades when the geographer Carl Sauer was hired by the university in 1923. Sauer had been trained at the University of Chicago in the tradition of **environmental determinism**, which at the time was the dominant theoretical approach in geography to cultural development. Kroeber was a proponent of the particularism of his mentor Boaz, and there is little doubt that Kroeber's ideas about how cultures originate and evolve had a profound influence on Sauer, who began to criticize and ultimately reject the deterministic philosophy of many of his teachers at Chicago. Sauer's adoption of the particularist approach in conjunction with elements of **possibilism** was quickly revealed in his concept of the cultural **landscape** and appeared later in the idea of **sequent occupance**. Eventually, Sauer's views led to the emergence of the fields of **cultural ecology** and landscape ecology. He suggested that cultures are the unique result of a collection of influences, but especially are shaped by those cultural patterns that were previously present in the landscape, and that these influences could be exposed and examined. He did not claim that the environment had *no* impact on the development and characteristics of a culture, only that it represented a secondary influence in comparison to preceding cultural attributes. Sauer became the most influential figure in American cultural geography, and his emphasis on a particularist approach to the study of landscapes remains at the heart of the discipline today.

Pastoralism

The practice of raising and husbanding domesticated quadrupedal livestock, typically sheep, goats, cattle, horses, or other animals. Pastoralism may be conducted either on a subsistence level, where stock are kept for the needs of the owner and family, or on a commercial scale, where large numbers of animals are herded on large tracts of land, eventually to be slaughtered and processed for the market. Many groups who are pastoralists are also nomadic and may engage in **transhumance**, a practice common in mountainous regions, in which herds of animals are moved between lower and higher elevations on a seasonal schedule. In other cases the **migration** of humans and animals involves a regular rotation among traditional grazing lands, which are abandoned once the animals have consumed most of the available vegetation. A key distinction between pastoralists and hunters and gatherers is that the animals associated with pastoralism are domesticated and live in close proximity to the herders, sometimes even sharing the same structure for shelter.

Non-commercial pastoralism is typically viewed in the context of **cultural ecology** as an adaptation to natural conditions that prohibit the development of complex agricultural systems. Conditions are either too dry (the Sahel region) or the growing season too short (northern Scandinavia, Siberia) to produce sufficient food for the population, and domesticated livestock then become the main source of calories in the diet. Moreover, the livestock when slaughtered may provide clothing, weapons and tools, and other items essential to the survival of the pastoralists. Pastoral groups may be classified on the basis of the dominant type of livestock they raise, or on the geographic **region** they typically occupy.

There are many pastoral societies remaining on the planet, although government modernization and settlement policies, in many cases forced on groups following a pastoral lifestyle, have reduced their numbers dramatically over the past two centuries. In Central Asia, for example, the Kazakhs and Kyrgyz were nomadic pastoralists for centuries, but were forced onto collectivized agricultural farms from 1928 to 1933, resulting in massive starvation and incredible losses of animals. Many herders in these groups simply slaughtered their livestock and left the animals to rot in the pastures rather than relinquish them to government ownership. Today few Kazakhs or Kyrgyz are truly pastoral in the way that their ancestors were, although many still keep some horses, cattle, or sheep. In eastern Africa, the Maasai are a well-known pastoral group who measure personal wealth by the number of cattle one possesses. The Maasai diet is also dependent on cattle, as their main source of protein is the blood of their cattle mixed with milk. The cattle are not slaughtered when bled; rather the blood is obtained by puncturing the animal's neck and draining a small amount of blood, without serious harm to the cow. The Maasai have retained their traditional lifestyle in spite of efforts by both the Kenyan and Tanzanian governments to encourage them to adopt sedentary farming.

Plate Tectonics

Plate tectonics is the notion that **Earth**'s crust is composed of a number of huge, interlocking pieces that slowly move. This knowledge is not abstract nor useless in that plate tectonics provides us with insight as to why continents and ocean basins are shaped as they are, why large **earthquakes** and volcanoes are focused in the geographic patterns in which they are observed, and why Earth has its mountains, folds and faults.

One of the great scientific triumphs of the latter half of the 20th century was the proof and fleshing out of the implications of plate tectonics. This is a concept so grand that it was difficult at first for most scientists to believe it. Indeed, some of the early proponents were ridiculed. Whereas the notion of plate tectonics is a unifying concept, the topic of continental drift is the part of knowledge from which plate tectonics sprang.

In 1598, the Flemish geographer/cartographer Abraham Ortelius noted the "fit" between the Atlantic-fringing continents. In 1620, influential Sir Francis Bacon echoed this as continental coastlines on maps became better defined. Both these scholars left the subject without proposing a causal mechanism. In 1858, Antonio Snider-Pelligrini drew "before" and "after" maps of continents and stated that the continents had moved apart because the biblical flood of Noah restored a lopsided single-continent Earth to a more even configuration with several continents; this reasoning became untenable as scientists found increasing evidence that such a worldwide flood never existed.

Continental Shelf

A continental shelf is the undersea extension of a continent, geologically related to the rocks of the continent and much shallower than most of the **ocean**. The shelf extends out to sea with a gentle slope toward the open ocean. Around the world, the outer edge of the continental shelf is 100 m–200 m (330 ft–660 ft), where the ocean bottom plummets along a steep incline called the continental slope. Geographically, continental shelves average 80 km (50 mi) and can be nonexistent along a coastline (e.g., the subduction zone off the Chilean coast) or as much as 1,500 km (930 mi) (e.g., the Siberian Shelf in the Arctic Ocean). Typically, the oldest rocks of the submarine continental shelf structure are overlain by sediments and sedimentary rocks eroded from the above-ocean parts of the continents. Continental shelves have become economically important because they are laden in places with oil and gas reserves and other minerals. Additionally, the world's richest fisheries tend to be on continental shelves. The United Nations **Law of the Sea** (1982) grants nations exclusive mineral rights for 200 nautical miles (370 km) away from their coasts. This works well in most places but nations arranged around gulfs and seas have had contending claims because of overlapping "exclusive" zones.

Increasing world knowledge and science led Alfred Wegener (the son-in-law of the climate classifier Vladimir Köppen) to publish papers and a 1915 book suggesting disparate evidence proved the existence of continental drift. He stated that fossilized remains strongly pointed to the existence of a supercontinent from which all the present continents evolved. Wegener named this supercontinent "Pangaea" and sought to prove its existence through the rest of his career. He died in a 1930 Greenland blizzard attempting to amass more evidence. Wegener is credited as the modern scientific father of continental drift, but he did no serious work on establishing the causal mechanisms and was harshly criticized by the majority of the scientific community. Through the 1940s, Arthur Holmes proposed that the causal mechanism was giant convectional currents within the Earth but he had no means to prove it.

Continental drift proponents were in the small majority until the early 1960s, when Harry Hess postulated that continental drift is a result of sea floor spreading. He came to this conclusion because increased knowledge of ocean bottoms showed complex underwater ridge structures that trailed around the planet. Hess had no direct proof of the spreading but that was quickly solved by Fred Vine and Drummond Matthews (1963) after they read Hess's work. They realized that their magnetic research along the Indian Ocean ridge was proof of seafloor spread. The Indian Ridge is composed of volcanic rocks, and volcanic rocks contain evidence of paleomagnetism. Iron in magma is aligned with the magnetic polarity

(N-S) of the planet at the time magma cools into volcanic rock. Earth scientists had previously realized that the polarity of Earth's magnetic field suddenly reverses from N-S and S-N. Reversals have been noted as long as 780,000 years apart and as short as 25,000 years. As new rock is created it records these polarity reversals. The stunning geography discovered by Vine and Matthews was that there are numerous stripes of matching polarity running in parallel out from the Indian Ridge. Indeed, the ages of these seafloor rocks increases with increasing distance from oceanic ridges. This was direct evidence of seafloor spreading and opened a cornucopia of research that is called plate tectonics.

There is now no doubt that Earth's crust is arranged in a series of a dozen major plates and many minor ones. The Pacific Plate is the largest at 103,000,000 sq km. Others are as small as 18,000 sq km and our knowledge of sizes and boundaries are often being refined. The size of a plate may not bespeak its significance in land-forming processes. For instance, Juan De Fuca is a modest size plate caught between the Pacific Plate and the North American Plate and is responsible for large earthquakes and volcanoes in the northwestern United States and southwestern Canada.

The vertical structure of Earth's crust has been described elsewhere in this book (see **Earth**). The plates themselves range up to 100 km in depth. In the horizontal realm, the plates are defined by their very definite edges. They have irregular, jagged shapes that have been likened to the pieces of a jigsaw puzzle. Some plates are almost entirely covered by ocean (e.g., the Pacific Plate) and some are covered entirely by land (e.g., the Turkish-Aegean plate). Most plates are covered by both land and ocean (e.g., the North American Plate).

The energy for plate motion is provided by slow convection within the mantle. The relatively cool and rigid crustal plates ride slowly on the hot, weak rocks of the asthenosphere, a part of the mantle usually 50 to 150 km deep. In the asthenosphere the hot temperatures of near 1600°C bring rocks to near-melting conditions. These rocks are plastic and capable of convectional flow.

Isostasy is a vital concept in understanding the behavior of plates. Because of materials and density differences, the crust floats on the asthenosphere. As materials are floated toward and away from a particular area, the altitude of the crust will adjust according to the amount and density of the materials. The compensation may be rapid over a few hundreds or thousands of years or slow over millions of years. Isostasy gives us the wherewithal to understand the grand movements at the **boundaries** of crustal plates.

Crustal plates move away from areas of seafloor spread in various directions. Motion ranges from 1 to 10 cm per year. Although these are slow in terms of familiar environmental events, they can cause massive crustal impacts. The Atlantic Ocean,

Many notable landforms result from plate tectonics. This view across Lake Geneva from southern Switzerland shows the titanic barrier of the French Alps, which top out at well over 3,000 m above sea level and were formed by the African plate colliding with the Eurasian plate. (Mihai-Bogdan Lazar/Dreamstime.com)

for instance, has opened up in the last quarter of a billion years at an average rate of the growth of a human fingernail. This rearrangement is enough to cause major deformations of the crust in the zones of seafloor spread and plate collisions.

Plate boundaries are where "the action is." They are the sites of large volcanoes, earthquakes, and intense crustal deformation. The boundaries are of various types: (1) Divergent boundaries are associated with seafloor spreading and are the locations of numerous shallow earthquakes because of the outpouring of lava to make the basaltic rocks of the seafloors. Divergent boundaries have constructed the midocean ridges. If the divergence happens under a continent the surface expression is volcanoes along and in continental rift valleys such as the East African Rift Valley. (2) If there is divergence in one area there must be convergence in another. Some converging boundaries bring together oceanic (dense) and continental (less dense) crust. The oceanic plate subducts ("slides") under the continental plate and the crustal rock melts as it reaches deeper than 100 km. The subduction is thought to pull the entire plate along and keeps the divergent boundaries active. Long, linear mountain ranges and oceanic trenches are found in **regions** of subduction. The subduction zones have a mixture of shallow and deep earthquakes caused by rearrangement of material around the margins of the plate. These materials can become the intrusive and extrusive forms of **vulcanism**. Perhaps the grandest example of subduction is the Andes of South America. Studded with many huge volcanoes, this 200-km-wide range extends the entire 7,000 km length of western South America over the collision zone of the Pacific and South American plates. They are paralleled by the Peru-Chile Trench offshore in the Pacific Ocean. (3) Another sort of convergence brings together oceanic plates underneath the oceans. The crustal expression is a deep trench with a paralleling island arc. A prominent example of this is the Mariana Trench and the volcanic arc of the Mariana Islands of the western Pacific. (4) Other convergences are continental-continental collisions. In this case, there is no subduction because of the relative buoyancy of both plates and a mountain range is built. The Himalayas are the result of 35 million years of the Indo-Australian plate colliding with the Eurasian plate. (5) Finally, some boundaries are transform in that they slide past each other. There tend to be some major faults and earthquakes created as the jagged pieces snag going by, but transform boundaries are conservative with no trenches or major mountain building.

Terranes are chunks of Earth's crust quite different in rocks and history from the plates around them. Sometimes they have been moved thousands of km by a drifting plate and geologically pasted to the edge of another plate. The complicated mountainous landscapes of the Canadian West (frequently known as the Canadian Cordillera) are due to subsequent accretion of terranes over hundreds of millions of years to the North American Plate.

Some places well within plate boundaries are associated with landforms resulting from plate motions. Worldwide, "hot spots" are located over more than 50 mantle plumes that are rising currents that remain stationary for very long amounts of time. Material is brought to the surface as the plate slides along overhead. There are all sorts of volcanics associated with hot spots and the progressively younger ages of the Hawaiian Islands from northwest to southeast can be pegged to the Pacific Plate passing to the northwest over a mantle plume. A wonderful aspect of knowledge of plate tectonics is found in understanding geographies much older than the current sequence of crustal movements. The study of large areas of rock deformation and age indicate the existence of several supercontinent formations and breakups during Earth history with time orders of (very approximately) 250,000,000 years. Quite a bit of climate change and history, including some evolutionary developments, can be explained through examining the changing locations of continents.

Population

A group of individuals sharing some characteristic or set of characteristics, and occupying a defined, common space. Geographers are primarily concerned with human populations at a variety of **scales**: global, regional, national, and local. However, biogeographers may study other populations, including various classes or species of wildlife. Demographers are scholars who study the dynamics of populations, including all the factors that direct the growth or decline of the number of humans. Population geographers are interested primarily in the spatial qualities of populations, especially as these are influenced by trends in population dynamics. Two measures of the spatial character of population are fundamental to the work of population geographers. *Population density* indicates the number of people in a given unit of a standard area. In much of the world population the unit for reporting density is a square kilometer, but in the United States and a few other countries an areal unit of a square mile is used more commonly. Maps may be produced showing population density based on either unit, so one must be certain to use the same units when comparing maps of population density. However, one of the shortcomings of examining human population using density is that the figure reported assumes an equal distribution across the entire area considered. If this is a large country, it reveals little about the actual spatial arrangement of people there. For example, the population density reported for the People's Republic of China in 2008 was 110 people per square kilometer. But it is far from the case that every square kilometer of territory in China has 110 people living on it, so the information is of limited utility. *Population distribution*, an indication of where people actually live in a **region** or country, usually shown on a **map**, often provides a clearer picture of how population and space are related there.

A number of metrics are used when assessing the dynamics of a population. The Crude Birth Rate (CBR) of a population measures the number of births in a given year, per one thousand of the existing population. Because this is a rate, it is reported as a decimal figure. For example, the birth rate in the Philippines in 2010 was estimated to be 26.01, meaning about 26 babies were born for every one thousand people in the country. The Crude Death Rate (CDR) indicates the number of deaths in a specified population per one thousand people, and is reported in the same fashion as the CBR. Considering these two figures together, one can quickly deduce the basic parameters of growth in a population for a given year. If the CBR exceeds the CDR, then the population is increasing; if the CDR exceeds the CBR then the population is shrinking, unless this is offset by additions to the population through **migration**. But these measures provide information only for the immediate year, and cannot offer much information regarding the long-term (meaning in most cases perhaps 20–25 years) trends. However, the Total Fertility Rate (TFR) can serve as a reasonable predictor of longer term trends in populations. The TFR calculates the number of children a fertile woman would have over the course of her reproduction years, based on the current year's average. A "fertile woman" is generally considered to be between the ages of 15 and 49, and all women in a population are assumed to be fertile. By examining the trends in the TFR for any population over time, conclusions may be drawn concerning population growth. The TFR for the entire world in 1950 was almost 5; by 2010 it had dropped to around 2.5, indicating that the growth rate for the planet as a whole is declining. Some demographic models established by the United Nations predict that by around 2040, global TFR will be 2.1, which represents the replacement rate for the existing population, or "zero population growth."

As of 2010, there are almost 7 billion people on **Earth**. This total is larger than at any other time in history, and each day brings a new global record for human population, since our numbers continue to increase, although the rate of increase, as shown above, is slowing. There are, however, some rather stark regional differences in population dynamics. Although the TFR for the developed world is now below the replacement value of 2.1, the TFR for the developing world currently approaches 4, meaning that virtually all current population growth is occurring in the developing countries. It is highly probable that this trend will continue for at least another generation. The **Infant Mortality Rate** (IMR) is also much lower among developed countries than in Africa, South Asia, and other lesser developed regions. The IMR for a country is calculated by counting the deaths of all infants in the first year of life in a given year, divided by every 1,000 live births. This figure again is reported as a decimal number. Infant Mortality Rates today range from 2 to 3 for Japan and some European countries, to above 100 for many countries in Africa. The IMR is not only useful for understanding various aspects of population structure and growth, but also serves as a general comparative measure of quality of life. For the most part, the IMR has been declining along with the general mortality rate in the developing world for the past 50 years, and this has resulted in continued expansion of the population despite a drop in birth rates in many regions.

The scientific study of population and the forces behind how it changes was initiated by Thomas Malthus at the end of the 1700s, and population growth has been a controversial subject ever since. Malthus developed his **Malthusian Theory** regarding the dynamics of human population growth, which suggested that humans reproduce in such a manner as to induce a recurring cycle of "feast and famine" in which periods of adequate food supplies are alternated with periods of shortages, because of a natural inclination for humans to over reproduce and exceed their ability to produce food. Offered at the beginning of the **agricultural revolution**, Malthusian theory has to date appeared to have little validity, but in the 1960s the neo-Malthusian school revised the theory, arguing that in fact the longer term trends in human population growth did follow the pattern described by Malthus. Most of the predictions of this school also have failed to come to fruition, but the basic question regarding how many people the planet can support has not been answered. Some scholars and policymakers hold that the "human family" has already reached its optimum size, and that only the approach of **sustainable development** will ensure that humanity will continue to have access to sufficient **natural resources** and food and maintain a reasonable quality of life for the great majority of people. Others believe that the planet has the capacity, and human beings the ability, to maintain a much larger population than even the nearly

7 billion people living on Earth today, perhaps 10 to 12 billion. As evidence they cite the fact that technological advances over the past two centuries have led to consistent advances in food production, and there is no reason to believe that this will not be true in the future. Regardless of which side is correct, the debate over population growth is certain to continue for many years to come.

Possibilism

A philosophy addressing the relationship between humanity and the environment that holds that humans have the adaptive ability to modify their responses to the physical world; in other words, that a wide range of possible responses to environmental stimuli exist. Possibilism stands in direct contrast to **environmental determinism**, which holds that the environment limits, shapes, and even controls the development of culture. Perhaps the foremost proponent of possibilism was the French philosopher Lucien Febvre, who argued against the **organic theory** put forth by Freidrich Ratzel and others of the German school in the late 19th century. Febvre maintained that there were "no necessities," and that humanity was in the "first place" rather than external, physical determinants. In the United States, possibilism clearly had an influence on the approach of Carl Sauer and the so-called Berkley School, who presented an alternative to the determinist position in the form of **sequent occupance** theory and other theoretical approaches that stressed the ability of humans to modify and adapt to physical conditions. Sauer and his followers developed a model of relations between humans and their environment rooted in **cultural ecology** and that was a direct refutation of the deterministic approach promoted by geographers at the University of Chicago, where Sauer had received his doctoral degree. In the United Kingdom one of possibilism's most eloquent promoters was Patrick Geddes, a biologist and urban planner, who held that the spatially constructed environment influenced cultural and social behaviors, an early form of "cultural determinism" in direct contrast to environmental determinism.

The theoretical approach of possibilism has had substantial impact on the philosophy of geography, especially in regard to the epistemological nature of the discipline. One can find the foundation of possibilism in the reformist movement known as humanistic geography that arose in the 1970s, whose most prominent proponent was Yi Fu Tuan. The **quantitative revolution** of the 1960s had redirected the discipline of geography toward the methodological and philosophical approaches of spatial science. Many human geographers were troubled by what they viewed as the rigid dominance of a positivistic, empirically based framework that assumed a static relationship between the observed and the observer, and which devalued the importance of spatial perception and "sense of place." Humanistic geography emphasizes the importance of human agency, especially in the construction of **landscapes**, which, it is argued, are socially constructed from a variety of possibilities and do not necessarily represent the same ideas, concepts, or even historical experiences of all observers. Moreover, borrowing heavily from the constructivist approach to literary criticism then emerging in the humanities, especially in art and literature, humanistic geographers evolved a perspective on cultural landscapes that viewed them as possessing a textual quality. The concepts of landscape and place could be interpreted as dynamic spatial entities that might be "read" in various ways, depending on a wide spectrum of factors influencing the reader: gender, social position, race, and others. More recently, possibilism has at least partially provided the theoretical basis of the generalized discipline of cultural studies, social and feminist geographies, and post-modernist geographical perspectives.

Precipitation

As part of the **hydrologic cycle**, water vapor is transformed into **clouds**. Clouds are common and, although in constant motion, usually cover about half of the planet at a time. Even so, precipitation is occurring over a small portion of **Earth** because most clouds do not produce precipitation. How and why precipitation occurs is vital in understanding the physical geography of Earth, because the precipitated water affects the surface lithosphere, biosphere, and hydrosphere.

Precipitation is triggered by one of two processes. The first is known as the warm cloud process and involves only water droplets (including supercooled water) above tropical locations. Cloud droplets collide and some of them coalesce to become large enough to fall out of the sky as rain. The cold cloud process (also known as the Bergeron process) takes place in middle latitude and polar clouds. The temperatures are below freezing and supercooled droplets and ice crystals coexist at temperatures between 0° and –40°C. The saturation vapor pressure is a bit higher over liquid water than over ice crystals and the net effect is that ice crystals grow by adding mass from the nearby droplets. The enlarged ice crystals become large enough to fall as snow.

The general term for water precipitated out of the sky is "hydrometer." There are several types of hydrometers and their occurrences are governed by moisture supply, rise of air, and the environmental lapse rate underneath the clouds. Snow starts as snowflakes in clouds and falls through below-freezing air to the surface.

Weather Modification

Societies have always had dreams of modifying weather to their advantage. Native Americans of the Southwest had rain dances designed to moisten their dry climates; weather modification prayers were common worldwide in early societies around the world. During the U.S. Civil War and World War I it was claimed that the roar of large cannon caused rainfall. More cloud-physics-based modification originated after World War II. Since World War II, the most common type of weather modification is based on the seeding of **clouds** with silver iodide. This substance mimics the crystalline structure of ice, helping initiate the cold cloud process of precipitation. Seeding has been accomplished with apparent success in some places, but there is always the question of how much precipitation would have fallen without cloud seeding. In terms of breaking droughts, one must consider that there must be a combination of adequate low-level moisture and rising air in existence before clouds can be seeded. Russia and other countries have attempted to suppress hail by seeding thunderstorms. There are no world standards for these efforts so it is difficult to assess success rates. Various modifications of **hurricanes** and tornadoes have been proposed, but the immense amounts of energy released in such natural events have made modification impractical. Finally, grand schemes of surface modification have been proposed to ameliorate regional **climates** (e.g., melt the ice of the Arctic Ocean), but these dreams have never gained serious traction.

Although the temperature at the surface can exceed freezing by a few degrees, the key is that snow does not melt on its journey. Rain starts as rain in tropical clouds and as snowflakes in middle latitude and polar clouds. Starting as snow, it must fall through a sizable above-freezing layer and melt to become rain by the time it reaches the surface. Sleet starts as snow and melts in an above-freezing layer and then freezes in below-freezing layer so it strikes the surface in pellet form (on the British Isles and some other places sleet is defined as a mixture of rain and snow). Perhaps the most pernicious form of **precipitation** is freezing rain. Freezing rain starts as snow, melts in an above-freezing layer and then comes through a shallow below-freezing layer so that it is slightly above freezing as it strikes the ground. Below-freezing ground causes this cold rain to instantly freeze into a layer of ice. Hail emanates from large thunderstorms and is the result of an ice embryo traveling through **regions** with supercooled water that freezes on contact with the ice embryo.

How much precipitation has fallen? This question is sometimes not easily answered because of complications due to reports from nonstandard gages, blockage by buildings and vegetation, and wind effects. A standard rain gage has a 20-cm mouth leading down to a narrow funnel that amplifies the catch and allows ease of measurement. A special measuring stick is placed in the cylinder and rain is measured to hundredths of inches in the United States and to millimeters elsewhere. Manual measurements are being replaced by automated gages. The most common type is a tipping bucket gage. Rain falls from the gage mouth into a small, conical bucket. When a certain mass of water (usually a millimeter of precipitation) fills the bucket, it tips to empty while its place is taken by an empty bucket. The tips register as pulses to an electronic device so that the amount and timing of the rain are registered. The rate of rainfall is measured in depth per hour or day. In the United States, inches per hour define rainfall rates with upper rates called cloudbursts and exceeding 4.00" (10" cm) per hour.

Measurement of snow is more problematic. The traditional method was to measure with a stick in places where there didn't appear to be drifting. The most modern method is to have the snow automatically melted into a heated gage containing a tipping bucket. Snowfall intensities are measured by noting runway visibilities in fractions of kilometers or miles.

The liquid equivalent (water equivalent) is the depth of accumulated water adding melted frozen forms to the measured rainfall. In the past this was not readily done, so the U.S. National Weather Service adopted a 10:1 ratio for sites where the snow was not melted for measurement. That is, for each 10" of snow 1" of liquid was added to precipitation totals. In reality there are "wet" (3:1) and "powder" (30:1) snowfalls that complicate matters. The water equivalent is of crucial interest in places like the mountainous American West where most precipitation comes as snow and melts over the warm season to supply water needs. No matter what the actual ratio, it bears noting that it takes large snowfall totals to equal the precipitation coming from one summertime thunderstorm.

Precipitation catches over Earth are highly varied. Mt. Wai'ale'ale on the island of Kaua'i, Hawaii, USA is officially the wettest at over 11,600 mm (460") a year, while worldwide there are several other mountainous locations that are similar but not well measured. There are some generalities that can be applied to world precipitation. The highest precipitation catches are in the tropical regions of the world followed by places that experience **middle latitude cyclones**. The wettest places have good water vapor supplies and consistent lift due to topography and passing low pressure disturbances. Dry places are defined by the relative lack of these above factors and the dryness is usually enhanced by one of several other factors. The Atacama coastal desert of western South America is said to be the driest continental area on the planet with rain occurring every few years at individual locations (see **Deserts**). Dry places tend to be dominated by subtropical highs or inland location or blockage by mountains.

Some authors refer to permanently frozen places as cold or polar deserts. These places have very little water vapor in the air so there is not frequently meaningful precipitation. For instance, in the snow and ice world at the South Pole, the precipitation averages about 25 mm per year. Melted down to liquid equivalent, polar precipitation totals rival the lowest ones in other deserts of the planet.

A concomitant feature of the amount of precipitation is how often it comes. Some places average less than one day per year of measurable precipitation while others garner more than 340 days. The world tendency is for the places with the least precipitation to have the most unreliable precipitation. London, United Kingdom has an expected precipitation of somewhat over 500 mm (20") per year and Londoners are shocked if the precipitation varies more than 10 percent from year to year. The arid parts of the planet have more than 50 percent interannual variability. At Luxor, Egypt, the yearly precipitation averages only 2.5 mm (.1") so any precipitation is unexpected.

Precipitation rates are also measured via **remote sensing** techniques. Remote sensing has great advantage in this realm because precipitation doesn't need to fall in a gage to be estimated and there is foreknowledge as to where the event headed. Radar uses microwaves to measure precipitation and modern Doppler radars can sense motions of precipitation within storms and pinpoint dangerous situations such as mesocyclones. Since the middle of the 20th century, precipitation has been artificially produced. This is usually accomplished by "cloud seeding," consisting of spreading silver iodide from aircraft. The silver iodide crystals act as nuclei around which water can gather, thus enhancing nature's methods. The precipitation processes described above need a significant supply of water vapor and a significant rise of air—factors that humans cannot control. Droughts occur when these factors are not present so that the production of precipitation by artificial means is relegated to possible enhancement of precipitation that might have occurred on its own.

Pre-industrial City Model

A concept in urban geography, sometimes also referred to as the Sjoberg model, and now disputed by many urban geographers, that all cities prior to the industrial revolution exhibited a common internal morphology and functionality. Prior to Sjoberg's work, several models of urban structure and development had been proposed in the first decades of the 20th century, including Ernest Burgess's Concentric Zone Model, and the **Sector Model**, proposed by Homer Hoyt. Several decades after the appearance of these models, the sociologist Gideon Sjoberg developed the pre-industrial city model, and laid out his argument in the book *The Preindustrial City: Past and Present*, published in 1965. The model emphasizes the role of social forces in determining the spatial characteristics of the urban place. Sjoberg's basic thesis was that virtually all cities that had emerged in the pre-industrial era, a stretch of somewhat over two millennia, shared common antecedents, functions, and internal organization, regardless of the **region** and cultural environment in which the city had developed. Those cities in the contemporary era that had yet to undergo industrialization also indicated this structure, according to Sjoberg. His urban model was constructed to be a partial refutation of the Burgess model of urban morphology put forth in 1925. Burgess had contended that cities are composed of circular zones around a central core. Sjoberg criticized the Burgess model on the basis that it applied only to the era of mass capitalistic production, i.e., the industrial age, and held that before the industrial revolution cities of the feudal period and before shared the same fundamental spatial plan, based primarily on social and political stratification.

The Sjoberg model offered an urban structure that emanated from the social class hierarchy of urban residents. The spatial arrangement of the city was directly related to the political power relationships between these groups. Sjoberg wrote that in general, there were three classes who lived in pre-industrial urban spaces. The most influential of the classes was an elite social stratum, comprised mostly of religious authorities, military leaders that of course held power through the threat of arms, and the political rulers. This class occupied the center of the city. Around this core, residential neighborhoods holding the lower class formed. Merchants and those engaged in some trade lived in this part of the city, with poorer classes living the furthest away from the center. This meant that commercial activities were not concentrated at the city center but further away, a spatial feature that distinguished the pre-industrial city from its successor in the industrial age. A third class, slaves and menial workers, lived interspersed with the poorer sections of the lower class, or at the very margin of the urban area. Many contemporary urban geographers find the pre-industrial city model overly generalized and simplistic, especially its questionable position of completely divorcing economic from political power. In addition, some scholars believe that the model suffers from dubious assumptions, such as the contention that levels of technological expertise were similar across the geographical and temporal spectrum of pre-industrial urban centers.

Primate City

A primate city is one that dominates the urban **landscape** of a country in terms of demographic weight and economic function. The term stems from an article published in the *Geographical Review* by Mark Jefferson in 1939. Jefferson described what he called the "law" of the primate city in countries where the largest city held a dominant position over the second- and third-largest cities by a constant ratio.

New York City is a classic example of a primate city. One in every 13 Americans lives in this metropolitan area. Here the southern end of the borough of Manhattan, one of the most intensively developed pieces of land in the world, is viewed beyond the Brooklyn Bridge. (Cpenler/Dreamstime.com)

The validity of this "law" has been greatly weakened over the intervening decades, but the general concept of a single urban place holding much greater significance demographically and economically than other cities in a state is still accepted and may be empirically demonstrated in numerous countries. If one extends the concept to include the phenomenon of the emerging **megalopolis**, found in many world regions as a single urban space, then these new primate "megacities" represent an urban dominance never before witnessed.

Many primate cities may be identified. Paris is more than eight times larger than the next largest city in France, and Mexico City is almost five times larger than Guadalajara, the country's second most populous city. In the United States, a country with many large urban areas, New York is more than twice as large as Los Angeles. Such cities typically contain not only a large share of the country's population, but also account for a disproportionate percentage of the country's economic production. For example, the metropolitan area of New York produces over a trillion dollars of goods and services in an average year, a figure larger than many entire states in the United States. Furthermore, the city plays a magnified role in the country's economy due to its hosting of the two largest stock exchanges in the country, and because an enormous number of corporations have their headquarters in the city. The city also functions as a cultural center, serving as a capital for art, design, music, and theater. New York is somewhat unusual as a primate city in that it does not function as the national capital.

In developing countries the overwhelming dominance of a primate city may lead to a clear **core-periphery** pattern and distortions in economic development. In addition, political power and decision-making may become highly concentrated in such urban leviathans. The concentration of wealth and power may lead to resentment in the country's **hinterland** and exacerbate centrifugal forces. On the other hand, the allure and opportunity represented by a primate city may also compound the challenges the city must face. Until recently, Mexico City had to absorb thousands of new residents every month as people migrated to the city seeking employment. These new arrivals placed enormous stresses on the city's infrastructure and facilities. Housing and services were in short supply, leading to the emergence of *barrios*—a ring of slums surrounding the main city. In spite of these difficult conditions, thousands of Mexico's citizens would migrate to the city if allowed, indicating that the dominant role of primate cities appears destined to become greater, as the **urbanization** of populations around the world increases.

Purchasing Power Parity

A concept employed when attempting a comparison of economic conditions across countries, which according to its proponents provides a more accurate assessment of relative wealth and economic conditions. A common way to compare economic conditions between countries is to use a measure of Gross National Product (GNP) per capita. This figure is usually reported in U.S. dollars—the GNP per capita in the United States in 2008 was approximately $47,000, while in Turkey it was approximately $10,000. The figure for Turkey is based on the current exchange rate between the American dollar and the Turkish lira. This implies that on average, an American is more than four times wealthier than a resident of Turkey. But this comparative measure also assumes that costs are the same for comparable goods in both countries, which in reality is rarely the case. The cost of basic necessities in Turkey may be much less than in the United States, meaning that the relative wealth of the average person there is actually greater, perhaps significantly greater, than is indicated in the GNP per capita comparison. Purchasing Power Parity (PPP) is a comparative measure that attempts to account for the difference in prices, inflation, variances in exchange rates, and other economic factors that are not included in other methods. There are two types of PPP commonly used:

absolute PPP, and relative PPP. Absolute PPP is based on the assumption that the exchange rates between two economies will be stable and this will result in equivalent purchasing power for each currency. Relative PPP attempts to account for differences in the relative inflation rates and exchange rates between the two currencies being compared.

When calculating PPP, economists will select a so-called basket of goods, or in some cases, a single good that is available in all economies in the calculation and that is produced using the same inputs. A famous example of the latter technique is used by the British news magazine the *Economist*, called the "Big Mac Index," which has been published yearly in the journal since the 1980s. This is in reality a calculation of absolute PPP and is based on the cost of the famous sandwich in various economies. Although there are some shortcomings to this methodology, it is a simplified way of analyzing PPP. Critics of the concept of PPP point out several weaknesses in the methodology and theory. First, the "basket of goods" must be comparable in the economies used in the calculation. Perhaps the most common criticism of the "Big Mac Index," for example, is that while a Big Mac may be commonly purchased in developed countries in the United States or Western Europe as "fast food," in other economies such as China or Mexico this purchase is considered more of a luxury and, accordingly, made much less frequently by the average consumer. It is essential that the basket of goods be composed of the same goods in each economy, and that these goods are similar in quality. Despite some flaws, many economic geographers regard PPP as the most accurate means of comparing economic wealth, development, and other characteristics among countries.

Push-Pull Concept

A theory that attempts to identify the motivational factors that drive **migration**. The concept was introduced by Everett Lee in his article, "A Theory of Migration," first published in 1966, in which Lee attempted to codify the various considerations that a potential migrant might take into account as a basis for relocation. Lee based his ideas on the earlier work of Ernest Ravenstein, a British geographer who in the late 19th century articulated a series of "laws" of migration. Ravenstein was the first scholar in the industrial age to attempt a systematic study of migratory flows, both within countries and across international **boundaries**. Ravenstein focused much of his study on the actual movement of people and less on the forces that convinced them to shift **location**, however, and Lee sought to address the causation of migration. He divided the array of pressures to migrate into two broad categories: those that would "push" an individual or group to migrate, and those that instead would "pull" migrants to a new place. Lee holds that these factors working either alone or in concert, and at various levels of intensity, are responsible for inducing the movement of human beings across space.

The "push" factors that Lee describes are all negative aspects of remaining in the home country that a migrant will seek to avoid and often flee from. Lee did not claim to have produced a complete listing of these, but his compilation contains the majority of common reasons behind the "push" to migrate. These range from life-threatening situations such as famine, threats to one's life, and **natural hazards** and disasters, to matters of personal life choices or to seek better economic opportunities, such as finding a suitable spouse or a higher-paying job. Other push factors are poor quality of life, especially low standards of housing and medical care, loss of farming opportunities due to desertification (indeed a major cause of migration in the Sahel region of Africa), actual or feared religious and political persecution, and other reasons. Any of these causes and others suggested by Lee are sufficient to convince an individual or group to move to another location, although the **friction of distance** may prevent migration until a certain threshold of tolerance is exceeded, at which point migration will take place.

Push factors by themselves will suffice to instigate a migratory flow, but pull factors also usually play a role in the decision to migrate. These are positive aspects of the target location, whether it is a foreign country or territory, or another place in the country of origin. These include better living conditions, better health care or treatments unavailable in the home country, political or religious freedom, greater opportunities of marrying, etc. Pull factors therefore frequently complement push factors. Some pull factors are related more to amenities; for example, moving to a **region** that has a more comfortable climate or beautiful scenery (a common factor in the migration of retirees in developed countries). Although rather general, the push-pull concept provides insight into why migration occurs.

Q

Quantitative Revolution

A shift in the methodological foundation of the social sciences and some humanities toward more "scientific" techniques and models, based on the increased use of statistical methods, a search for empirical, natural "laws," the widespread development and application of hypothetical models and theory, and the adoption of a positivist-based emphasis on scholarship grounded in the "scientific method." In American geography, this "revolution" was initiated in the late 1950s by a group of emerging scholars who incorporated analytical techniques into their work that relied heavily on mathematical proofs and who wished to push the discipline in the direction of what some termed "spatial science." Many of those who considered themselves "quantifiers" felt that the discipline of geography had stagnated, remaining essentially devoid of a body of general theory that had emerged in other social sciences in the early 20th century, was overly dependent on simple descriptive narratives that offered little more than (in their opinion, at least) superficial accounts of topographical and cultural features encountered on the landscape, and provided little in the way of application to other fields of inquiry, or even other closely related social sciences. It should be pointed out that in fact mathematical modeling and sophisticated, abstract theoretical approaches were not unknown in geographical thinking at all, as witnessed by the promulgation of the **von Thunen model** in economic geography in the early 1800s, Walther Christaller's work on urban hierarchy presented in **Central Place Theory**, as well as the research of many other urban geographers in the first decades of the 20th century, along with other examples. But much of this tradition lay in European geography, while such methodologies were more rarely encountered in geographical research in the United States.

At the time the quantitative revolution began to influence geography, the discipline in the United States was experiencing something of an identity crisis. Related subjects like geology, political science, and even anthropology and sociology were firmly rooted and defined as academic disciplines, but geography had only emerged as a separate subject at the university level in many cases in the early 20th century. The first academic department of geography in the United States was founded at the University of Chicago in 1903. Ironically, the integrative nature of the discipline actually worked to undermine its status as a distinctive intellectual

Professional organizations

The National Geographic Society dates from 1888 when it was formed as an elite club for explorers and others with geographical interests. The society has since broadened to include all with geographic interests, most of whom are not professional geographers. The Society publishes *National Geographic Magazine*, which is a popular journal of 50 million copies per month. The Society is not the most important professional organization for American geographers. As in other disciplines, geographers are actively involved in lifelong learning under the aegis of professional organizations. These organizations provide venues for meeting with other geographers and published outlets for scholarly research. The Association of American Geographers (AAG) was formed in 1904 and membership was, at first, gained only with the recommendation of a member of the organization. Today, with an open membership, the AAG is the predominant scholarly geographic organization in the United States with over 10,000 members. It hosts large yearly meetings and is proactive for the growth of geography. Its most respected journal is the *Annuals of the Association of American Geographers*. The National Council for Geographic Education, founded in 1915, is focused on geographic teaching and learning, particularly in pre-college geography. Its main journal is the *Journal of Geography*. There are many geographical organizations in other countries. In the United Kingdom, the Royal Geographical Society was founded in 1830 and supports research, fieldwork, education, and public interest in the discipline. Its flagship scholarly journal is *Transactions of the Institute of British Geographers*.

perspective. The new quantitative methodology of the 1950s and 1960s was embraced by those who sought to address this issue by establishing a scientific basis for the subject. In addition, the revolution coincided with the appearance of new technology that complemented and enhanced the mathematical methodologies of the quantifiers. Early computers were already in use in some geography departments by the late 1950s, and the innovative technological advancements often meant additional resources for those departments and scholars who employed them in their research. Young scholars who led this charge included Brian Berry, Waldo Tobler (who is credited with formulating Tobler's "first law of geography"), and Arthur Getis, all of whom received doctorates in geography from the University of Washington between 1958 and 1961. For approximately two decades Berry was the most cited scholar in American geography, and his influence in propelling the methods of the quantitative revolution toward the mainstream in geography in the United States was unmatched.

Many geographers initially welcomed the shift in methodologies and philosophy brought by the quantitative movement, even if they did not incorporate the new techniques into their own scholarship. However, as the 1960s progressed, those who continued to produce descriptive, qualitative work found that many of the scholarly journals and academic departments now under the influence of the quantifiers rejected their research as "unscientific" and lacking intellectual rigor. This resulted in a counter reformation in the form of new approaches to the study of human geography, accompanied by a vigorous critique by some geographers of what they viewed as an infatuation with numbers and methods, at the expense of the human dimension of geographic inquiry. These scholars, who believed there was more to geographic study than **locational analysis**, began to develop intellectual linkages to disciplines in the humanities like art, history, and literature, using the conceptual framework of the **landscape** as the basis of their analysis. Eventually this school came to be known as "humanistic geography," which generated new directions within human geography like behavioral and social geography, the examination of spatial dimensions of race and gender, and other innovative lines of investigation. A resurrection of the descriptive, qualitative tradition of geographic research ensued, because these new areas in human geography were based more on perception and subjective interpretation of the dynamic between humans and space, and did not lend themselves readily to quantification or mathematical measurement. By the late 1970s, the humanistic school was solidly established, led by scholars like Yi Fu Tuan and Donald Meinig.

In retrospect, the controversy brought on by the quantitative revolution in geography engendered substantial change to the discipline that reached beyond analytical techniques. Fifty years after the new paradigm was implemented, it seems clear that the discipline benefited from a renewed examination of its very nature and purpose. The question of whether geography is an art or a science, exemplified in an extended and active debate in the literature over the course of two decades, forced geographers to reconsider their intellectual position, as well as the place of the discipline in the academy. The emphasis on statistical methods and theoretical modeling prepared the way for a technological transformation in the field of **cartography** through the application of computer-generated **maps**; the development of **geographic information systems (GIS)** utilizing the comparative analysis of data bases; and the evolution of new theoretical directions. Qualitative human geography underwent its own "revolution" of sorts, finding linkages with other intellectual fields that previously had been largely ignored by geographers, a process that resulted in an expansion of both theory and application in human geography. Many geographers today cultivate a familiarity with both quantitative and qualitative methods, resulting in a greater appreciation of all manifestations of spatial inquiry.

R

Rain Shadow Effect

A rain shadow is a **landscape**-induced, long-term relative lack of **precipitation** of an area compared to its geographic surroundings. Rain shadows exist because the landscape "gets in the way" of the horizontal flow of air. Air cannot reverse direction when it encounters significant topography. Thus, it must stop, flow over, or flow around the barrier; this has a significant influence on patterns of precipitation. The orographic effect is an enhancement of precipitation because of the rise of air over the topography. A rain shadow can be viewed as the muting of precipitation because of the landscape. Frequently, the orographic effect and rain shadows occur in tandem.

Precipitation occurs because air containing water vapor *rises* and cools. The air cools until it becomes saturated, clouds are formed, and then precipitation processes cause some of the water in clouds to fall toward **Earth**. One type of rain shadow occurs because of the strong *descent* of air. If stable air has been topographically lifted to the crest of a mountain, it will attempt to sink back to its original, unlifted altitude after it crests the mountain. It will undergo adiabatic warming at 10°C per every kilometer of descent because, even if originally saturated at the mountain crest, the air quickly becomes unsaturated. Adiabatic warming increases the air's capacity to hold water vapor and decreases its relative humidity sharply away from saturation. Without a significant rise of air, the potential for precipitation is smaller and the area's dryness enhanced by greater solar radiation due to the relative lack of clouds.

A second type of rain shadow is formed as moist air encounters topographic barriers but does not have the dynamics to rise over the barrier or divert around it. In this case the air pools in front of the barrier and the land in the lee of the barrier does not receive the moist air.

A third type of rain shadow is formed when air encounters higher terrain but is able to divert around it at low elevations. For instance, Pacific air encounters the Olympic Mountains in Washington State and is able to flow around them through the Strait of Juan de Fuca to the north and a near-sea-level lowland to the south. The result is that east of the Olympics—near the western reaches of Puget Sound—the precipitation is significantly less than in the mountains. Embedded within a regional climate of many cloudy days with light precipitation, the locals jokingly label this part of Washington "the banana belt." Port Townsend directly downwind of the Olympic Mountains receives a bit over 65 cm (26 in) of precipitation. To the east of Puget Sound, Seattle receives nearly 100 cm (40 in) of precipitation because the air streams that have flowed around the Olympic Mountains are able to converge and cause the rise of air.

The largest rain shadows occur in locations where topographic blockage forces air to sink through distances of thousands of meters so that the adiabatic warming is pronounced. In the United States, the crest of the Cascade range of the Pacific Northwest is oriented at right angles to the moist westerly flows coming from the Pacific Ocean. Air must ascend over 3,000 m in some places and then sink over 2,000 m into the continental interior. The maritime air ascends the windward sides of the mountains and produces copious amounts of precipitation in some places. On the leeward side of the mountains the strong descent of air inhibits precipitation so that the climate is markedly drier. Paradise Ranger Station on the windward slopes of Mt. Rainier in Washington State averages more than 1,700 cm (675 in) of snow a year, while Yakima, Washington—some 100 km (60 mi) downwind—averages a mere 50 cm (20 in). In practical terms this means the difference between a relatively moist climate capable of supporting a significant forest cover and a climate supporting a semiarid grassland.

There are numerous major rain shadows on the planet. Some notable examples include:

- The Patagonian Desert in Argentina east of the Andes Mountains.
- The southwestern (leeward) slopes of Haleakala volcano on Maui, Hawaii are semiarid compared to the rainforest on the northeast (windward) slopes.
- The western side of Madagascar is shadowed from moist air from the Indian Ocean to the east.
- The Thar Desert of eastern India made dry from mountain ranges to the northeast west and southeast.
- The presence of the Gobi Desert of Mongolia is partially attributable to blockage of moist tropical air by the Himalayas and Tibetan Plateau.

Although it is tempting to conceive of a rain shadow location as having an arid or semiarid climate, this is not always the case. In Great Britain, the mountains in Wales and western Scotland put leeward areas to the east in rain shadows, yet these rain shadows had natural forest cover in the lowlands because there is still enough precipitation to support a forest **biome**.

Unquestionably, the largest of the rain shadows increase the difficulty of productively using the landscape for human habitation and agriculture. However, the rain shadow areas are not without some positive aspects. In the American West, the orographically induced snow packs of the winter contain significant water. During the warm season these snow packs melt and water is concentrated into streams. These streams are dammed and otherwise utilized to supply water for urban and agricultural purposes in the numerous rain shadows.

Another positive benefit is the sunnier climates of rain shadow areas as compared to their windward counterparts. For instance, the city Bend is some 35 km (22 mi) alee of the crest of the Oregon Cascade range. Lowland Bend is considerably sunnier than any location west to the Pacific coast. This sunny climate has led to Bend's emergence as a recreation and retirement destination. Bend averages over 250 days a year of sunny and mostly sunny weather while some of the windward slopes of the Cascades have over 300 cloudy days per year. Bend is proximal to the forests and ski slopes of the Cascades so its residents use the rain shadow to advantage.

Region

A region is a defined territory that indicates some common characteristic or feature, or that is structured or organized in such a way that it is distinguishable from the surrounding territory. Any territory that is enclosed by a formal boundary is a region, so all **nation-states** and their administrative subdivisions are considered to be regions, but these by no means are the only examples. The concept of dividing the **Earth** into constituent components called regions is the basis of the branch of geography known as regional geography. This way of approaching the study of Earth dates at least to the ancient Greek geographer and historian Herodotus (died circa 425 B.C.), and geographers have widely utilized the concept of the region ever since. The "region" has proven to be such an enduring and useful device for the study of geography precisely because Earth is marked by a high degree of variation, and identifying zones of commonality, i.e., "regions" allows the geographer to organize and categorize this variation. Nearly every subfield of geography utilizes the concept of the region. It is not the case that all regions are clearly and uniformly defined, or understood, in the same way by everyone who uses them. Several examples would include the Midwest and Deep South in the United States, the Outback in Australia, and even international regional designations like the Middle East. Yet even these somewhat imprecise regions continue to be used as a means of mentally organizing and viewing the world across a wide spectrum of **scales**. Regions may also be used in a generalized conceptual sense and do not require a stark, exact demarcation. The regionalized concept of **core and**

Terra Nullius

A legal principle often invoked during the era of **imperialism** to justify the acquisition and settlement of new land. In Latin "terra nullius" roughly translates to "unclaimed territory," and colonizing powers often employed the doctrine when discovering and settling overseas lands from the 1500s to the 1800s. In the case of Australia, terra nullius was used to effectively claim the entire continent for British settlement, because according to British legal standards, the local inhabitants had no official rights to ownership. In recent decades native people in Australia have challenged existing ownership rights on the basis that the doctrine of terra nullius was illegally or unfairly applied and deprived their ancestors of property rights. The fundamental precepts of terra nullius were followed during the colonization of Africa and North America. In the latter case the concept of "manifest destiny" and the Turner Thesis are both intellectually rooted in the doctrine. Antarctica remains terra nullius according to international law. Some countries have made claims to sections of the continent, but because there is no permanent human habitation associated with these parcels, the land remains "unclaimed," and ownership is not officially recognized by most of the world's countries.

periphery is an example of this type of regional approach, where there is never a clear delineation identifying where the core begins/ends, and the periphery ends/begins. The regional designation of **hinterland** is a similar concept.

Geographers frequently draw a distinction between an area and a region. The latter term is more precise than the former. All regions are areas, but not all areas are regions. An area may be simply a portion of space that lacks any cohesive quality that differentiates it from the surrounding space. A region, on the other hand, is distinguished by a boundary of some sort, although the boundary may not be apparent, or even be physically manifest. A region, unlike a simple area, does hold certain unifying characteristics that contribute to a spatial uniqueness, although it might be the case that just the presence of a legal boundary is all that makes the region unique. A region that indicates a significant level of uniformity across its extent is called a *formal region*. Such a region might be defined on the basis of a physical characteristic that is present, or it could be identified by a cultural trait or set of traits that predominant there. For example, a space exhibiting a specific soil type that is different from surrounding **soils** could be identified and bounded on a **map**, resulting in a formal soil region. In the study of **linguistic geography**, linguistic regions are mapped and analyzed based on a common language that is spoken or understood there. Of course, it is rarely if ever the case that a regional quality occurs across 100 percent of the region's spatial dimension, and thus those defining the region must set the criteria for uniformity, i.e., 90 percent of the region are native speakers of French, etc. But some spaces are integrated not by a feature or set of characteristics that are homogenous there, but by a certain level of functionality across the space in question. This also is a type of region, typically called a *functional* or *nodal* region. In this type of region the common quality is a system that integrates the space represented by the region. An example is the region serviced by the subway system of a large city.

What is the rationale for regionalization? Both physical and human geographers use the concept of the region, and it remains the foundation of **locational analysis** along with other geographical methodologies. Recent theoretical innovations in geography focused on "place" or "locale" are merely refinements of the regional approach. Despite the alleged homogenization brought on by **globalization**, it remains true that one of the more obvious qualities of the Earth's surface is variation, both in the naturally occurring environment, and in the features created by human activity. To understand this variation, geographers seek out patterns of commonality, and after identifying *where* these may be found, attempt to explain *why* they appear as they do. The spaces marked by this variation become the regions that geographers discover and analyze. One regional pattern within a region may directly influence another, either overtly or in a less obvious fashion. For example, the pattern of soils in a region may have a direct bearing on the resultant patterns of vegetation there, and this in turn may have substantial influence on the patterns of settlement by humans. Moreover, the relationships between regions are of great interest to geographers, and the study of these may be quite complex and challenging. Two or more regions may share **complementarity**, meaning that a partnership exists in which one region is able to supply the needs of the other, while also receiving benefits in return, usually in some kind of economic relationship. Mercantilism, developed under the system of **imperialism** in the 19th century, although exploitative, was an example of complementarity.

Regional geography is the branch of the discipline that directly studies regions. Regional geographers typically are broadly trained, have even broader interests, and much if not most of their scholarly work is descriptive. Often a regional geographer develops an expertise in a specific region, frequently defined by political boundaries, economic linkages, or some other unifying feature, and then studies the manner in which various aspects of the geography there are integrated to provide coherence to the region as a whole. A regional geographer interested in contemporary Europe might focus on the trend toward **supranationalism** there, the economic region represented by the European Union, and the relationships between the nation-states that comprise the organization. The perspective of regional geography perhaps is not as scientific or as technical as other fields of specialization within the discipline, but in fact regional geography is the branch of geography that non-geographers most frequently study, and a knowledge of regional geography is essential to diplomats, policymakers, tourists, businesspeople, and many others. It is the type of geography every well-informed person should carry with them in the form of a **mental map**. This point was perhaps made best by the writer James Michener in his essay *The Queenly Science*:

> If I were a young man with any talent for expressing myself, and if I wanted to make myself indispensable to my society, I would devote eight to ten years to the real mastery of one of the earth's major regions. I would learn languages, the religions, the customs, the value systems, the history, the nationalisms, and above all the geography, and when that was completed I would be in position to write about that region, and I would be invaluable to my nation, for I would be the bridge of understanding to the alien culture. We have seen how crucial such bridges can be.

The "region," one of the oldest spatial concepts in geography, was also the first to attempt to organize the world into logical patterns that may be studied, analyzed, and perhaps to some degree, understood. At the fundamental level, this understanding is what all scholarship in geography continues to strive for today.

Religious Syncretism

Religious syncretism occurs when the characteristics of two or more faiths blend to alter one of the faiths, or to create a completely new religion. This is usually the result of **cultural diffusion**, when a new religious worldview encounters the established set of sacred precepts already present in a given **region**. The process of syncretization, or the incorporation of new rituals, dogma, or concepts, may be voluntary or enforced. The process of religious syncretism frequently leads to alterations in the **sacred space** of a location, as new religious structures may be built, or existing ones changed to suit the changes brought about by the integration of new beliefs. This may involve the establishment of entirely new sacred locations, or it may mean the use of previous sacred spaces, but with a modification of their former characteristics, relevance, or importance in the new religious system. Religious syncretism is often viewed by at least some believers as a threat to the "purity" of religion, and an attack on orthodoxy and tradition. Those who support the syncretic process are sometimes labeled heretics and accused of undermining the principles of the faith. However, most faiths indicate at least some ideas or practices that have been borrowed from other religious traditions, and the process of religious syncretism may be so subtle and lengthy that it is not

Santeria is a blend of Roman Catholic and traditional African belief systems. Here a Santeria believer participates in a ritual in Cuba. (AP/Wide World Photos)

recognized by believers. On the other hand, some faiths have been specifically created on the basis of incorporating the "best" elements from others.

In Latin America, when Christianity was imposed on the indigenous population in the 16th century, Roman Catholic officials frequently employed parallels between native beliefs and Christian traditions to facilitate conversion. The outcome was the retention of some elements of the non-Christian faiths within the context of Roman Catholicism, resulting in a syncretic version of Christianity that is quite different in some respects from that practiced in Europe. When millions of African slaves were introduced to the region, additional religious perspectives were added to the local culture in some areas. An example of a syncretic religion emerging from this process is Santeria, which originally developed among the black population on plantations in Cuba. Santeria combines the saints of Roman Catholicism with the status of *orishas*, or gods worshipped by the Yoruba people of western Africa, many of whom were transported as slaves to the New World. A good example of religious syncretism from Asia is Lamaism, the variation of Buddhism observed in Tibet, Mongolia, and some other parts of eastern Asia. Some of the traditions of Lamaism are derived from the Bon faith, a polytheistic religion that existed in Tibet prior to the widespread adoption of traditional Buddhist beliefs. The deities of Bon were incorporated into Lamaism in the form of spiritual guardians of the new faith, resulting in a syncretic melding of the Buddhist and Bon faiths. Many additional examples may be identified, and few religious movements exist today that do not show some effects of syncretism.

Remote Sensing

Geographers are keen in using fieldwork as part of their professional toolkits for understanding the world. Fieldwork can be quite laborious, costly, and—sometimes—dangerous. Since the middle of the 20th century when aerial photography became widely available, geographers have made frequent use of remote sensing.

Remote sensing is a term that refers to a set of techniques by which the environment is studied. Electromagnetic radiation is sensed by devices as it is reflected, refracted, scattered, transmitted, and emitted by matter in the environment. Each feature of **Earth** and the **atmosphere** has a "spectral signature," that is, the interaction of the various electromagnetic wavelengths with that feature. Thus, we are able to identify, measure, and model features that are not in physical contact with the features under study. There are many measurement devices that are in direct contact with what they are measuring and so, by definition, are *not* remote sensing devices. A thermometer is *not* a remote sensing device for air temperature because its temperature readings are dependent on the collision of atmospheric molecules with the device. A thermal radiometer, however, can be mounted in a satellite and measure the temperature in the atmosphere it is not touching. Table 3 illustrates the types of remote sensing commonly used by geographers, primary wavelengths involved, and typical uses.

The most well-known and commonly used form of remote sensing is photography. A camera contains film sensitive to light. The film is kept in the dark until the lens is opened to expose the film. The image is made permanent by developing it with chemicals. Aerial photography is of oblique and vertical types. Oblique photography is the perspective we see looking out of an aircraft window. Much more useful to the geographer is vertical photography in which the camera is pointed underneath the aircraft and features can be mapped and measured.

Color infrared photography is sometimes mistaken for thermal infrared sensing. Color infrared photography uses color film sensitive to most visible wavelengths but also sensitive to the shorter (non-thermal) infrared wavelengths that are plentiful in Earth's environment. These near or reflective infrared wavelengths cannot be perceived by humans. In color infrared film emulsions, reds are assigned to strong near infrared reflectivities. Living green vegetation is very reflective in the near infrared and appears very red on color infrared film. Water absorbs near infrared wavelengths so that lakes, streams, and soil moisture appear dark.

Photography has been the mainstay of imaging for the better part of two centuries. Yet, photography has inherent disadvantages in its need for film and its limitation to visible and near infrared wavelengths. Thus, non-film sensors have been developed.

Thermal infrared sensing does not use film because there are no film emulsions sensitive to these wavelengths. All objects *emit* in the thermal infrared portion of the spectrum and knowing the relationship of the most plentiful emission wavelengths to **temperature**, we are able to determine temperatures from afar.

At wavelengths much longer than most of the natural electromagnetic energy from the sun and Earth, geographers use microwave and radar; these remote sensing types use identical wavelengths and are particularly sensitive to all the states (phases) of water. Microwave sensors are passive sensors because they do not generate the energy they are receiving from targets. Microwaves naturally occur but are neither powerful nor plentiful. Therefore, microwave systems usually have relatively poor spatial resolution when mounted on airplanes or satellites. Radar (an acronym for radio detection and ranging) generates its own, concentrated microwaves and measures the strength of their return as they are backscattered off a target.

Sonar (an acronym for sound navigation ranging) is another active sensing technique. Sound is propagated and then the nature of its backscatter is measured. In the case of geographers, it is the distance to the bottom of the water that is being sensed. Sonar senses through water, which is much denser than air, so its speed is considerably slower than analogous radar pulses in the air, and sonar wavelengths are on the order of several meters. Sonar has revealed the nature of sea and lake bottoms and has been key in proving **plate tectonics**.

Lidar (an acronym of light detection and ranging) is based on the technology of lasers, which have beam coherences and densities not achievable with sonars or radars. These are active systems that use a variety of wavelengths much shorter

Table 3. Remote Sensing Techniques

Name	Wavelengths	Geographic uses
Photography	.4 μm–.7 μm	Mapping, measurement of cultural and physical features
Color infrared photography	.7 μm–.9 μm	Vegetation type and health, surface soil moisture
Reflective infrared sensing	0.7 μm–3.0 μm	Vegetation type and health, surface soil moisture
Thermal infrared sensing	3.0 μm–14 μm	Temperature differences in rural and urban systems, "night vision," cloud temperatures
Microwave sensing	0.1 mm–1 m	Soil moisture and temperature
Radar	0.1 mm–1 m	Clouds, storms, landscape roughness
Sonar	3 mm–3 m	Ocean bottom features
Lidar	250 nm–10.0 μm	Nature of vegetation canopy, aerosol, and cloud studies
Multispectral sensing	Various	Land resource analysis, change detection

than radars and sonars and so are sensitive to smaller Earth features. Indeed, the spatial resolution of lidar from aircraft platforms can be less than a fifth of a meter.

Multispectral remote sensing is a clever way of overcoming the limitations of use of various wavelengths. Multispectral remote sensing devices break bands (continuous intervals) of wavelengths into several smaller bands of simultaneously acquired data. Each band has its strengths and weakness for study of geographic phenomena. For a given purpose, the most meaningful of these bands are numerically comanipulated so that they maximize information extraction from a satellite or an aircraft sensor array. In recent years, work has been accomplished in hyperspectral remote sensing that simultaneously acquires data from tens of bands.

Rimland Theory

A perspective on **geopolitics** and international relations, presented as a critical response to the **Heartland theory** proposed by J. Halford MacKinder. The Rimland theory was the brainchild of Nicholas Spykman, a political theorist and professor of international relations at Yale University. Spykman was a proponent of *realism* in international relations, a view which holds that political states act solely to promote their own interests and agendas. Writing in the 1940s at the height of World War II, Spykman suggested that the Heartland theory put forth by MacKinder several decades earlier was flawed, in that it overemphasized the role of the Heartland in determining the balance of power in global relations. Spykman also drew heavily on the writings of the American naval strategist Alfred Mahan in constructing his theory. Although his academic training was not specifically in geography, Spykman considered the discipline to be of the utmost importance in analyzing and understanding international relations and the dynamics of global power.

The Rimland theory appeared in Spykman's 1944 book *The Geography of the Peace*, published posthumously the year following his untimely death. He adopted the basic spatial framework of the Heartland theory, but made some changes in terminology. Spykman retains the concept of the Eurasian landmass representing a "Heartland," but calls the region that MacKinder labeled the "inner" or "marginal" crescent the "Rimland." He rejects MacKinder's characterization of North and South America, Australia, Japan, and Great Britain as lying in the "outer" or "insular" crescent, and instead simply terms this region the "off-shore islands and continents," although he agrees with MacKinder's view that for these countries sea power is of paramount importance, and represents the main means of projecting power. In the Rimland theory, the Heartland does not represent the pre-eminent seat of power that it symbolizes in Heartland theory. Rather, it is the Rimland that is the foremost seat of power and is the key to dominating the Heartland. Paraphrasing MacKinder's summary of the Heartland theory, Spykman offered his own summary:

> Who controls the Rimland rules Eurasia;
> Who rules Eurasia controls the destinies of the world.

Rimland theory differs from the Heartland theory in two major aspects. First, Spykman believed that MacKinder had greatly exaggerated the prospects of the Heartland as a base of power. Having the advantage of writing four decades after MacKinder's original thesis appeared as the *Geographical Pivot of History* in 1904, Spykman noted that the development of railway infrastructure in the Eurasian core area had not progressed in the way that MacKinder had expected in his original thesis. Indeed, the quality and quantity of transportation linkages in MacKinder's Heartland, especially in Siberia and western China, had remained at quite a low level, and could not compete with the sea transport of the inner and outer crescents. In addition, Spykman criticized MacKinder's conceptualization of the Heartland as a repository of vital resources, especially because its agricultural potential was much less than in the surrounding regions.

Second, Spykman pointed out that the Heartland theory oversimplified the historical relationship between sea power and land power in regard to controlling the heartland. Spykman admitted that the location of the Heartland provided a defensive depth and presented daunting challenges to an invading adversary. But this was well-known from a cursory review of Russian history (Russia being the country occupying most of the Heartland), and this seemed to be almost the only real geographical advantage the location enjoyed, vis-à-vis the Rimland. Spykman revealed that in fact the history between the Heartland and what he called the Rimland was not based exclusively on the sea power of the Rimland versus the land power of the Heartland. He cogently pointed out that various alliances involving the Heartland power (he used Russia as an example) and countries in the Rimland had emerged over the course of modern international relations, and that in fact Russia had on occasion joined with one or more Rimland powers to counter the ambition of an aggressive Rimland country—World War I itself was a case in point.

For Spykman, the key region for global control was the Rimland. This was because the Rimland had to function as both a land power and a sea power, and it also functioned as a **buffer zone** between the powers of the Heartland and the naval powers of the outer islands and continents. It is strategically more important than the Heartland due to its greater **population** and more advanced economy, two geographic factors that he believed MacKinder had completely failed to consider. Furthermore, the history of the balance of power on the Eurasian landmass, and the numerous conflicts associated with maintaining that balance, were about limiting control of the Rimland, not about controlling the Heartland.

Spykman's revision of the Heartland theory, appearing in the last year of World War II, had a deep influence on post-war geopolitical strategy in the United States. His argument that the power of the Heartland could be held in check by controlling the Rimland attracted the attention of western policymakers in the late 1940s, who were concerned about the possibility of Soviet expansion. Spykman advocated reconstructing Germany as a counterweight to Soviet ambitions in the **region**, a strategy realized in the Marshall Plan and other efforts to rebuild Europe. The formation of NATO in 1949, an organization consisting primarily of American allies located in the Rimland, may also be regarded as an outgrowth of Spykman's influence. Even more influential was the notion of limiting the expansion of power from the Heartland by using the Rimland as a barrier—this was the basis of the Western policy of containment, pursued in response to the Soviet occupation of Eastern Europe and the spreading of Soviet influence via "wars of liberation" in other parts of the world. "Containment policy" guided the foreign relations of the United States and its allies for the second half of the 20th century, until the collapse of the Soviet Union in 1991. In spite of what his critics suggest is an inflated estimation of the strategic role of the Rimland in global power politics, Spykman's theory continues to influence the debate over foreign policy in the United States and elsewhere.

Rural Settlement

In general, a rural place or **region** is identified by a low density of **population**; a predominance of land used for agriculture, ranching, or that is undeveloped; and relative lack of urban places, especially those that occupy a large area whose economic activities and employment are concentrated in the secondary and tertiary **sectors of the economy**. The territory in a country that is precisely designated as rural often depends on the standards and definitions used by the government agencies of that specific country, just as is the case when considering urban geography. Moreover, even countries that show a high level of economic development and hold large populations may also be predominantly rural—the United States Bureau of the Census identifies 95 percent of the United States as rural, and the proportion of rural territory in Canada and Australia is similar, although as in the United States, a large majority of the population living in these countries is urbanized. On the other hand, many lesser-developed countries have a majority of their people living in rural districts, with only a relatively small portion of the population considered residents of cities. Regardless of the world region in which they are located, most rural **landscapes** show distinct patterns of settlement, which are typically determined by environmental and economic factors, cultural practices, social relations, and sometimes the type of crop produced.

The presence or absence of water may influence the spatial nature of rural settlement. In **deserts** and other regions where rainfall is scarce and agrarian production is dependent on surface water sources for irrigation, or on underground supplies from wells or springs, settlements are centralized around or near the source of water. These may be gathered at an oasis, where a single freshwater spring provides water for the human population, livestock, and usually a limited amount of cultivation. If local farm production is dependent on a lake or river to supply a system of irrigation canals, rural settlements tend to be located adjacent to the water source or to the central irrigation works. The construction and maintenance of irrigation canals, along with the equitable allocation of water, requires significant oversight and collective effort and results in a settlement pattern that is closely tied to the availability of water.

The predominant settlement pattern encountered in most of Europe, East and South Asia, Africa (among indigenous groups), and most of Latin America is a clustered formation of dwellings, in the form of a farming community or village. Here farmers live in a group, which gives advantages of security, shared resources, and social contact. Farmland may be held in common or individually, and fields usually are adjacent to the farmhouses or only a short distance away. In some parts of the world, such villages may be centuries old, and as a result may show little structural organization, having grown and changed with no central planning. Such clustered settlements probably emerged quite early in the history of agriculture, because they allowed for more effective defense against nomadic peoples who attempted to raid the food surpluses produced by the agriculturalists; they also provided a pool of collective labor that could be tapped at harvest time or at other times when required, such as for barn raisings, construction of irrigation works, or communal buildings.

In other regions, farm villages reveal at least a rudimentary level of planning, with homesteads aligned along a central avenue that runs the length of the settlement. These are found quite commonly in Russia and much of Europe, parts of the Middle East, and in some regions of Asia. The land under cultivation lies along either side of the village, often immediately adjoining the residential property of the farmers, a geographical feature that gives ready access to the farmland and that minimizes time and transport costs for cultivation. Geographers call a small settlement of this sort a "street settlement," because of the single thoroughfare that linearly connects the residences. In parts of Europe, North America, and Australia, the spatial character of rural settlement follows the pattern of a "commons" or "green" village. Here the homes of the settlement are arranged around a central open "commons" area (sometimes referred to as the "greens") that is owned by the local community, and which may be used for a variety of general purposes, although crop production is rarely allowed. The commons is frequently laid out as a square or rectangle, with streets radiating away from it at right angles. Larger planned rural communities may take on a so-called checkerboard structure, in which the homes are spaced at regular intervals with property lines meeting at right angles, forming a grid pattern.

Some patterns of rural settlement follow a more dispersed spatial pattern. These may appear as semi-clustered settlements, or as isolated, individual homesteads. The latter pattern is common in areas of plains that were developed by European colonists during the past two centuries and is frequently encountered in the interior plains regions of the United States and Canada, much of the outback of Australia, and in the veldt of southern Africa. Settled by pioneers, or in the case of South Africa, Boers, these areas were claimed from native peoples and opened to individual farmers and ranchers, who were provided land grants by their respective governments. The result is an isolated pattern of settlement, often with several miles separating each farmstead. In hilly or mountainous terrain, smaller semi-clustered settlements are frequently the rule, and these may be linked by religious or familial connections. In some cases, the entire settlement, consisting of several dozen family units, may all be related through blood and marriage ties. "Row villages" occur in rugged or marshy landscapes in central Europe, Latin America, and North America. In this type of settlement houses are spaced farther apart than in a typical street village, but generally follow a linear feature, often a river, railway, or canal. The agricultural land extends away from the homes in an elongated rectangular pattern, forming what rural geographers call "long lot" holdings, and this type of settlement may be called a "long lot village."

S

Sacred Space

A sacred space is a location that has acquired an elevated, often spiritual status for a specific group. This status may truly be religious, or it may be more connected to **territoriality** and **cultural identity**. This is in contrast to profane space, which does not carry the same status, and which is the location where everyday activities take place. Generally speaking, a place identified as belonging to sacred space requires a special reverence and behavior of those who recognize it, which is usually expressed in the form of ritual or sanctity. Sacred space can be created by human activity, or it can consist of a natural setting that acquires mystical, religious, or nationalistic importance.

There are thousands, and probably millions, of examples of sacred spaces that have been constructed by human activity. These range in size and significance from a single tomb or gravesite to entire cities, such as the city of old Jerusalem, almost all of which may be considered sacred either by Jews, Christians, or Muslims (or in some cases, by all three faiths). Conflict can arise when more than one group claims a space as sacred. The Temple Mount of Jerusalem is a prime example, which both Jews and Muslims claim to be a vital part of their exclusive religious tradition, but other conflicts abound. In 1992 in the Indian city of Ayodhya a mosque was destroyed by a mob of Hindu believers because they believed that the building occupied land on which the Hindu god Rama was born, and thus considered the Muslim structure to be intrusive and offensive. Before the mosque was built in the 1500s, a Hindu temple had stood on the site. In some cases rather than being destroyed, the sacred space of one faith may be adopted and converted by another. A good example is the Haghia Sophia in Istanbul, which functioned as a Christian church for over 1,000 years, and since 1453 has served as a mosque.

Locations may be considered sacred that are tied to historical events that are especially important to the identity of an ethnic group or country. Serbians, for example, have a special reverence for the region of Kosovo, because they suffered a dramatic loss there in 1389 at the hands of the Turks. For Serbians, the plain of Kosovo represents a tragic, costly, but heroic sacrifice, and similar places may be found across the world. In the United States, Gettysburg, Pennsylvania, the Alamo in Texas, and Pearl Harbor in Hawaii would all qualify in a similar way as sacred space for many Americans. Natural features may also function as sacred space.

The Temple Mount in Jerusalem is an example of contested sacred space. This location is held sacred by Jews, Christians, and Muslims. (Lucidwaters/Dreamstime.com)

In Australia, the rock known as Uluru to native Australians, and as Ayers Rock to others, represents a religious place vital to the culture of the native people. In Shinto, Mount Fuji and many other natural features are considered to be the residence of *kami,* or spirits, and are thus part of the sacred landscape of Japan. Indeed, natural features were likely the first sacred spaces claimed by human beings.

Satellite Images

Since the late 1950s humans have constructed and launched artificial satellites for a variety of purposes. These satellites employ **remote sensing** to explore the **Earth's** surface and atmospheric phenomena by using the electromagnetic properties of those phenomena. Usually, satellites are capable of multispectral remote sensing to simultaneously assess electromagnetic properties in a variety of wavelengths.

Of considerable importance in the current era is the fact that satellite data are widely available through the Internet.

Some satellites are in polar orbits completing an orbit roughly every hour and a half at altitudes between 700 and 1,700 km. That is, they orbit south to north or north to south passing near the poles. The Earth rotates underneath the satellites so that each orbit brings the satellite over longitudes west of the previous orbit. These are sun synchronous satellites in that they pass over the same latitudes at the same time each day so that solar illumination differences are minimized. Their relative low orbits make them relatively "inexpensive" to launch. Many weather satellites and all the Earth resources satellites are in polar orbits. The longest continuous Earth resources satellite series is the Landsat series first launched in 1972. These polar orbiters image large swaths of Earth in each scene. Current scene sizes are about 170 km in latitude by 183 km in longitude with spatial resolutions of 30 m across in some of the wavelengths. Yet, a drawback is that the current versions recapture the same scene area once every 16 days. This is problematic if the goal is to do some monitoring of short-period events such as floods or crop harvests. Another limitation is Landsat's inability to sense the surface through **clouds**; thus some passes are useless because of cloud cover. New American satellite series such as the Earth Observation System satellites are able to cover the globe in a day or two with somewhat lesser resolution but can image in three dozen different spectral bands. The new instruments allow specialized work such as three-dimensional profiling of atmospheric features. Also, other countries (e.g., SPOT from France and DAICHI from Japan) have launched their own Earth resources satellites.

Geosynchronous satellites are placed in much higher orbits and their orbital speeds around Earth's axis match the rotation of the Earth in a way such that they remain over the same Earth location. Geosynchronous orbits are roughly 36,000 km above Earth's surface and there is considerable expense in placing satellites in such orbits and keeping their orbits synchronous. As a result, there are considerably more polar-orbiting satellites than geosynchronous satellites. GOES (Geostationary Environmental Satellite) is a long-running satellite program of the U.S. National Oceanic and Atmospheric Administration that has two satellites positioned to monitor the United States. GOES East is positioned over the equator at 75°W; it views the eastern United States and can monitor Atlantic **hurricane** threats to the western Atlantic Ocean basin. GOES West is positioned over the equator at 125°W; it has a view overlapping that of GOES East and monitors the western United States and the **middle latitude cyclones** approaching from the Pacific Ocean. Other countries have deployed geostationary satellites around the globe so that virtually the entire Earth is under continuous monitoring (e.g., the European Community's METEOSAT, India's INSAT and Japan's GMS).

Geographers make frequent use of satellites. These are of two varieties: weather satellites and Earth resources satellites. Weather satellites have been in orbit since 1960 and geographers have used them to study all manner of phenomena ranging from severe storms to seasonal mountain snowpacks to the monitoring of El Niño. Earth resources satellites are polar orbiters and have been used for purposes such as monitoring land use change, identifying prehistoric habitation sites, and estimating city **populations**. The large area views afforded by satellites have aided in refining our knowledge of the physical and cultural features of Earth and discovering spatial relationships between seemingly disparate elements. Some uses depend on the "picture" qualities of the satellite data. Yet, it must be remembered that these are "images" with no film involved. The data are used in digital form and, as such, the numeric data generated can be used in sophisticated modeling such as observing the components of Earth's energy balance or automated detection of change in Earth surface properties over time. The various satellites can detect energy in both visible and invisible wavelengths so that combined information is able to make more assessments and environmental models more certain.

Satellite State

A satellite state is a country that lies in the "orbit," metaphorically, of a larger, more powerful country, meaning that it owes political allegiance or economic tribute to the more powerful state. Before the 20th century, such states were typically referred to as "vassal states," "tributary states," "puppet states," or "client states." The governance of such states is frequently characterized by authoritarian or totalitarian political structures, and satellite states may compose a **buffer zone** between empires; thus historically they have played a role in **imperialism** and **geopolitics**. In some cases, a satellite state functions in a fashion similar to a colony, in that its foreign policy aligns closely with that of the more powerful state to which it is subordinate. However, a satellite state differs from a colony or a dependency in that it is a de jure sovereign power, despite the fact that such sovereignty is compromised by its satellite status. Furthermore, satellite countries may form an economic bloc with the dominant power, based on a shared ideology or simply to enhance the development and authority of the dominant polity. The term was widely employed by both politicians and scholars during the Cold War, especially in relation to the communist states of Eastern Europe. "Finlandization" is a related term also used during the same era. In this case, Finland was not subject to the same degree of domination as most of the Eastern European countries, but nevertheless adopted political self-restrictions in deference to the military might of its large neighbor, the Soviet Union. For example, Finland followed a foreign policy of strict neutrality, and the Finnish press engaged in self-censorship, rarely criticizing Soviet actions or even providing much coverage of aggression against other states, such as the 1968 invasion of Czechoslovakia by the Soviet military.

The Warsaw Pact and COMECON represented official organizations composed of Soviet satellite states. The official document that formed the Warsaw Pact was the Treaty of Friendship, Cooperation and Mutual Assistance signed in 1955. The members of the Warsaw Pact stretched through Eastern Europe from Poland in the north to Bulgaria in the south, sharing their eastern **boundaries** with the USSR. Initially only Yugoslavia, which did not have a common border with the Soviet Union, remained outside the organization although Albania, a charter member, later withdrew its membership. The geographical proximity of the Soviet Union played a key role in maintaining the satellite relationship. Those countries that occasionally deviated from the expectations imposed by Moscow could be readily brought to heel by the Soviet military, as happened in Hungary in 1956 and Czechoslovakia in 1968. When the Communist Party in Hungary attempted sweeping reforms in 1956, Soviet troops invaded the country and after heavy loss of life, crushed the reformists, replaced the regime with officials who supported retaining close ties to the Soviet government, and continued to follow the dictates of Marxist-Leninist ideology. This tragic episode made clear that Moscow had no intention of allowing any of the cluster of satellites it had secured along the USSR's western margin to remove themselves from its orbit. A similar series of events transpired in Czechoslovakia in 1968 during the so-called "Prague Spring," although the level of violence and loss of life was much lower than in Hungary in 1956.

The countries of the Warsaw Pact were not the only Soviet satellites during the era of the Cold War. Mongolia, a state occupying a strategic position between the People's Republic of China and the USSR, closely aligned its foreign policy with that of the USSR after the Sino-Soviet split in 1960. Moreover, satellite states were established in **regions** well beyond the borders of the Soviet Union, the most prominent being Cuba. The proximity of Cuba to the United States and its role as a Soviet satellite generated serious and dangerous tensions after 1959; the Cuban Missile Crisis of 1962 brought the United States and the USSR to the brink of war. The crisis was set in motion when the Soviet government placed offensive nuclear missiles in Cuba that could strike the entire continental United States, an escalation of tensions that the American administration of John Kennedy was unprepared to accept. For ten days in October 1962 the United States and the Soviet Union teetered on the precipice of war, a conflict that almost certainly would have resulted in the annihilation of both countries and their allies. Eventually the crisis was defused, but the situation had highlighted Cuba's role as a Soviet satellite in the Western Hemisphere, and how the role of such client states could quickly disrupt the balance of power between the two superpowers.

In the post-Cold War era, there are few countries that may be identified as satellite states in the same way as the examples given above. A handful of states are closely aligned with a more powerful neighbor, and a case may be made that these countries represent satellite states in the contemporary context. Some commentators have argued, for example, that Belarus for much of the post-Soviet period has functioned as a satellite of Russia, due to its dependency on energy supplies from its large neighbor and its consistent and vigorous support of Russian foreign policy positions. The small country of Bhutan, situated in the Himalaya, could be considered a satellite of India until 2007, when a new treaty redefined the relationship between them, asserting Bhutan's independence in crafting foreign policy. However, a case might still be made that Bhutan, due to its almost complete economic dependence on India and its close and consistent support of the policies of the Indian government, remains in fact a satellite of New Delhi. A final example of a contemporary satellite state might be Lebanon, a country long under the influence of its more powerful neighbor, Syria. Both countries emerged from the French Mandate established after World War I, but Syria in recent decades has played a major role in Lebanon's politics, both directly and through proxies like the Islamic extremist organization Hezbollah.

Scale

The term "scale" can have several meanings in a geographical context. In a precise usage in **cartography**, scale refers to the proportional relationship between space represented on a **map** and that which occurs in reality. Scale is an important consideration when constructing a map, because the level of detail displayed on the map will change with the scale, and some features that are evident at one scale may disappear from view at another. There are several possible ways to indicate the scale of a map, which is often shown in the map legend. A verbal scale is expressed as a simple statement: "one inch equals ten miles," which of course cannot be taken literally. This sentence means that one inch of distance on the map is the *equivalent* of ten miles in the real world. This is a commonly used scale on road atlases, as it is easily understood even by those with little knowledge of cartography, and makes the approximation of simple distances on the map quite easy. The verbal scale may be accompanied by a bar scale, which appears as a graphic representation, typically a small bar with intervals marked off indicating a specific distance between the intervals. Using a ruler or virtually any device that corresponds to the interval on the bar scale, a map user can then apply the distance on the bar directly to the map and calculate the distance between two **locations**. A third method of indicating scale on a map is the representative fraction (RF), which is reported as a simple ratio or fraction. A typical RF would be 1:250,000, which might also be written as 1/250,000. This simple ratio informs the map reader that one unit of distance on the map is equivalent to 250,000 in reality. An advantage to using an RF to express scale lies in the fact that since it is a ratio, any units may be employed, as long as they are consistent across the ratio. In other words, a British scholar using the map might utilize miles in considering distances and locations on the map, but her counterpart in France could use exactly the same map and work in kilometers! Maps on which the scale is expressed as a representative fraction therefore have a wider applicability, because someone using the map can use any units to which she is accustomed.

Scale can also be used in relation to maps in a general sense. One may speak of a "small-scale map" or, conversely, one may encounter a "large-scale map." A small-scale map is usually a map showing a relatively large area, perhaps a large country, a continent, hemisphere, or even the entire world. The map is able to show little detail, due to the scale at which it has been constructed. A small-scale map also has a relatively small RF. A typical map of the world in an atlas might have an RF of 1:100,000,000—this is clearly a small scale map, but an RF of 1:250,000 might be considered small scale as well, depending on the size of the area portrayed on the map. A large scale map is a map of a relatively small area that shows a considerable level of detail. The RF of a large-scale map is relatively large, perhaps 1:50,000 or larger. A typical large-scale map might be a street map of a small city, a map of a college campus, etc.

The concept of scale is also applied to the magnitude of spatial consideration a given geographical study or argument may assume. This is sometimes called the "geographical scale" of a study or research project, and may be conceptualized at a variety of levels. For example, various scales of approach may be employed in a comparative study—one might examine a specific geographical feature at the global, regional, and local scales. A political geographer could investigate the expression of political cooperation in a **federation**, a geographical alliance that operates at a regional scale, while another political geographer might conduct research on precisely the same topic but at the local scale, perhaps in the government agencies of a single state or province. Geographical scale is an important consideration when engaging in the study of any spatially expressed phenomenon, because the scale at which research is performed often directs the way in which a geographical space and its features are perceived and analyzed. Scale affects how one views the world, engages the world, and understands the world.

Seasons

Rhythmically varying angles of the sun in the sky are a basic feature of physical geography because they represent systematic changes in the amounts of **solar energy** received in various locations. It is the appearance of the sun in the sky that separates day from night and winter from summer. From time immemorial humans recognized the latter, more subtle, effect and attached great societal importance to it. For instance, a millennium before the Egyptian pyramids were built, Celtic tombs overlooking Ireland's River Boyne were outfitted with portals over the entrances. During a few sunrises in the third week of December the sun shines through the portal and illuminates the tombs' central chamber. This event marked the time of year when the sun was lowest in the sky. Clearly, the builders understood there was a relationship between the path of the sun in the sky and expectations of **temperature** and **precipitation** patterns. Worldwide, societies have incorporated such notions into religious and social rites and planned planting and harvesting. Today, the reasons for seasonality are well understood and seasonal effects are still vital to life on the planet.

Earth orbits the sun approximately once per year and it is the position of Earth in its orbit that causes changes in seasons. Earth's orbit is elliptical with about 5 million km of difference between Earth's closest approach to the sun (perigee) and furthest swing from the sun (apogee). This orbital shape affects seasons in only very minor ways. It is *not* changes in Earth/sun distance that causes seasons. Two observations prove this: (1) summer in the Northern Hemisphere occurs during the apogee of the orbit and (2) there are opposite seasons in the Northern and Southern hemispheres; that is, when one is experiencing winter the other is in summer.

The reason for seasons is rooted in the fact that Earth's axis is tilted at an angle of 23.5° away from perpendicular to the Plane of the Ecliptic. The Plane of the Ecliptic is the imaginary surface that bisects the sun and Earth and is the orbital plane of Earth. As Earth orbits the sun the angle of tilt is maintained at 23.5°. The north end of Earth's axis always points in the same direction. In our part of Earth history the north end axis always points to Polaris, the North Star. (In fact, there is a 26,000-year axial wobble that can be ignored on a year-to-year basis.) However, the angular relationship between the axis and the position of the sun systematically changes during the course of a single orbit. In June the north end of the axis points toward the sun, in December the north end is pointed away from the sun, and in March and September it is in an intermediate position. This configuration causes warmer temperatures in whichever hemisphere is pointed toward the sun.

The progression of seasons is controlled by the **migration** of the solar path in the sky. On any date, there is a single latitude at which the sun can be viewed straight overhead and where solar energy entering the top of the **atmosphere** is most intense. This latitude is known as the solar declination. Declination progressively changes because of the consistent tilt of Earth's axis. In June, the north end of Earth's axis is pointed toward the sun and the south end is pointed away from the sun. This causes the sun to appear at a higher angle and have a longer path length in the sky in the Northern Hemisphere compared to the Southern Hemisphere. Solar declination varies between 23.5° north latitude and 23.5° south latitude. The "tropical" year is the time it takes for the sun to make one round trip between these two latitudinal extremes. The northern extreme is also called the Tropic of Cancer and the southern extreme the Tropic of Capricorn. The equator at 0° latitude functions as the middle of the range of solar declination. These lines are frequently drawn on **maps** and globes. One complete orbit of the sun is a revolution and causes the migration of the vertical rays of the sun. Revolution is not the same as rotation, which is the movement of Earth about its axis. Whereas the former is related to seasons, the latter is related to day versus night.

Temperatures at individual locations are typically highest just after the time of highest solar angles because a high-angle sun imparts much more energy than a low-angle sun. So, timekeeping has long been based on the solar path through the sky. The starts and ends of seasons are reckoned from knowledge of the solar path. The third week of December is a solstice at which time the sun's vertical rays are overhead at the Tropic of Capricorn (23.5° south latitude). This marks the beginning of winter with the shortest days of the year in the Northern Hemisphere. By the third week of March, the declination has progressed northward to the equator and the day is known as an equinox—the start of spring in the Northern Hemisphere. The third week of June holds another solstice and the beginning of summer in the northern hemisphere as the vertical rays of the sun reach the Tropic of Cancer (23.5° north latitude). Then the declination progresses southward and crosses the equator in the third week of September to start the northern fall with another equinox. A round trip of the declination is completed as the vertical sun reaches the Tropic of Capricorn. Inconveniently, this round trip takes approximately 365 ¼ rotations of the Earth on its axis so that the modern calendar has added a leap day every four years.

It is important to understand that the delineation of major lines of latitude are a direct function of the tilt of Earth's axis. The tropics include the latitudes the solar declination crosses. In that the sun is always close to being overhead at midday, the tropics are always warm. The polar latitudes are circumscribed by the Arctic and Antarctic circles that are the equatorward limits where the sun can remain above or below the horizon for more than a day at a time depending on the season. In that the sun is either low in the sky or not above the horizon the polar regions are cold to frigidly cold depending on the season.

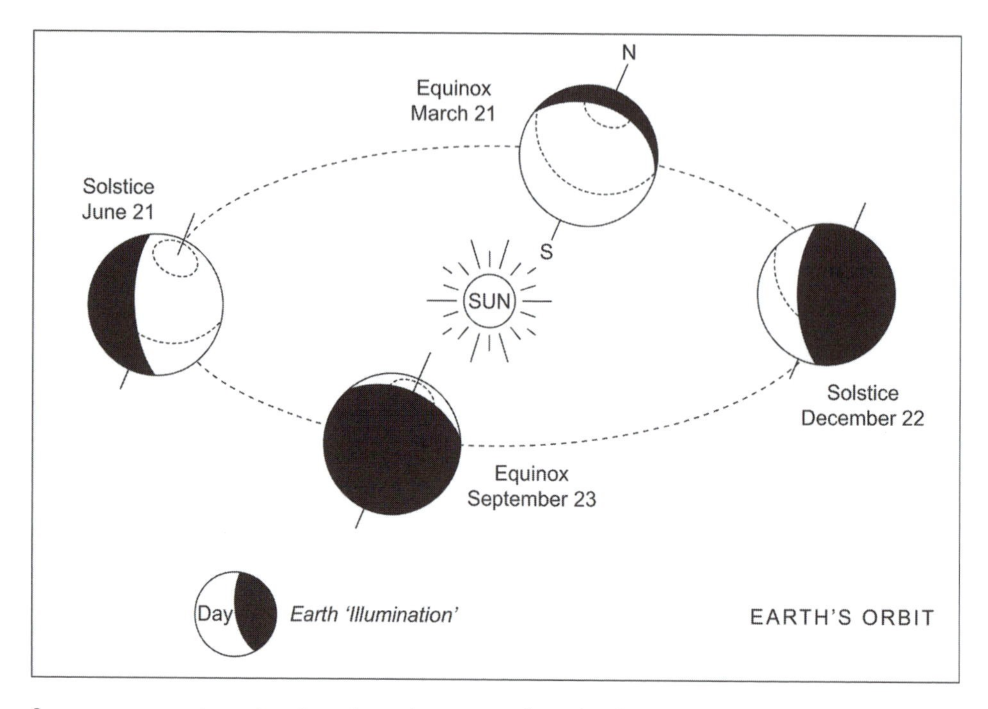

Seasons are experienced at the polar and temperate latitudes due to the tilt of the earth's axis as it orbits the sun. The seasons are reversed beween the Northern and Southern hemispheres. (ABC-CLIO)

Seasonality is intimately related to day length. Instantaneously, half the Earth has daytime while the other half has nighttime. Yet, the tilt of the Earth's axis ensures that day lengths vary by latitude. At the equator, there are 12 hours of day and 12 hours of night (excluding twilight) throughout the year. If it is the time of an equinox, all places on Earth have approximately equal day and night periods. At other times, there is a systematic variation of day length with distance from the equator. In the winter hemisphere, day lengths get shorter and shorter with increasing latitude and, poleward of the Arctic or Antarctic Circles there is no direct sunlight. In the summer hemisphere, day lengths become longer and longer with increasing latitude and, poleward of the Arctic or Antarctic Circles the sun does not set for increasingly great amounts of time for up to six months.

Though seasonality is based on Earth/sun relationships described above, it is expressed quite differently over the globe. Latitudinally, there are three seasonal temperature zones. In the deep tropics there it is always "summer"—warm to hot. In the middle latitudes there is a striking winter/summer temperature difference. In the polar latitudes, temperatures range from winter frigid to summer cold. Precipitation is also a part of seasonality. Most locations have their peak precipitation during or just after the time of the year the sun is highest in the sky. There are some notable exceptions in the middle latitude caused by the presence of summertime subtropical highs. Finally, there is a land/sea geography to seasonality. Locations within continents tend to have much greater seasonal temperature and precipitation swings than do oceanic locations.

Sector Model

A model of hypothetical urban morphology proposed by Homer Hoyt in 1939. The Hoyt model, as it is sometimes called, was a revision of the Burgess Concentric Ring model that had been developed in 1925 by the sociologist Ernest Burgess of the so-called "Chicago School." Hoyt saw little evidence that cities evolve and grow in the form of rings around a central zone, and instead offered a schema that divided the typical city into sections clustered around a central business district. The sectors were differentiated on the basis of the fundamental economic activity each would be expected to contain along with the residential characteristics, if any, found there. The model is theoretical and thus does not match any real city perfectly but offers a generalized explanation of the urban structure. Hoyt's model places much more emphasis on the role of transportation in developing the land-use pattern in urban areas than the Burgess model. He conceived of the city's expansion following the lines of transportation, primarily railroads and street car routes, and this resulted in a more linear pattern of development, resulting in blocks of similar spatial functionality, rather than the circular bands of the Burgess model.

In the Hoyt model, the Central Business District, or CBD, lies at the heart of the city (see diagram on page 111). It is characterized by high land values, resulting in tall buildings and congested development. The transportation lines that service the metropolitan region all converge on the CBD, and thus in the Hoyt model most of the sectors of development adjoin the CBD to some extent. Adjacent to the CBD is an area of degenerated industrial production and residences where the unemployment rate in the city is typically at its highest. This is identified as sector "B1" in the model and represents the old industrial core of the city. This is a sector in need of urban renewal, and in many contemporary cities would be experiencing **gentrification**, although Hoyt did not use this term in relation to his model. Located on either side of the old industrial sector are sectors of poor-quality housing, usually of tenement-style apartment buildings with little open space. These sectors are labeled "B2." Sector "C' in the model represents middle-class housing, which according to the Hoyt model is constructed on estates of various sizes. Sector "D" is a commuter zone that follows some line of transport and is composed of more expensive residences with more parks and open space, because land values are lower here than closer to the CBD. The model predicts that at the outskirts of the urban area a sector of low-density residential settlement will develop, marked by large open spaces, smaller hamlets and communities, and suburbs. The sector model fit many cities in the industrialized world reasonably well in the mid-20th century, but innovations in technology and economic changes have reduced its applicability for many metropolitan areas today.

Sectors of the Economy

The term "sector" when applied to an economy may have several meanings. In general usage, any modern economy may be split into a "public sector" and a "private sector." The public sector is represented by all those economic activities that are funded and supported by government sources through the collection of tax revenue and public bonds, or which are granted special status and do not compete with similar businesses in a defined market, like public utility companies. The private sector encompasses all economic and business activities that are privately owned, funded, and administered. This includes a vast **scale** of economic endeavor: everything from a small business employing only one individual to huge corporations with tens of thousands of employees. Investment capital in the private sector is not generally derived from tax revenue, but rather from private investment in the form of loans, personal savings, offerings of stock (somewhat confusingly, since this is called an "initial public offering" in investment parlance), and other sources. More precisely, economists divide any economy into "sectors" based on the predominant types of activities and products found in the sector. In economic geography, these are the primary, secondary, tertiary, quaternary, and more recently, the quinary sectors of the economy. These sectors are typically presented as though they represent clear, discrete packages of economic activity, but in reality there are many businesses that might be considered to be in more than one sector, and indeed, economists are not always in complete agreement on the precise characteristics that define each sector.

The primary sector of economic production generates goods that are produced or extracted directly from the **Earth's** land and **oceans**. They are taken and used in the condition they are discovered, or they may be refined or concentrated into more useful forms. The simplest forms of primary activity would be hunting and gathering, or fishing for one's immediate food requirements. Economic activities that fall within the primary sector are typically those that seek to discover and exploit **natural resources**. Mining, forestry, fishing, and agriculture are the main commercial activities in the primary sector. All of these activities are extractive, meaning that they withdraw economic value from the land or water of the Earth. The primary sector is crucial to the economy because it produces materials for the secondary sector and also provides basic necessities such as food. Before the industrial and **agricultural revolutions** of the 18th and 19th centuries, the primary sector was the largest sector in most economies, both in terms of employment as well as for volume and value of production. In most modernized economies the primary sector accounts for only a small percentage of employment, but this is not universally true. In the United States, for example, the primary sector accounts for about two percent of total employment, and the figure is similar to that in most countries of the European Union. On the other hand, China, a country with a large economy and significant industrial capacity, still has approximately 50 percent of its workforce employed just in agricultural production.

The secondary sector consists of those economic activities that utilize the products and materials generated in the primary sector to make more complex kinds of goods. This sector is frequently characterized in a broad sense as the manufacturing sector, but this label includes several processes that do not typically result in a manufactured product, such as electrical power generation, smelting of metallic ores, sawmilling, refining, and processing of some foods. The economic processes in this sector are multilayered and complex, as many of the items or substances created in one secondary activity are often used in a subsequent activity to produce yet another, more sophisticated good. For example, crude petroleum, a commodity obtained through primary sector activity (mining) can be manufactured into synthetic rubber as part of the secondary sector. The rubber obtained via this procedure may then be used to make tires, which in turn are incorporated into a yet more complex and expensive good, an automobile. Parts of the secondary sector that utilize large quantities of bulky raw materials from the primary sector, such as iron and steel manufacture or petrochemical refining, sometimes referred to as heavy industry, frequently locate in close proximity to the sources of their primary inputs, or at break-of-bulk points. The early iron and steel industry in Great Britain, for example, was situated adjacent to deposits of iron ore and coking coal, the two most important ingredients in steel manufacture. This **location** lowered the transportation costs of the ore and coal. In some industries **agglomeration** is common, which leads to lower costs and greater efficiency.

Most economic activity in so-called post-industrial economies takes place in the tertiary, or service sector. In the United States' economy, the largest in the world, approximately 80 percent of the labor force is engaged in service activities. Just as in the secondary sector, a large range of activities falls within this sector. Retail stores, personal services like barbers, tanning salons, restaurants (in the

United States, the fast food portion of the tertiary sector accounts for a large percentage of income), as well as more sophisticated and expensive services such as health care, legal advice, financial investment assistance, and others all form this branch of the economy. Transportation activities are typically included with the tertiary sector and account for a large percentage of economic value in many advanced economies. In addition, all types of entertainment are also considered tertiary activities, so music, film, and sporting events all contribute to the tertiary sector's percentage of GNP. Tourism and the so-called hospitality industries, like hotels, food services, recreation, entertainment facilities, etc., make up another large segment of the tertiary sector, and these businesses represent one of the fastest-growing components of this portion of the economy. Globally, the service sector accounts for more total value than any other sector, and continues to expand faster on average than either the primary or secondary sectors in the economically advanced world. Advances in technology, especially the adoption and expansion of the Internet, has greatly accelerated the growth in tertiary activities in the past two decades.

Many economists distinguish a quaternary sector, which may be best approached as a distinct component of the tertiary sector, only in this case quaternary activities are those that are derived from intellectual activity, and generally involve the transfer of specialized information. Examples would include research and development, consulting services, financial planning and advice, education and training, and others. Most of the investment in these activities comes from businesses and corporations, which seek to ensure their future prosperity by gaining some competitive advantage from improved skills and knowledge of their workforce, more efficient means of offering their services or products, or detecting trends in the specific markets they service. Quaternary activities are also unusual from other services in that there is frequently no immediate result from investment; rather, the quality of the information or advice procured may not become apparent for months or even years. This sector is found only in advanced economies, because its development is dependent on the ability to obtain and process large and diverse amounts of information quickly, and also on a significant labor force that has a high level of analytical expertise. Universities, private research institutes and think tanks, consulting firms, and other similar businesses are the key units driving the quaternary sector's expansion. In some industries like health care, the activity represented by the quaternary sector is vital, because the time required to move an innovative type of equipment, a new drug, or even a pioneering treatment technique to market may be extensive, and mistakes are not only extremely costly, but may even result in the financial collapse of the company.

The four sectors identified above all represent a kind of continuum in which value is added at each stage, from extractive activities that focus on obtaining raw materials, to refining those materials into more sophisticated products, to various levels of services that require some skill, from the relatively simple (the shining of shoes) to quite complex (research and development of a new cancer-fighting drug). An emerging school of scholars argues that yet a fifth sector must now be considered: the quinary sector. This school of thought holds that there are "hidden" economic activities of considerable value that remain unaccounted for in the traditional four-sector model. These include non-profit activities and public services funded by governments, such as police and fire protection, public education, and others. In addition, domestic work performed by homemakers, activity that typically does not generate profit but has value and is ubiquitous, is also considered by many economists as belonging to the quinary sector. In some cases it is difficult to assign a monetary value to the quinary sector, because the "products" of this sector are not necessary subject to the law of supply and demand, yet many economists have come to recognize that economic value and output do not always conform to market dynamics.

Segregation

The spatial division of a **population** according to racial or ethnic characteristics. Since the 1950s many studies in social geography have been conducted on the residential racial segregation present in most major cities in the United States. Before the 1920s, many cities in the northern **region** of the United States had a relatively small black population, but a large **migration** of Southern blacks in the 1920s and 1930s northward greatly increased the percentage of African Americans living in urban places like Detroit, Chicago, Milwaukee, Cleveland, and New York. By the 1950s sociologists of the so-called "Chicago School" were conducting research on the spatial separation of racial groups in Chicago, and this research stimulated similar studies of the racial segregation patterns found in other cities. Segregation in residential areas in these urban environments followed more or less the same geographical arrangement—an inner city that was predominantly black, with smaller clusters of other ethnic or racial minorities; and a surrounding fringe of suburbs that were overwhelmingly populated by white residents. In the 1960s these spatial patterns appeared to become even more distinct, and were accompanied by a decline of the inner-city area. Property values declined, the local tax base eroded, and the quality of services provided to residents suffered, including law enforcement, fire protection, and education.

The term "white flight" was coined to describe the process whereby cities became racially segregated. Whites began leaving inner-city neighborhoods in the

1930s as greater numbers of black families moved in, but this process dramatically accelerated after World War II. Several factors contributed to the movement of whites to the suburbs. The construction of the interstate highway system in the 1950s allowed whites in the suburbs to live a considerable distance from the inner city, but still commute there on a daily basis relatively easily. Low land values in the countryside, relatively low mortgage rates, and the application of mass production techniques to housing construction after the war enabled many whites to purchase homes in the suburbs, but blacks were unable to follow because incomes in the black communities were lower, and few blacks could afford to relocate. Moreover, discriminatory practices in lending prevented blacks from purchasing property in predominantly white areas. "Redlining," for example, was a tactic mortgage lenders used to maintain the segregated character of housing in the large cities. When a residential area was redlined, mortgage lenders would deny mortgage applications of blacks and other minorities for property within the area, effectively prohibiting them from living in that part of the city. This was done primarily to protect property values, because the mortgage holders felt that the presence of minority residents would depress the value of existing residences. Such discriminatory practices are now illegal, but their effect remains, as most urban areas in the United States remain quite segregated. However, the U.S. Bureau of the Census has declared in recent census reports that the level of segregation in American cities is declining, as white neighborhoods have gradually become more integrated.

Sequent Occupance

The American geographer Derwent Whittlesey is credited with first articulating the concept of sequent occupance, although he drew heavily on the theoretical work of others in the field of cultural geography, especially the ideas of Carl Sauer (see sidebar). Sequent occupance is the notion that **landscapes** are shaped over time by the sequential settlement, or at least sequential use, of that landscape by various groups who occupy the land. Thus, according to the proponents of this approach to landscape study, a place can be understood only through an examination of the historical impact of such occupation, and a comprehension of the nature of the culture at each stage of occupation. The theory had a deep influence on the approach of historical and cultural geographers in the United States in the 1920s and 1930s. The concept of sequent occupance has its roots in Social Darwinism, in that the landscape is seen as evolving over time due to the various influences upon it, and the condition of the present landscape contains elements from all previous occupations. Most of the proponents of sequent occupance emphasize the necessity of fieldwork to identify the ways that the land had been utilized by past cultures. Past occupancy is reflected in the presence of structures from previous cultures, alterations of the natural environment, etc. Such a reading of the landscape reveals the practices and processes that combine to form the landscape in its present condition. Sequent occupance as a theory of geographical development focuses on the relationship between human activity and the natural environment, although there is no implicit assumption that the sequence is necessarily continuous, or that successive occupations and uses are superior to those that came previously.

The theory is in almost direct opposition to the intellectual position of **environmental determinism**, because sequent occupance suggests that how cultures use the landscape is not predominantly determined by the physical environment, but rather is a product of the culture itself. Various cultures, in other words, interact with the landscape based on their technological capacity, and it is that capacity, not the environment, that dictates the human-landscape dynamic. Sequent occupance is a conceptual component of many contemporary studies in **cultural ecology**, and both emerged from an emphasis on human ecology in the work of some geographers in the early 20th century. In some cases the historical patterns of settlement of a place may be partially revealed through a study of the local **toponymy**, which frequently carries clues to the history of previous cultural occupants as well as to a place's former function on the landscape. Sequent occupance continues to have a strong influence on studies in historical geography, and most contemporary historical geographers engaged in holistic landscape studies would subscribe to the basic principle of the theory, i.e., a landscape cannot be completely understood in a vacuum of time, and only a consideration of all past cultural uses will result in a full picture of the relationship between human activity and the physical environment.

Shatterbelt

A region of persistent political fragmentation due to **devolution** and centrifugal forces. The term has been applied by political geographers to a number of places since the Second World War, especially East Central Europe, but also Southeast Asia, the Middle East, and Africa. A synonymous phrase is "crush zone." Shatterbelts sometimes serve as **buffer zones** between hostile states or empires, and historically have played an important role in **geopolitics**. Shatterbelts in the Middle East and Africa emerged in the latter 20th century due to the collapse of **imperialism**, and the subsequent devolution of structures of governance to multiple independent states, most of which had never existed prior to the collapse of colonial authority. Regions classified as shatterbelts are characterized by states or territories that have a large degree of ethnic, linguistic, and/or religious diversity, and a history of antagonism and hostility between the groups living there, and can result from the **balkanization** of larger political entities. **Boundaries** in shatterbelts tend to be fluid and often contested, due to the fact that such political divisions frequently cross cultural regions, splitting ethnic groups between two or more countries. Although the term itself did not come into wide use until after World War II, the general concept of a shatterbelt appeared in the writings of political geographers several decades earlier. In a more modern context, shatterbelts often hold what Samuel Huntington has described as "civilizational fault lines," a key concept in his "clash of civilizations" thesis.

The classic example of a shatterbelt is southeastern Europe, especially the Balkan Peninsula. This region has been functionally a shatterbelt for at least 500 years, as it has been geographically sandwiched between more powerful states that attempted to control part or all of the territory. From the 16th century to the early 20th century, the Ottoman Empire controlled large swaths of the Balkans at various times, and imposed Islamic culture on many of the peoples under its rule. Almost the entire population of what are today Albania and Kosovo were converted to Islam, and significant numbers were converted in portions of modern-day Bosnia and Bulgaria as well. A majority of the Slavic-speaking peoples in the region retained their Christian religious identity, but they too were divided into Roman Catholics (Slovenes, Croatians) or Orthodox Christians (Serbs, Montenegrins, Macedonians, Bulgarians) whose relationships were frequently antagonistic. Compounding this complex ethnic geography was the presence of large non-Slavic Christian minorities like the Hungarians and Germans, especially in those portions of the **region** that fell within the borders of the Austro-Hungarian Empire, and many other groups scattered throughout the peninsula. The collapse of the Ottoman Empire after World War I, the subsequent creation of Yugoslavia, and four decades of the Cold War only temporarily subsumed the turbulent character of this shatterbelt, as witnessed by the violent birth of seven new countries, most of which had never been independent previously, between 1991 and 2008. Kosovo's independence in the latter year indicates that shatterbelts will remain volatile regions well into the 21st century.

Sjoberg Model. *See* Pre-industrial City Model.

Social Distance

A phrase used to describe the degree of interaction between various social groups divided by socioeconomic status, race, gender, sexual orientation, or other factors. It is distinct from physical distance, which represents the actual locational space between groups. The process of **segregation**, which determines physical distance between groups, may contribute to greater social distance by reducing the frequency of contact between groups. In theory, and frequently in practice, the physical distance separating groups compounds their social distance, due to **friction of distance**. Social distance may be measured and analyzed through several theoretical approaches. One method is to question a defined group about attitudes and views of other social groups and with the resulting data create a scale or index of interaction. For example, a social geographer who wishes to gauge the social distance between blacks and whites in a large urban area might construct a questionnaire for whites that requires them to rank the level of social interaction they would find acceptable with blacks. A widely used instrument among sociologists and social geographers that follows this methodology is the Bogardus Social Distance Scale. The Bogardus scale assigns a numerical value to differing levels of social interaction, from marriage to excluding a group from entering the United States. Social distance measured in this manner is a concept that is a function of how receptive a group is to interacting with another. The Bogardus scale has been used in research since the 1920s and over time indicates that social distance in the United States as a whole is decreasing.

Some social scientists contend that social distance may take other forms besides that measured on the Bogardus scale. An alternative is to consider social distance as a social "norm," meaning that it is established as a part of one's social and **cultural identity**. In this sense, social distance is a part of group cohesion, because it serves to separate those who "belong" to a group and those who remain external to the group on the basis of some characteristic, for example, race or ethnicity. Because group identity is dynamic, social distance conceived in this way can change over time and either become greater or be reduced. This "normative" social distance between Caucasian groups in large American cities has diminished since the 19th century, when the social distance between Italian, Irish, Polish, and other ethnic groups who had recently arrived via **migration** was larger and significantly restrictive. Social distance can also be perceived as the level of social interaction between groups, with the underlying assumption that the greater the magnitude of interaction, the smaller the social distance between them. Critics of this approach argue that in many instances, greater levels of interaction do not necessarily result in a reduction of social distance, but point to the persistence of distinct social and religious groups even under conditions of frequent and open social interaction. On the other hand, greater interaction appears to have reduced the social distance separating racial groups in the United States since the 1950s; more recently, the social distance between groups of different sexual orientation seems to have diminished.

Soils

Soils are at the interface between the lithosphere and the **atmosphere** and cover a large majority of **Earth**'s land surfaces. Incredibly variable over the planet, they are natural and, most critically, capable of supporting life in ways impossible in solid bedrock. They are able to store water and provide vital links in the **biogeochemical cycles**.

What makes up soil? About half the mass of most soils is composed of highly weathered remains of rocks. Usually, soil contains a bit of decayed and semi-decayed plant and animal material (about 5 percent of a good agricultural soil). The remainder of the soil is pore space, which is occupied with soil air and soil moisture.

In an examination of a pit dug through the soil and known as a soil profile, there are usually major differences in soil characteristics with depth. Most soils can be conceived of as having horizontal layers called horizons. These horizons can vary in color, fertility, texture depth, structure, and water percolation rates. Moreover, these main horizons can be subdivided into subhorizons. The O horizon is composed of organic material and is on top of the mineral soil. The A horizon is the topmost layer of the mineral-dominated soil, and is known as *topsoil*. The B horizon is below the A horizon and is where materials from above and, sometimes, below have accumulated. Together, the A and B horizons are known as the *solum* (true soil). The C horizon resides beneath the solum and consists of unconsolidated materials beneath the weathering and most of the organic material. Not all soils have horizons (see Table 4) and not all soils with horizons have subhorizons.

The interaction of organic and inorganic features of the physical environment vis-à-vis soils is frequently illustrated by a pseudo-equation:

$$\text{soil} = \int (\textit{parent material, time, climate, topography, and biota})$$

This means that the soil observed in the field is the product of a combination of mass and energy exchanges associated with these five factors.

Parent material is the matter forming the bulk of mineral soils. The parent material can be rock, weathered remains of rocks (like clay, silt, or sand), organic materials, previously existing soils, or organic matter. Parent material has a large

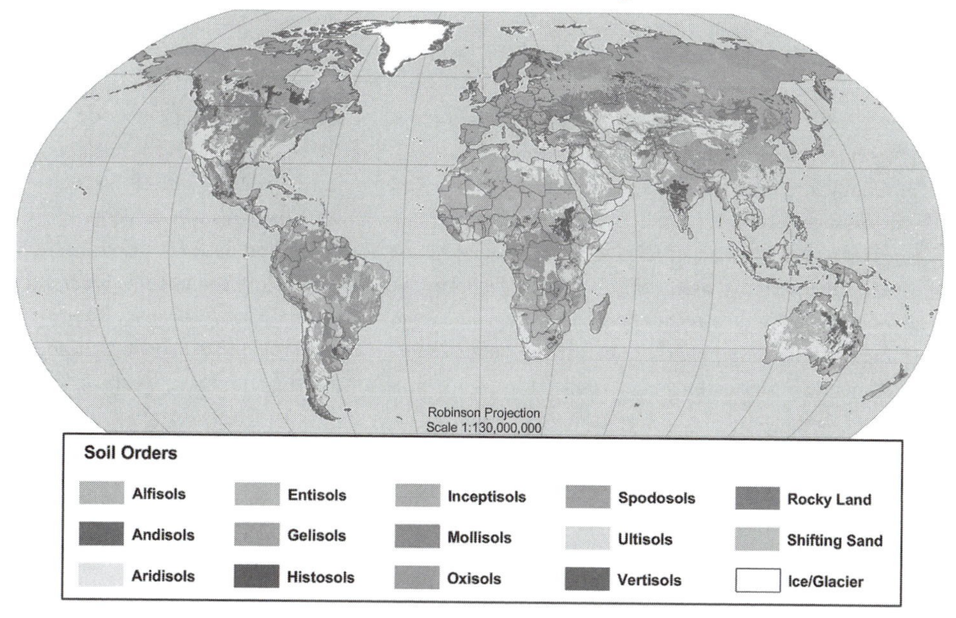

A map showing the global distribution of soil types, based on the USDA soil taxonomy system. (U.S. Department of Agriculture)

role in determining mineralogical, chemical, textural, and structural properties of soil. For instance, sandstone of bedrock usually produces relatively infertile, sandy-textured soils.

One of the hallmarks of Earth's physical system is the incomprehensible scope of time. Parent material is morphed into soils because of availability and exchange of mass and energy between Earth's surface and its atmosphere. Physical and organic weathering agents penetrate the parent material. With great amounts of time there can be significant remineralization and translocations of materials up and down through the soil profile. Given enough time, the soil can become physically unlike the parent material. As with rocks on continental surfaces, soil ages do not approach the age of Earth. However, there is a wide range of ages. Some soils have had hundreds of millions of years to evolve, while some can be dated to the withdrawal of continental glaciers a few thousand years ago, and others are merely tens or hundreds of years old.

Climate is the most important factor in a worldwide sense. Witness the slow chemical processes of tundra soils that are palpably dissimilar from those creating deep, infertile soils found in the tropical rainforest. Climate exchanges of mass and energy with underlying soils determine the rates of physical and chemical

Permafrost

Permafrost is soil that is frozen except for a brief period during the summer. In much of Russia in the region known as Siberia, and in Alaska and northern Canada, vast stretches of permafrost are present. It may also be encountered in alpine **locations** around the world. In Siberia and North America this frozen zone can extend to enormous depths. In some places in Siberia, the permafrost may extend over a thousand feet into the soil and subsoil. Such an environment presents huge difficulties for permanent human settlement, as well as significant engineering and construction challenges. Buildings must be elevated on pillars in permafrost areas, otherwise the weight of the building generates heat that melts the soil underneath and causes the structure to become unstable, lean, or shift, and potentially collapse. Roads, railroads, water lines, and other infrastructure is also extremely difficult and expensive to build and maintain. Agriculture is possible for only a few months in the summer in permafrost regions, and then the crops produced must mature quickly to be harvested before the winter sets in and the soil freezes again. Such regions typically have quite low population densities because of the harsh environmental conditions found there.

processes. Given sufficient amounts of time to evolve, it can be said the nature of the soil is highly reflective of climate.

Topography can speed or slow soil evolution. Soil depth and chemistry are profoundly affected by slope and there is a vegetative feedback in that various plant communities are optimized at various elevations. The topography does not need to be extreme for there to be significant soil differences. Modest hillslopes make for major soil differences on scales of tens to hundreds of meters.

Biota are the plant and animal forms associated with the soil. Plants range from microscopic to the largest trees. Their importance includes anchoring of soils via root systems, storage and exchange of energy and nutrients, and the increase of infiltration of rainwater. Animals enrich soils through their surface and below surface droppings (especially earthworms) and stir nutrients through the digging of burrows. Clearly, the nature of the biota influences the development of soils.

Even though soil formation is slow compared to human lifetimes, soils can be seriously degraded or lost in a matter of a few years. Societies have declined with concomitant degradation of their soil resources. The story of European and Asiatic ports becoming silted and unusable because of erosion from agriculture is all too common in historic times. **Desertification** implies the degradation of soils and is of great concern in the Sahel, south of the Sahara. The Dust Bowl of the U.S. Great Plains is a prima facie example of the interaction of overuse of the soil resource with natural drought.

It is well appreciated that knowledge of soils is of intense agricultural interest. Yet, the world is increasingly urbanized and in this realm, too, soils can be critical. For instance, soil characteristics are key in causing foundation cracks, potholes in roads, and corrosion of pipes. Unfortunately, most of the world population has little knowledge of the nature and fragility of the soil resource and the word "dirt" is used as a synonym for "soil," whereas dirt is technically soil misplaced from its natural setting.

Pedology is the term denoting soil science. The roots of pedology are quite old. Soil classifications have been performed for more than two-and-a-half millennia. Chinese and other early civilizations worked out simple systems of taxation based on agricultural productivity. In the late 1800s V. V. Dokuchaev and colleagues studied soils using science and was the first to geographically consider soils over large areas through the invention of soil mapping. C. F. Marbut and others of the U.S. Department of Agriculture (USDA) presented a genetically based soils taxonomy in the 1930s and a greatly revised scheme was released in 1960, which is in use in the United States today. No soil classification is universally used, although in 1998 FAO (UNESCO) and the International Union of Soil Scientists published the World Resource Base classification through international collaboration.

The U.S. Soil Taxonomy (1960) is a system based on the observable and testable properties of the soil. That is, it is an objective classification based on many features. It is highly useful in that it allows positive identification through the examination of soil profiles and their chemical and physical properties. Prior to the Soil Taxonomy system, U.S. soil classifications were based on problematic inferences as to the origins of a particular soil. This caused significant problems in situations such as soils that had evolved during times of climate change. Whereas the Soil Taxonomy system precisely identifies each soil by observable characteristics, it should be noted that there are some geographic concerns not resolvable. For instance, different combinations of the soil-forming factors explained above can produce virtually identical soils under widely disparate physical conditions.

As in the classification of animals, plants, or climates, soils are classified at various levels, from general to specific. These levels are soil orders, suborders, great groups, subgroups, family, and series. Table 4 shows the characteristics of the dozen soil orders of Earth. The USDA has classified over 20,000 soil series in the United States, and there are an indeterminate number for the world because many **regions** have not been classified to the series level. The geography of world soil orders is shown in the accompanying map. It should be noted that this map represents areas dominated by each soil order but that larger scale maps of counties, states, or countries would show considerably more complexity.

Table 4. USDA Taxonomy World Soil Orders

Order	General characteristics
Alfisols	Mature soil, extensive in low and middle latitudes, transitional environments, subsurface clay horizon, plentiful nutrients
Andisols	Volcanic ash as parent material, not highly weathered, minimal profile development, dark color, usually fertile
Aridisols	Light colored, in arid and semiarid climates with little water to leach soluble minerals, sandy, thin, productive if irrigated
Entisols	Least well-developed order, horizons sometimes nonexistent, thin and relatively infertile except on floodplains
Gelisols	Recent soils associated with permafrost, found in the arctic and subarctic, mixing of materials because of freeze/thaw
Histosols	Least extensive of any order, organic soils, always saturated, lack of oxygen slows decay, acidic, briefly fertile if drained
Inceptisols	Immature but possessing faint horizons, parent materials "young," steep slopes and floodplains, various climates
Mollisols	Surface horizon dark, thick, and fertile, typical of grasslands, most fertile of all orders, most extensive order on Earth
Oxisols	Deep, warm, and wet climates, highly altered from parent material, tropical, not usually fertile
Spodosols	Cool, moist climates, coniferous forest cover, upper layers light in color and heavily leached, tend to be infertile
Ultisols	The most heavily weathered soils of middle latitude soils, reddish, subsurface clay accumulation
Vertisols	Alternating wet and dry climates of tropics and subtropics, large amounts of clays prone to swelling and shrinking

Solar Energy

It is no understatement that solar energy is the lifeblood of our planet. Weather, oceanic circulation, and life owe their existence to the energy output of our sun. Although it is true that solar energy is responsible for heating up our planet, there is a common misconception that solar energy and thermal infrared energy are synonymous. In fact, only a small proportion of the direct solar beam is composed of thermal infrared energy.

The sun is a star that has natural variation in output but the variations are currently insignificant enough to maintain life on **Earth**. The evolution of the sun and certain prehistoric output changes has been known to force **climate** change, but the present status of the sun is not such that we do not expect large changes in the next thousands of years. Sunspots are dark features on the sun's radiating photosphere some 1,500°C cooler than their surroundings. They are caused by disturbances in the interior solar magnetic field. These disturbances are tube-shaped and sometimes pop above the photosphere. Sunspots occur where the magnetic field lines leave and enter the photosphere. These areas can grow until some reach sizes exceeding 30,000 km with individual sunspots usually lasting a few days to a few months. Numbers of sunspots cycle over 11–13 years, and increased numbers of sunspots are associated with increased solar activity. It has been conjectured that solar cycles influence weather and climate but this relationship is unclear because of all the other factors influencing Earth's **atmosphere**. Prominences, coronal mass ejections, and flares are other output variations related to sunspots. However, the measured year-to-year solar variability at the top of Earth's atmosphere is about 0.1 percent of the average.

The sun is able to emit huge amounts of energy through space because its gases are heated as a result of fusion. The hotter a substance, the more energy it radiates because of the greater intensity of molecular vibration. The vibrating electrons of the gases' molecules emit electromagnetic energy, and this energy is propagated through the void of space and through our atmosphere. In space, the speed is about 300,000 km per second with the speed of transmission through our atmosphere being negligibly slower. Solar energy can penetrate some Earth materials.

Solar output is a staggering 3.31×10^{31} joules per day. Of course, Earth is small and 150 million km distant and the sun radiates in all directions, so we receive only about one two-billionths of the total solar output. The mean instantaneous solar input is 1.74×10^{17} joules per second at the top of the atmosphere; compare this figure with the "measly"1.5×10^{13} joules per second consumed by human activity on the planet.

Solar energy can be understood as having two physical realities. On one hand, radiant energy is composed of streams of particles called photons. Photons are the best way to explain the manner in which energy interacts with plants to cause photosynthesis. Photons incident on plant leaves energize the photosynthetic process.

On the other hand, streams of photons are arranged in waves, and the waves can be described by the frequencies by which their crests pass or by the lengths of the waves; most geographers use the latter. There is an exceedingly wide range of wavelengths in the universe; however, solar energy is fairly well focused around its most plentiful wavelengths near 0.5 micrometers (1 micrometer = 1 millionth of a meter). This peak energy is visible blue light, and it is not surprising that photosynthesis has evolved to capture the solar wavelengths most available. Various wavelengths of solar energy act quite differently when encountering matter (see **Heating and Cooling**). Of immense importance in the Earth system is the propensity for the particular mix of Earth's atmospheric gases to be relatively transparent to solar energy and allow about half of that energy to penetrate directly to Earth's surface.

The solar beam reaching Earth is well described as "sunlight." Because the sun's radiating photosphere is at approximately 6,000°C, visible wavelengths are the most plentiful comprising about 44 percent of the solar output. Seven percent of solar energy is invisible ultraviolet energy mainly absorbed by the atmosphere and 11 percent is thermal infrared energy. About a third of the energy (37 percent) is in the near infrared portion of the spectrum and not sensible by humans. The sun also emits microwaves, television waves, and radio waves and these corporately account for less than 1 percent of the solar energy output. Significant amounts of solar energy are transformed into thermal infrared energy on interaction with Earth's atmosphere and surface, and it is this transformed energy that is largely responsible for heating of the planet (see **Heating and cooling**).

The penetration of solar energy deep into the atmosphere and onto Earth's surface directly forces winds to blow and **ocean currents** to flow. That is, solar energy forces the circulation systems of the planet. When examining **seasons**, it is clear that the systematic changes in the latitudinal receipt of solar energy as caused by the revolution of the Earth about the sun make for systematic variations in the amount of air and water circulating in any geographic region.

Of considerable concern is the increasing amount of ultraviolet radiation reaching Earth's surface. In recent years the ozone of the stratosphere has been eroding, particularly in the far southerly latitudes. Unmitigated solar ultraviolet energy is fatal to land life, and any increase in present amounts reaching Earth's surface will bring an array of unwanted effects. These include increased incidence of human skin cancer, DNA modification in plants and animals, and breakdown of exposed metals and fabrics.

Solar energy has been used in myriad ways to aid human activity. Drying foods, passive heating strategies for traditional housing forms, and distillation of salt water to fresh water are simple historic examples of our conscious manipulation of energy from our local star. Modern humans have begun to use large amounts of solar energy to make electricity and warm water for household purposes. Such uses can only grow as Earth's fossil fuels become more expensive.

Spatial Inequality

A spatial differentiation of abundance of any measurable quality or characteristic. No geographical feature is evenly distributed across any **landscape** of the **Earth** except at a very small scale, so every spatial characteristic is marked by spatial inequality to some degree. Spatial inequality is primarily of concern to scholars who study the **geography of economic development**, and is related to the concept of **segregation**, a condition in which spatial inequality is established in an urban environment on the basis of race or other considerations and is reflected in the **social distance** between groups. Spatial inequality is often the motivating factor behind **migration**, because economic opportunity, wages, and amenities are never evenly distributed, and people tend to relocate to those places that offer the greatest advantages. Economic geographers may analyze and explain spatial inequality at any **scale**: global, regional, national, or local. **World Systems Theory** is an attempt to identify the factors and conditions that contribute to global spatial inequality of income and wealth. At the regional and national scales, concepts of **core and periphery** are sometimes employed to explain the presence and effects of spatial inequality. Many countries possess an economic core **region** that is characterized by an underdeveloped **hinterland**, resulting in an unequal distribution of economic development, income, and standards of living.

The spatial inequality of wealth and income are of interest because other spatial inequalities are often connected to this specific pattern of inequality. For instance, the spatial distribution of the quality or accessibility of health care in many countries is directly correlated with levels of personal or family income. **Infant mortality rates**, longevity, and other factors influencing the quality of life are also typically geographically associated with the spatial characteristics of monetary income. If it is assumed that such inequality is an undesirable geographic pattern, then a comparative, reasonably precise means of measuring such disparity is useful, as countries and regions may then be evaluated relative to one another. A common method used to accomplish this is the Gini coefficient. The Gini coefficient is a mathematical index that ranges between a hypothetical value of zero to a maximum of one. A Gini coefficient of zero indicates a situation in which there is complete equality, in other words all individuals or family units receive exactly the same level of income or wealth. A coefficient of one indicates that all wealth is concentrated with one person or household, with all others receiving nothing, resulting in the maximum degree of inequality. Frequently, the coefficient is multiplied by 100 for reporting purposes, as this produces whole integers and is somewhat easier to use when making comparisons. In this system the maximum coefficient is 100, while the minimum remains zero. In reality, for countries the range of Gini coefficients is between 20, signifying low spatial inequality of income, to around 70, indicating a relatively large inequality. As a general rule, developing countries tend to have the largest Gini coefficients.

Regional spatial inequality may be influenced by a number of factors. **Globalization** may either exacerbate or ameliorate spatial inequality between regions, depending on the circumstances of the regional relationship. The era of European **imperialism** unquestionably increased spatial inequality between the industrialized, developed countries of Europe and North America, and the colonized regions of Africa, Asia, and Latin America. Ironically, globalization may be lowering spatial inequality between countries like India and China, where average incomes have risen in recent decades, and Europe and North America, where the average increase in incomes has been proportionally less. On the other hand, globalization has seemingly increased the degree of spatial inequality *within* these emerging economies. In China, for example, the development of the Special Economic Zones (SEZs) along the coast has resulted in a dramatic rise in China's integration with the global economy. As a result, personal incomes in most of the SEZs are much higher than in many other parts of China, a fact of China's economic geography that has led to greater spatial inequality in the country. A similar increase in spatial inequality has occurred in India, where incomes in urban areas have risen much faster than in **rural settlements**, partially as a result of outsourcing and the emergence of high-tech industries in many Indian cities. Some economic geographers argue that such increases in spatial inequality are not necessarily a negative development, because while urban incomes have increased most rapidly, *all* incomes have been rising, increasing living standards as a whole.

Economic spatial inequality is also a phenomenon of developed countries, and in fact has been studied by scholars longer in these regions than in the developing world. In developed countries spatial inequality is frequency a function of differential rates of economic growth between regions, due to a variety of factors. For example, in the United States in the early 20th century, many urban areas in the upper Midwest, located in what today is called the rust belt, were growing rapidly. Many of these cities had developed clusters of heavy industry due to transportation advantages or through **agglomeration**. A **location** adjacent to the Great Lakes, providing cheap water transportation, was a distinct benefit to the growth of Chicago, Milwaukee, Detroit, Cleveland, and other cities. Other parts of the region, however, remained badly underdeveloped, like the Appalachian Highlands of West Virginia and eastern Kentucky and Tennessee. The latter regions were some of the poorest sections of the country, with much lower income levels than surrounding areas. Much of the back country of the Appalachians was not even serviced by electricity in the 1920s, and standards of health care and education there were well below the national average. During the 1930s the U.S. government pursued policies to raise living standards in the highland region and reduce the spatial inequality in the eastern section of the country. Ironically, the old manufacturing belt today is losing its status and falling behind some of the other regions of the country. Spatial inequality is dynamic and can quickly shift in the modern globalized world.

Spatial Interaction Models

Geographers who study the flow of phenomena across space, especially if such flows are organized and occur via networks, are interested in conceptualizing the ways such flows transpire and the factors that interrupt or facilitate them. Scholars of transportation and economic geography are especially focused on such movement, but other geographers may also study flows. Movement occurs across space because resources, people, and cultural attributes are unevenly distributed on the **landscape** of the **Earth**, and two **locations** may share **complementarity**, meaning that exchange is advantageous to both. However, this movement is subject to **distance decay**, which means that less interaction takes place as the distance between two locations increases. This is due to the effect of the **friction of distance**, a concept that holds that costs, both material and in terms of time invested, increase with distance. The larger the distance that separates two points, the greater the costs associated with interaction between those points. Yet countless flows are transpiring at any given moment on the Earth. Many flows are interrelated, and in turn stimulate new movement. Trade flows, for example, may result in economic expansion and greater employment opportunities in a location, and thereby generate **migration**, a flow of labor. Spatial interaction models provide a means of representing and explaining the movement of people, commodities, technology, and many other things through space. Many of these theoretical constructs were developed in the 1930s and were elaborated on during the **quantitative revolution**. Christaller's **Central Place Theory** can be considered a spatial interaction model, and a number of others are in common use.

Perhaps the most frequently employed, and one of the best known, of the spatial interaction models is the gravity model. This model applies the basic concept of Newtonian physics regarding the force of gravity. Newton stated that the gravitational force between two objects was equal to the mass of the first object multiplied by the mass of the second body, and then divided by the distance between them squared. This model has been widely applied because it has been shown that friction of distance between two places does not function in a linear way; rather, like the physical force of gravity, it is inversely proportional to the square of the distance separating them. In the gravity model employed in the social sciences, Newton's "force" is represented by the degree of flow, or interaction, between two places, which are assigned a respective "mass" based on the **population** of each, or perhaps some other measure. The formula is:

$I=P_1 \, xP_2/d^2$ where I is the degree of interaction;
P is the respective population at each place;
and d is the distance separating the two places.

The gravity model is highly useful in the study of transportation geography and economic geography because it can be used with reasonable accuracy to predict the degree of flow of traffic, goods, migration, and other items. Used in combination with various statistical techniques, the model provides substantial insight into the workings of networks through space and what factors affect their effectiveness. It assumes that the closer places are to one another, the greater the level of interaction between them, an assumption repeatedly borne out in studies of actual flows.

An adaptation of the gravity model is Reilly's Model of Retail Gravitation, a somewhat more sophisticated spatial interaction model that attempts to establish **boundaries** of interaction between **regions**, rather than the interaction between two places. Reilly used retail sales data from counties in Texas to develop his model. He was interested in discovering the breaking point, or boundary, between two sales regions. Reilly assumed that the larger the population in a city, the greater the retail sales area it would command. But where would the boundary be between two cities of unequal population? Reilly's model predicts the location of this border through the equation:

$$BP = (\text{distance between cities})/1+\sqrt{P_2 \div P_1}$$

Where: BP is the breaking point measured from city 1.
P_1 is the population of city 1.
P_2 is the population of city 2.

Here again, population is the equivalent of mass in Newton's theory. In the case where two cities would have the same population, the breaking point would appear exactly half way between them. Reilly's model has been used not only by economic geographers in academic studies of sales areas and interaction but has applications in business and marketing. Furthermore, it may be applied beyond retail sales, to find boundaries between media markets, zones of loyalty to sports teams, and many other phenomena.

Squatter Settlements

Areas of dense, low-quality housing around many large cities in the developing world, usually formed by illegal occupation, or "squatting," of unclaimed land. Squatter communities are created by the **migration** of poor, rural residents to cities who are drawn there by the **push-pull concept**, especially the allure of higher-paying jobs. Over the past forty years many urban areas in the developing world have witnessed a surge in new arrivals from the **hinterland** they serve, and the supply of local housing has simply been unable to keep pace with the increase in population. New residents have therefore been forced to build ramshackle sheds on unclaimed or public land. Globally, an enormous number of people live in squatter settlements, with some estimates running as high as one billion, and in some cities the residents living in such conditions may approach one-quarter of the **population**, although exact figures for the number of people living in local squatter neighborhoods are frequently difficult to obtain. In the Spanish-speaking countries of Latin America squatter settlements are locally referred to as *barrios,* while in Brazil a community of illegal residents is called a *favela*. The city of Sao Paulo, Brazil's most populous city, has more than 600 of these settlements, and the United Nations estimates that about 26 percent of Brazil's total population lives as squatters. In South Africa the local squatter settlements are called shanty towns. The shanty towns found around South African cities have expanded exponentially over the past 20 years, due to the elimination of the laws that prohibited blacks from settling in many urban areas. Under the system of apartheid, the white-dominated government had created some squatter communities by forcefully relocating illegal residents to maintain racial **segregation** in urban areas; but as apartheid was dismantled, many rural blacks migrated to the cities and established new shanty towns.

Regardless of where the squatter settlement is located, all such developments share some basic characteristics. Housing is inferior, often built haphazardly on unstable ground like the sides of hills and ravines. Building materials are usually scrap wood and sheet metal, cardboard and plastic. Most of the structures lack flooring, and frequently thousands of shacks, perhaps homes to ten thousand people, are constructed without any plumbing. Because the settlement is illegal and unplanned, in most cases streets are unpaved, narrow, and lack any naming system or other means of identification, just as individual residences have no numeral system of location. Running water is often available only at some distance, because water lines are usually not supplied to squatter settlements by municipal governments, and communal spigots are frequently the only source of water.

The lack of sanitation is a major problem in squatter settlements. Trash and debris may accumulate in large piles, attracting rats, and residents who attempt to burn such refuse risk burning the entire community down, because fire protection is seldom provided to the squatter areas in many cities. The most serious threat to public health is the lack of proper sewage disposal. Outdoor toilets that are improperly situated and may contaminate the local water supply are sometimes the source of epidemics that may then spread to other parts of the city. Outbreaks of cholera have occurred in many squatter settlements in tropical **regions**, and because most residents of the squatter neighborhoods cannot afford medical care, and in fact local clinics or other health care facilities are rare there, almost any disease tends to spread rapidly in such congested, unsanitary conditions.

High crime rates, fire, and **natural hazards** like **earthquakes** represent other challenges to residents of the squatter settlements. Earthquakes and landslides are a constant threat to many residents of squatter settlements, especially in Latin

America where barrios are often built on steep slopes or ravines. The soil in such locations is frequently unstable, and during heavy rains or tectonic activity can slide downslope, taking the entire community with it, resulting in heavy loss of life. Fires are a common source of destruction in squatter communities for several reasons. Because utilities like natural gas and electricity are seldom available, the ramshackle houses, made of highly flammable materials, are heated with open fires, and these are used to prepare meals as well. Houses are constructed with little if any open space between structures, allowing fire to spread quickly throughout an entire neighborhood. A second frequent source of fires is the practice of stealing electricity by throwing exposed copper wiring across a live electrical line, and thereby tapping into the electrical grid. Although this is obviously a dangerous practice because of the danger of electrocution, it also causes fires because the exposed wires frequently short out and shower sparks onto the roofs of surrounding homes. Crime is often rampant in squatter communities, because little if any police protection is provided to the residents, and a considerable number of people living as squatters may be engaged in illegal activities, from petty crimes like pickpocketing, to more serious violent criminal acts related to drug trafficking and armed robbery. Women in particular are vulnerable in squatter settlements, not only because of the lack of police protection, but also because social and family support networks are often much weaker than in the rural villages from which they migrated.

In the 1970s and 1980s, many governments attempted to eliminate squatter settlements, sometimes by forcibly removing or relocating the residents. In recent years, however, some urban leaders have recognized that squatter settlements in many cases are the only realistic solution to the chronic housing shortages their cities face. Moreover, the residents of these communities contribute to the local economy by contributing to the supply of labor and to the local tax base, often without benefiting from many of the services the city provides to other neighborhoods. In some cases, city leaders have attempted to regulate the squatter settlements and provide some basic services like trash collection and general health care to limit the spread of disease and improve the quality of life there.

Straits, Passages, and Canals

Straits and passages are relatively narrow, navigable waterways lying between two points of land and connecting two navigable, and usually larger, bodies of water. In large archipelagoes like Indonesia and the Philippines, the waters between islands are typically called "passages," or sometimes "channels." Straits and passages are terms that generally refer to naturally occurring navigable bodies of water. A canal is an artificial strait or channel, constructed to connect two bodies of water. Canals may be quite small and lie entirely within the **boundaries** of a state, but a canal that joins two large bodies of water such as seas or **oceans** typically has great commercial and military importance. Straits and passages also frequently have strategic importance and form **choke points**. Historically, access to these stretches of water were a major element of the **Law of the Sea**, as they are easily controlled by the countries that hold the land on one or both sides of the strait or passage, and fall within the territorial waters of the adjacent countries. Blockading of the waterway can have serious strategic and/or commercial impact on states whose economies are dependent on transit through the waterway, and this action is often considered an act of war.

Possibly the most important strait in the world is the Strait of Hormuz, linking the Gulf of Oman and the Persian Gulf. The strait is about 30 miles wide at its narrowest point, and separates Iran from Oman. Several small islands occupy the strait, and it represents one of the busiest stretches of water supporting commercial shipping in the world. The strait controls access into and out of the Persian Gulf, and thus is a key artery to the global market for the world's most vital oil-producing region. Approximately 40 percent of all petroleum transported via water in the world passes through the strait, making it a key location in the **geopolitics** of the Middle East. Although the strait lies within the territorial waters of Iran and Oman, ships are granted the right of innocent passage according to international agreement. So many large tankers ply the waters of the strait that zones have been designated for use when either entering or exiting the Persian Gulf, to lessen the chance of collision. Tensions in and near the strait have been high for almost 20 years, due to the antagonistic relationship between the Iranian regime and Western powers, especially the United States. When Iranian naval forces mined portions of the strait in 1988, resulting in severe damage to a U.S. vessel, the United States responded by destroying Iranian oil platforms and gunboats. Iranian leaders periodically threaten to close or limit access to the strait, a threat that Western powers take quite seriously.

For centuries the Turkish Straits have been two of the most coveted and vital waterways in the world. The Bosporus connects the Black Sea with the Sea of Marmara, a small body of water lying between the Bosporus and its partner strait, the Dardanelles. The latter connects the Aegean Sea and the Sea of Marmara, allowing passage into the Mediterranean Sea and onward to global shipping lanes. The Bosporus is the smaller of the two passages, at less than a half-mile wide and about 14 miles long. The Dardanelles, called in ancient sources the *Hellespont,* is about 30 miles long and on average is approximately three-quarters of a mile wide. The Turkish Straits were the focus of the so-called "straits question" in the 19th

One of the most strategic straits in the world is the Bosporus, shown here as it passes through the city of Istanbul, Turkey. This narrow stretch of water controls access between the Black and Mediterranean seas, and carries a large amount of the region's goods to the remainder of the world. (Tomas Sereda/Dreamstime.com)

century, when European powers sought to establish a strategic balance on the continent that would limit the possibility of war. The London Straits Convention of 1841 resulted in the de-militarization of the Turkish Straits, a move supported by most major European powers in the hope that the prohibition of warships from the straits, under any flag, would lend support to a weakened Ottoman Empire. The Convention had the effect of denying Russian warships access to the Mediterranean Sea, and ultimately increased tensions between the Russian and Ottoman empires. Jurisdiction of the straits is now governed by the Montreux Convention Regarding the Regime of the Turkish Straits, a series of agreements concluded in 1936, which granted Turkey full military control over the Bosporus and Dardanelles.

Other notable strategic straits include the Strait of Gibraltar, and the Strait of Malacca. The Strait of Gibraltar is a narrow band of water only about eight miles wide that separates Europe from Africa, or most specifically, Spain from Morocco. The strategic value of the strait is the primary reason the British Empire took control of the Rock of Gibraltar in 1713, at the conclusion of the War of Spanish Succession. Control of the Strait of Gibraltar allows a state to control access from the Mediterranean Sea to the Atlantic Ocean, and vice versa, making the Strait of

Gibraltar one of the world's most important choke points. Likewise, the Strait of Malacca, which separates Malaysia from the Indonesian island of Sumatra, is a vital economic conduit that carries much of the oceangoing traffic between the Indian Ocean and the South China Sea. For countries as far away as Japan and South Korea, the strait is a lifeline to the global economy, and a primary source of energy from the Middle East, although the largest oil tankers cannot use the strait due to its relatively shallow depth. By some estimates, a third of the world's trade by value passes through the strait in an average year, making it arguably the most important economic connection in the world. As a result of the enormous wealth traversing the strait, it is also plagued by modern piracy, a growing problem that Malaysia and Indonesia have struggled to contain in recent years.

Canals that connect international bodies of water are strategic bodies of water by design. The two most important canals in the world are the Panama Canal and the Suez Canal, both of which allow for the transport of billions of dollars of goods in an average year. The Panama Canal is a locked canal that connects the Pacific Ocean to the Gulf of Mexico. A locked canal is one where locks, or sealed compartments, allow the vessel traversing the canal to be lifted by flooding the compartment and raising it to the next level, much like a series of steps on a staircase. This has to be done when the terrain the canal crosses is rugged or elevated. Such a canal is much more difficult and expensive to build than one where there is no intervening topographic barrier. The Panama Canal, begun by the French and completed by the United States, carries thousands of ships per year and billions of dollars of goods. Since 1999 it has been under the control of Panama. The Suez Canal, linking the Red Sea and the Mediterranean Sea, is a somewhat different type of canal from the Panama Canal. Because there is no intervening topographic barrier along the western side of the Sinai Peninsula between the terminal ports of Suez and Port Said, the Suez Canal does not require locks. The modern canal was opened in 1869, but Egypt, the country which now controls the canal, did not gain sovereignty over the Suez until 1956. The strategic importance of the canal may be seen in events such as the Suez Crisis of 1956, and the blockade of the canal in 1967, which led to war between Egypt and Israel.

Stream Erosion and Deposition

Streams are predominant shapers of the terrestrial **landscape**. Main streams and their innumerable tributaries are responsible for much of the landscape we see. Even in arid areas, stream water can do significant work when available because of the lack of protective vegetative cover fostering increased surface erosion and

Equilibrium

The concept of equilibrium is of great consequence in geography as in other modern sciences. Systems are explained by studying a factor (or a combination of factors) over time and describing its (their) changes. From this comes a sense of the average condition of a system that can be conceptualized as a balance of factors. In that **Earth** is an open system, mass and energy enter and leave the Earth's system so that external factors disturb equilibrium conditions. For instance, the Earth's energy balance depends on the amount of solar radiation emitted from the sun. If this amount were to increase, Earth's **temperature** would tend to increase. Other perturbations of equilibrium begin within the system. **Plate tectonics**, for example, sometimes cause mountain building, which sets off series of changes in erosion and regional weather. Earth's subsystems are dominated by negative feedbacks that slow and dampen changes away from the habitable equilibrium of current Earth conditions. In the case of increasing solar radiation, there would be increasing evaporation from **oceans**, possibly causing more **clouds** and greater albedo, thus minimizing the change in temperature. The most fruitful geographic application of equilibrium theory has been in **geomorphology** in the explanation of landforms. In this instance, equilibrium theory holds that slope form is shaped by a balance of processes such that an energy balance is maintained over the long term.

runoff. In the absence of significant tectonic disturbance, the overall effect over very long amounts of time is to make the higher places lower by erosion and lower places higher thus "smoothing" the landscape. Within the immense time **scale** related to smoothing it is common for streams to produce hills, valleys, and canyons by lengthening, deepening, and widening their valleys

The landscape has been given a simple classification vis-à-vis stream erosion. The surface area drained by a stream is a drainage basin (or a watershed) and its **boundaries** are drainage divides. Drainage basins exist on vastly different scales and, because streams have tributaries, main streams are associated with the multitude of tributary drainage basins that are nested at various scales one within another. It is through the organization of drainage basins that the flows of mass and energy associated with stream erosion and deposition can be understood. Stream-produced landscapes are segmented into interfluves, valley sides, and valley bottoms. High ground dividing adjacent river valleys is known as an interfluve. At the highest elevation along an interfluve is a drainage divide. Interfluves can be many kilometers across or nonexistent in the case of drainage basins edged by steep mountains. The valley side is the increased slope down to the valley bottom. Valley sides can have slopes of only a few degrees from horizontal or be bluffs or canyon sides. The valley bottom is the surface along which a stream is eroding or depositing and can be negligible such as on a canyon bottom or many kilometers wide.

Streams gain flow in two ways. One is by the organization of runoff water (called overland flow) from thin sheets running off of upland surfaces. Invariably, the water moves downhill via gravity and becomes channelized as the water increases its turbulence with increases in speed on its downslope run. Organization into channels is simple but profound because it allows the multiplication of erosional force many times that of overland flow. The second way in which streams gain flow is through the intersection of their channels with the water table (top of the **groundwater**); a stream might flow year round even through dry seasons.

Erosion within streams takes place in four ways. The first is by direct assault by force of the moving water (hydraulic force). Moving water is quite powerful as can be attested to by those who have toppled attempting to wade through a fast-moving mountain creek in shin-deep water. The second type of erosion is caused by the propensity of streams to erode laterally. The erosion on the outside of meander bends undercuts the banks, the bank materials fall into the stream and are transported away. The third type of in-channel erosion is performed by the materials in transport. Analogous to sandblasting performed by sediments carried during wind transport, stream materials are able to abrade the channel sides and themselves through myriad collisions. The net result is that materials become smaller and smoothed and, ultimately, are mostly silt. Even a cursory examination near streams shows this evidence. The fourth type of channel erosion is corrosion—the chemical assault on rocks. Chemical weathering is elaborated on in the **weathering and mass wasting** article. Stream corrosion occurs primarily via solution and hydrolysis. In most situations, erosion via corrosion is minor compared to the other three erosion mechanisms.

Streams have widely varied rates of erosion depending on the stream's dynamics. Increased erosion is caused as water turbulence increases as a result of either roughness of the channel or increase in the flow velocity. Of course, the total amount of material removed by stream erosion is also related to the volume of water.

Of considerable interest is the erosional profile of a stream. Given long enough, and in the absence of significant tectonic forces, a stream will form a concave longitudinal profile. That is, the slope of the stream with the horizontal is steeper near the headwaters than near the mouth. This results in more erosion near the headwaters and more deposition near the mouth. A smooth profile is known as a graded profile and, before a stream achieves a graded profile, it erodes away the breaks in profiles represented by falls and rapids.

Even without rocky channels, flow within streams is anything but smooth. Channel sides and bottoms are not perfectly smooth and this unevenness causes water to be turbulent and flow in swirls known as eddies. The noses of these eddies concentrate kinetic energy and provide increased force for the accomplishment of erosion.

This small river in southern Utah illustrates the erosive power of water in an arid environment. (Pmccall/Dreamstime.com)

When the speed and direction of eddies are averaged over minutes, streams are known to exhibit helical flow, which is the corkscrewing of water downstream. Helical flow leads directly to the meandering of streams.

The largest materials that can be transported at a particular time and place are pushed along the bottom via traction. Slightly smaller pieces skip along the bottom in parabolic arcs (saltation). The smallest pieces are transported relatively rapidly and for longer distances by suspension. Finally, the corroded materials are moved in solution. Of these modes, the suspended load is the largest transported component in most streams. The transportation of eroded materials in streams is analogous to the processes of wind erosion except that wind is not capable of transporting dissolved materials.

Especially when streams flow over their own deposits they are actively changing their channels over time because helical flow does not allow water to flow straight in a channel for more than a dozen or so times its channel width. The fastest, deepest flow in a channel is known as the thalweg and swings from side to side in the channel. The thalweg is usually near the outside of stream curves and this concentration flow helps to erode the stream bank in that direction. Over time, the meanders themselves migrate, increase their amplitudes, and cut themselves off, and new meanders form.

Sometimes streams flood sending water out of the channel. The water flowing over the general landscape is not as fast as the in-channel flow and the water loses some of its capability to move materials. So, materials are deposited near streams and given the general name of "alluvium." Many times alluvium is composed of silt. The flat areas along channel peripheries are floodplains and they are depositional features resulting from repeated flooding. In a sense, floodplain deposits are alluvial materials waiting to be transported by a larger flood at some later time. As streams meander and flood along floodplains there is a large variety of landforms created out of the depositional materials. These include natural levees caused by plentiful sediments as they depart the channel during floods and oxbow lakes that represent the cut off former stream meanders.

In locations actively generating large amounts erosional materials compared to the amount of stream flow, the flow becomes braided. Braided streams are analogs to the interwoven paths of braided hair. Instead of a single stream channel there are many stream channels that merge and diverge and readily change pattern with flow increases. Noteworthy examples are meltwater streams transporting the prolific erosive debris of glaciers and rivers flowing in drier climates such as over the Great Plains of North America, where incomplete weathering makes for sandy stream beds and limited flow is able to transport the material only intermittently.

Material carried along by a stream is frequently deposited as the stream nears its mouth emptying into a lake, an ocean, or a dry basin. This is due to the slowing of the water due to concave longitudinal profiles as explained above and results in a delta that has distributaries branching from the main stream in channel forms that fit within a shape fancied to resemble the Greek letter Δ. Consider the gigantic deltas of the Mississippi, Ganges, and Nile rivers. The materials deposited by these rivers cover hundreds of square kilometers.

Supranationalism

A form of political organization that partially transcends the sovereignty of its constituent units, increases the permeability of their **boundaries**, but that falls short of becoming a complete **federation**, in that the states so incorporated retain a significant measure of political independence. A cluster of states that joins together to promote trade and economic development, or to enhance their collective security, is not necessarily a supranational group. To reach this distinction, a greater degree of integration beyond simply promoting mutual interests must be sought and achieved. Some commentators consider supranationalism to be a form of political **globalization**. The first use of the term in a legal document appears to be in the Treaty of Paris in 1951, which formally established the European Coal and Steel Community (ECSC), setting the stage for the creation of the European Economic Community (EEC) with the promulgation of the Treaty of Rome in 1957. After adding members steadily for several decades, the EEC became the European Union (EU) under the Masstricht Treaty in 1994. Other international organizations may be "supranational" to a limited extent, but to date the European Union remains the most ambitious example of supranationalism.

The motivation behind the drive toward supranationalism in the European **region** has several points of origin. In the aftermath of World War II European countries sought some mechanism beyond simple treaties to ensure a lasting peace on the continent and to avoid a repetition of the massive destruction of the first half of the 20th century. Economic integration was viewed as such a mechanism, because by tightening economic (and ultimately, political) ties, the basis for conflict would be undermined, a notion that is the foundation of *economic peace theory* in the field of international relations. This is made quite clear in the famous statement by the French politician Robert Schuman, in which he held that economic integration would make war "materially impossible" in Europe. In addition, some of the smaller European countries were at a disadvantage in competing against the larger economies and sought to cooperate as a larger, single economic space to be more competitive. This was the case with the Benelux customs union, formed in 1948 between Belgium, the Netherlands, and Luxembourg. The reduction of customs duties and other impediments to trade between the Benelux countries led to rapid economic growth and other benefits, and the Benelux union along with the ECSC provided the basis for further integration. The three Benelux states then joined with the larger economies of West Germany, Italy, and France in forming the EEC. The EEC was much more than a simple customs union, because from an early point it attempted to coordinate economic policy among its members, some of whom initially resisted what they viewed as efforts to erode their national sovereignty.

The EEC, or Common Market as it was labeled in the British media, had as its central goals economic advancement, but it also acquired political dimensions early in its development. For example, the Treaty of Rome, which formed the EEC, allowed for additional members to join the organization, but no new members were accepted until the United Kingdom, Denmark, and Ireland were added in 1973. The United Kingdom had applied for membership as early as 1960, but the French government under Charles de Gaulle had consistently blocked British membership, fearing that granting the British a place in the organization would allow for too much Anglo-American influence in formulating EEC policies. It was also the case that only those states that were governed according to democratic principles such as universal suffrage, free media, and other democratic institutions were welcomed into the EEC. In the 1980s European countries that had suffered under dictatorships, some for decades, were allowed to join the EEC because the dictatorships had collapsed and been replaced by representative democracies—these included Greece, Spain, and Portugal. Moreover, the Single European Act promulgated in 1986 set the stage for the most significant advancement in cooperation among the EEC's economies in decades—the creation of a single European currency backed by a central banking system. For the British and Danes, this step represented too great a loss of national identity and sovereignty, and those governments opted out of the monetary union, and to date retain the use of their national currencies.

The Maastricht Treaty created a truly supranational organization in which member states have relinquished a significant amount of control over their borders, and for those in the monetary union, over their economic policy as well. The so-called "Schengen area" allows for a borderless region across most of Europe, including several countries that are not members of the European Union. A handful of EU member states are not party to the Schengen agreement but are expected to implement its provisions at some point in the future. In effect, virtually the entire European region now functions as a single entity in terms of border security, representing a population in excess of 400 million people, or only slightly less than the population of the United States and Mexico combined. In addition to a common currency, the euro, the EU has developed other characteristics of a **nation-state**, such as an anthem, a flag, and a number of governing institutions, including a bicameral legislature composed of the European Parliament (EP) and the European Council. The members of the EP are directly elected by European voters. Thus, in a number of areas the EU appears to have acquired the powers of a **federation**, but there have been signs in recent years that the public in a number of member states is unwilling to grant supreme sovereignty to the organization. In 2005 both the French and the Dutch rejected ratification of the European Constitution, resulting in a setback for those who wished to invest and centralize more political power in the EU political institutions. Nevertheless, the EU represents a special accomplishment in terms of political and economic integration, a feat acknowledged in the organization's self-description on its official Web site:

> The European Union (EU) is a family of democratic European countries, committed to working together for peace and prosperity. It is not a State intended to replace existing states, but it is more than any other international organization. The EU is, in fact, unique. Its Member States have set up common institutions to which they delegate some of their sovereignty so that decisions on specific matters of joint interest can be made democratically at European level (capitalization as in the original).

The European Union's experiment with supranationalism is a work in progress, and few observers, both within Europe and outside the region, would attempt a prediction as to how broad the geographical sweep of the organization will ultimately be, or how successfully future challenges to even greater integration will be met. One of the more vexing problems the EU faces is the question of Turkey's ascension to full member status. The Turkish government first applied for full membership in the late 1980s, and after 20 years it appears Turkey remains at least a decade away from joining the organization as a full member, although the country has had associate membership for decades. Turkey's turbulent relationship with

Greece and the continued uncertain status of Cyprus remain major obstacles to Turkish ascension. Another challenge lies in the realm of legal and social policy. There is considerable variation across the EU countries regarding legal codes and social standards, and it remains unclear how the organization will reconcile these differences, or even if the establishment of a uniform legal code is desirable. In some European countries, laws governing so-called "victimless crimes," like drug use and prostitution, are quite liberal even by Western standards. For example, in the Netherlands and Germany prostitution is legal, and in a number of EU countries the limited use of certain drugs for personal use is tolerated, if not technically legal. Yet in other EU members, such activities are illegal. Standards and laws governing environmental degradation also vary widely in many cases. For the time being, laws regulating these activities remain within the jurisdiction of individual member states; but as integration proceeds, it may become necessary to codify across the EU region laws dealing with social behavior.

Although the EU currently stands as the supreme example of a supranational polity, it is quite possible that other organizations will attempt to replicate, to one degree or another, many of the achievements of Europeans. The advantages of a regional bloc of states all using a single, centrally backed currency have become obvious with the advent of the euro, which now vies with the dollar as an international currency. The newly formed Union of South American Nations (UNASUR) clearly aspires to emulate many of the aspects of the EU, and although a monetary union has not been explicitly articulated as a goal as of 2010, the member states of UNASUR appear committed to dismantling many of the barriers that impede the movement of people, goods, and capital across the South American region. There is potential for several such organizations emerging, and political and economic cooperation on the supranational level may be a trend of the future.

Sustainable Development

A philosophy and set of behaviors that strive to find methods of maintaining current quality of life while simultaneously minimizing the degradation of environmental and social systems to perpetuate these systems for future generations. The term "sustainable development" is sometimes generally associated with the stewardship of **natural resources** and preservation of wildlife habitats. Certainly wise and judicious use of energy and other products, and maintaining a balanced approach to the exploitation of ecosystems, is one of the core goals of the philosophy, but the concept is considerably broader than just these objectives and extends beyond just environmental concerns. Indeed sustainable development is a holistic

Green Revolution

The term "green revolution" applies to the funding and development of advanced agricultural technology by Western agencies, and its subsequent transfer to the developing world. The "revolution" began in the 1940s, when the Rockefeller Foundation established a research center in Mexico devoted to developing high yield varieties (HYVs) of wheat and corn. The new hybrids were genetically engineered to store more calories in the grain and to produce larger grain heads. In the early 1960s the focus shifted to rice production in Asia, and a "miracle rice" hybrid emerged from research that dramatically increased yields in India, the Philippines, and other rice-growing countries. While Green Revolution programs are generally acclaimed as successful and have certainly enabled countries with high rates of population growth to feed their people, critics argue that there are also many negative side effects. For example, the HYVs typically require large amounts of water, fertilizer, and pesticides, which tend to damage the local ecological system, especially in the case of rice production. In addition, proponents of **Malthusian Theory** claim that while the new technologies have staved off famine currently, they cannot do so indefinitely. Some scholars have noted that the hybrids frequently contain less nutrition than traditional varieties and are susceptible to crop failure.

notion designed to achieve a state of equilibrium between development and conservation across the spectrum of human experience. Advocates of sustainability seek to avoid the pitfalls and degradational consequences of the **tragedy of the commons**, a situation in which a resource is mismanaged and potentially lost because of a lack of communal oversight and concern. Both raising awareness and implementing practice are vital to achieving sustainability, and the United Nations has been a strong proponent of the concept, especially through the vehicle of the UN Earth Summit meetings held in Rio de Janeiro in 1992 and Johannesburg in 2002. The meeting in Rio generated Agenda 21, a detailed set of objectives meant to promote and highlight the potential benefits of sustainability, as well as to articulate mechanisms for reaching an equilibrium of production and consumption economically, ecologically, and socially.

The use and status of "capital" lies at the heart of sustainability. There are at least three types of capital that are important when considering sustainability. *Economic capital* represents capital in the standard way that economists use the term; that is, something of economic value that serves as the means of production. This might mean financial capital, or it could indicate the physical attributes of production like machines, factories, tools, etc. *Social capital* generally refers to the social connections and shared communal values found in any group of significant size. This capital has value because the interrelationships it represents provide cohesion and integration to the social structure of a community or organization, without which the group tends to fracture and perhaps disperse. The sources of such capital are many and include marriage and family relationships, religious organizations, philanthropical and service groups, and many other platforms that individuals might use to establish relationships with the larger community that contribute to the formation of civil society. *Natural capital* is the entirety of the naturally occurring systems that provide support to all forms of life on the planet. Any and all ecosystems are the basic components of natural capital, which not only provide natural resources for use by humanity, but also are the "means of production" for the products of the global "natural economy," commodities like air, pure water, productive soils, etc. The system of natural capital functions as a whole, is deeply integrated and interdependent, and is dependent on diversity to maintain its value.

The notion of sustainable development emphasizes the valuation and preservation of the three types of capital, and seeks to integrate this approach through modification of lifestyles, cultural values, and activities. Protecting and conserving the natural environment is obviously part of maintaining natural capital, but sustainable activities that contribute to this objective also may contribute to the preservation of one or both of the remaining forms of capital. For example, recycling of paper, plastic, aluminum, and other materials clearly is a means of protecting natural capital—by recycling paper, less wood is required to manufacture new paper products, meaning that fewer trees must be cut to provide wood pulp, a basic ingredient of paper production. But recycling paper products also contributes to the preservation of economic capital, because it means energy will be saved by avoiding the harvesting of additional trees, fewer pulp mills will be required to process trees (which saves not only the economic capital of energy and construction costs, but makes another contribution to natural capital by lowering levels of air and water pollution), and fewer vehicles will be manufactured for transport. Recycling also benefits the accumulation of social capital, because civic groups (local Scout troops, high school ecology clubs, religious organizations) may "adopt" a local recycling drop-off site, thereby constructing social support networks and encouraging the community at large to engage in the process, which may lead to additional accumulation of forms of social capital. In this manner the practice of sustainable development, according to its supporters, moves humanity toward a more balanced, measured means of development.

Those who criticize the sustainable development approach find weaknesses in the way the philosophy is presented and structured, and believe that economic and social policy cannot be developed on the basis of a philosophy that remains vague and, to some critics, based on a utopian vision of economic motivation. According to this view, sustainable development is too reliant on altruistic behaviors that may be fashionable and affordable in affluent countries where economic standards are high and basic needs are easily and readily satisfied, but are not likely to be popular in those societies attempting to rapidly raise the average standard of living, and where population pressures place a strain on available resources. They point to the difficulties of convincing residents of marginal regions like Africa's Sahel not to engage in practices that exceed the **carrying capacity** of the local environment as one example among many from developing countries. A few commentators even suggest that the supporters of sustainable development hold concealed agendas designed to monopolize and control use of the planet's resources and perpetuate the dominance of the economically developed economies of the world over those located in poorer regions. They believe that the focus on cataloguing and documenting resources, an important process for establishing and managing an inventory of natural capital, is merely a cover for advanced **nation-states** to gain control of resources, and that sustainability is simply a mechanism for preventing the world's poorer states from eventually challenging the existing global economic order.

T

Temperature

All matter contains heat (whether or not it is hot to the touch). How much heat is present relates to a combination of the mass of an object, its temperature, and how that mass is arranged. "Heat" and "temperature" are commonly used as synonyms. Instead, they are related but separate concepts. Heat is the aggregate internal energy of an object and is transferred by conduction, convection, and radiation; these mechanisms are examined in the **heating and cooling** article. The rate at which heat is transferred is expressed in watts (joules per second) and is maximized as temperature differences between objects become large. The net transfer of heat is always from the hotter to the colder object.

All molecules in the **Earth** system vibrate and as they gain heat they vibrate more rapidly. This energy of motion is kinetic energy and is responsible for radiation of electromagnetic energy. Temperature is an expression of the average kinetic energy of a substance and this measurement provides a ready determination of how much and which wavelengths of energy are being emitted.

There are three common ways by which temperature is measured. The first dates back to Renaissance Europe and is the liquid-in-glass thermometer. In this device, a liquid is placed in a sealed glass tube with a bulbous reservoir on the bottom. The liquid volume visibly expands on heating and contracts on cooling. The first thermometers contained water, but now have liquids with freezing temperatures below freezing. Mercury was used for many years but its toxicity and cost has caused it to be largely replaced by red-dyed alcohol. The other two devices became quite common in the late 20th century because of their abilities to measure temperature without a human being present. The second device is a thermistor, which is based on precise calibration of electrical resistance in a metal probe. Small—typically a centimeter or less long—thermistors are frequently made of parallel platinum alloy wire heads together with semiconductor paste. A small current is passed from a source to the thermistor and the resistance measured. The amount of resistance decreases considerably over small increases in atmospheric temperatures so it is a simple matter to calibrate the thermistor to known temperatures. The third type of device is a radiometer, which is a **remote sensing** device that detects invisible thermal infrared energy. By noting the most plentiful emitted wavelengths from a target, its surface temperature can be determined. Satellite radiometers are able to "take the temperature" of clouds, Earth's water and land surfaces, and vertical atmospheric temperature profiles.

There are three scales of temperature in common use. They are Celsius, Kelvin, and Fahrenheit. The Celsius scale has 100 degrees between the freezing and boiling of pure water at sea level pressure. This is the scale in the most common use in the world and is the International System of Units (metric system) official scale. A Kelvin is the same "size" as a degree Celsius but the Kelvin scale has different zero point. Kelvins are based on absolute temperature. Every molecule vibrates until it is as cold as absolute zero (–273°C), which is defined as 0 Kelvins. Average room temperature is 293K (22°C). Kelvins are used heavily in the sciences, especially physics, chemistry, and the atmospheric sciences. The Fahrenheit degree is about 5/9 the "size" of a Celsius degree and there are 212 degrees between freezing and boiling with 0°F being considerably colder than 0°C. Very few countries besides the United States use Fahrenheit as their primary temperature scale.

"Official" air temperatures are usually measured in an instrument shelter that is at about eye height and shades the measurement device from the sun and **precipitation** while allowing the free circulation of air around the instrument. If, for instance, a measurement device is exposed to the sun, it is measuring the air temperature *plus* absorbing solar radiation increasing the heat content of the device.

The lowest official air temperature near Earth's surface was –80°C in Antarctica and the hottest was 58°C in Libya. Not all temperature change is related to **solar energy**. Some large temperature changes are associated with the rapid onset and ebbing of chinook winds; Spearfish, South Dakota experienced a 27°C temperature rise in two *minutes*.

Air temperatures are governed by a variety of factors. Although it is true the amount of solar radiation (controlled by latitude) is key, a number of other factors need to be considered. They include a **location's** position in the global circulation system, altitude, surface material properties (specific heat, albedo, etc.), topography (roughness of **landscape**), aspect (direction of slope with respect to the sun), and vegetation. Land surfaces heat and cool much more rapidly than water in daily and seasonal cycles so that the most notable temperature extremes are associated with land surfaces.

Where are the coldest places on the planet? In a simple sense, the answer is "the poles," but there is quite a difference between the ice-covered Arctic Ocean at the North Pole and the 2.8 km of glacial ice at the South Pole. Some energy is able to reach the atmosphere from the Arctic Ocean. Thus, temperatures are significantly warmer in the North Polar region. In fact, the coldest temperatures in the Northern Hemisphere are over land in Greenland and northern Siberia. These places are known as the "cold poles" and have significance in the way wind systems are generated.

Where are the average hottest places on the planet? The equator is the latitude with the highest average solar angle, but cloudiness prevalent in the intertropical convergence moderates equatorial latitudes. The warmest places on the planet are associated with the land areas of the subtropics. The combination of relatively high solar angle and cloud-free conditions generated by the subtropical highs make the region of 20° to 30° north and south of the equator the hottest latitudes.

Considering daily and seasonal cycles, temperatures are quite variable over the planet. However, Earth temperatures (except near volcanism or fires) are far below boiling and, in most latitudes, above freezing, thus allowing photosynthesis and a plentiful chain of life.

Territoriality

The claiming and controlling of space by animal species, including humans. The space is defined by **boundaries**, and such delimitations may be clearly indicated, or may be indistinct and alter with changing circumstances. The dimension of territorial space ranges from personal space claimed by most individuals to sovereign political space represented by the **nation-state** or even supranational organizations. The dominance of territoriality in the organization of human political space is clear from the fact that almost all of the **Earth**'s land surface, and at least a portion of the **oceans** and seas, are separated by legal borders into territorial states, all of which claim complete sovereign authority over the space contained within those limits.

People encounter the concept of territoriality on a daily basis in the form of personal space. Everyone possesses what political geographer Martin Glassner labels an "envelope" of territory immediately around his or her person, which varies in size according to culture. Latin Americans, for example, tend to have smaller personal spaces and stand much closer to one another when communicating than North Americans typically do, and females in most societies generally establish smaller personal spaces than males, especially when interacting with other females. This personal territory may be violated only by friends, relatives, and lovers, and strangers who intrude into an individual's personal domain will be judged to be aggressive and possibly threatening, often resulting in either flight or a hostile response on the part of the "owner" of the space. The concept of "private property" extends territorial space further, and this form of territorial claim is supported by law in most societies. Unlike personal space, the boundaries of our claims to private ownership of space are supported by clearly demarked divisions shown on property deeds, and often reinforced by structures that signal the limits of the space and inhibit violation of the territory, such as fences or walls. Unauthorized and deliberate trespassing into this territory is considered a criminal act in many countries, another indication of the serious nature of territorial control in human culture.

The origin of territoriality has been a topic of intense debate among political geographers, anthropologists, and sociologists. Some argue that territorial behavior is genetically encoded in humans and other species, and therefore humans are "programmed" to establish and defend territorial claims. The best known proponent of this view was the anthropologist Robert Ardrey, who argued in his seminal work the *Territorial Imperative* that humans are inescapably territorial like many other animals, and that such behavior is innate. In Ardrey's view, the array of nation-states that now occupy most of the Earth's surface are not only a necessary organization of political space, they are inevitable. Many others take issue with this position, and hold that the desire to claim and defend territory is not instinctive but rather a learned behavior. Regardless of whether territoriality arises via "nature or nurture," few social scientists would suggest that such behavior is inconsequential in human relations or history. Indeed, Robert Sack has proposed that territoriality is pervasive in human experience, and is found on some scale in all cultures, and he suggests that territoriality is a strategic behavior designed to exercise control over a given space. As such, territoriality may function to reinforce **cultural identity** as well as nationalism in some groups, and the desire to control a specific space, perhaps an historic "homeland," becomes a key component of belonging to the group.

The expression of territoriality in space has a profound impact on human relationships. The features that delimit territoriality, such as international borders, serve to reduce the influence of **cultural diffusion** and other types of interaction, limiting to some degree contact with, and understanding of, people who reside outside of one's own territorial space. Of course, violent conflict over territory is a recurrent theme in human history, and the desire to control larger spaces has led repeatedly to warfare and **imperialism**. Emotional attachment to a specific territory, especially one that acquires the status of a **sacred space**, can extend across many generations and many thousands of miles, resulting in seemingly intractable and endless struggle. An example is the "holy land," where conflict from the first Crusade in 1095 to the contemporary hostility between Israel and its neighbors indicates the resilience and significance of territoriality in this region. Whether the drive to control space is part of our genetic code or a behavior we involuntarily acquire from social conditioning, it nevertheless is an inherently geographical phenomenon, and one that shapes virtually all aspects of the human experience.

Thunderstorms

A thunderstorm is an intense convectional storm with accompanying thunder, lightning, and **precipitation**. At any one time there are, perhaps, a couple of thousand thunderstorms in progress over **Earth**. Although thunderstorms can be dangerous, it is important to note that these small disturbances help maintain the energy balance between the surface and the upper troposphere. Moreover, there are many **locations** where thunderstorms are the primary mechanism causing precipitation.

Most thunderstorms are the result of localized instability. The most common type of thunderstorm is called an air mass thunderstorm. These are formed in warm, moist **air masses** and frequently occur in places dominated by maritime tropical and equatorial air masses. Scattered afternoon thunderstorms are quite common in the southeastern United States in the summer because of surface heating causing unstable maritime tropical air to rise.

There are three phases to thunderstorms. The first is the cumulus stage. In this stage, cumulus and towering cumulus clouds are formed by condensation and sublimation in the strong rise of surface air through the unstable air mass. The change of phase out of water vapor releases latent heat retarding the rate of cooling of the rising air, keeping it warmer and spontaneously rising high into the **atmosphere**. Individual small clouds eventually merge to form a cumulonimbus cloud that is the thunderstorm. In the cumulus stage, all air motion is upward. The second phase is the mature stage. Here, the air has risen to the top of the layer of instability producing cloud droplets and ice crystals along the way. The ice crystals and cloud droplets accumulate enough mass via the precipitation processes to start to fall. The falling precipitation is the start of downdrafts that entrain surrounding air. In that the entrained air is relatively colder and dense it accelerates the rate of motion of the downdraft. The thunderstorm enters the mature phase when it spawns lightning and thunder. In the mature phase, updrafts and downdrafts are approximately equal; this is a tenuous equilibrium usually achieved for only a few minutes. The final phase of the thunderstorm life cycle is the dissipating stage. This occurs as the updraft fails to keep up with the flow of the downdraft as the downdraft entrains air. The updraft weakens to the point where the entire storm becomes downdraft. Tops of tall storms can literally collapse and downdrafts can reach speeds in excess of 160 km per hour. Air mass thunderstorms are usually short-lived. Individual cells last on the order of 30 minutes. However, as one thunderstorm dissipates others pop up as long as there is great instability.

Lightning is an electrical discharge that defines the presence of a thunderstorm. As ice and liquid water is quickly transported through the cumulonimbus cloud there is a net transfer of ions from warmer to colder pieces creating a large place-to-place electrical potential. When electrical potential exceeds 3,000,000 volts over 50 meters there is an electrical discharge that averages on the order of 10,000 amperes. This is lightning that "steps" 50–100 m at a time through a narrow channel about the diameter of the human thumb. The lightning stops for a few ten-millionths of a second each few meters and then steps again. Only about 20 percent of discharges reach the ground with the remainder being cloud-to-cloud lightning. A ground strike establishes a connection with Earth's surface and there are usually multiple round trips transferring large numbers of electrons back and forth. Thunder is the explosion of atmospheric gases as the lightning passes through. **Temperatures** instantaneously reach several times the surface temperature of the sun and cause an explosion of gases. The gases become so excited that they emit intense light for a fraction of a second. The explosion triggers a sonic shock wave that is heard as thunder. Thunder is not usually heard more than a few kilometers from the storm. At night, "heat lightning" is sometimes visible; this is regular lightning but the observer is too far away to hear the thunder. Lightning strikes are dangerous and kill more people per year in the United States than **hurricanes** or tornadoes. The combination of explosions and electrical charges are frequently fatal to living organisms. However, the atmosphere is an exceedingly poor conductor of energy so that damage is minimal with any distance from the strike.

There are several types of significant thunderstorms that can be differentiated from air mass thunderstorms. The National Weather of the United States defines

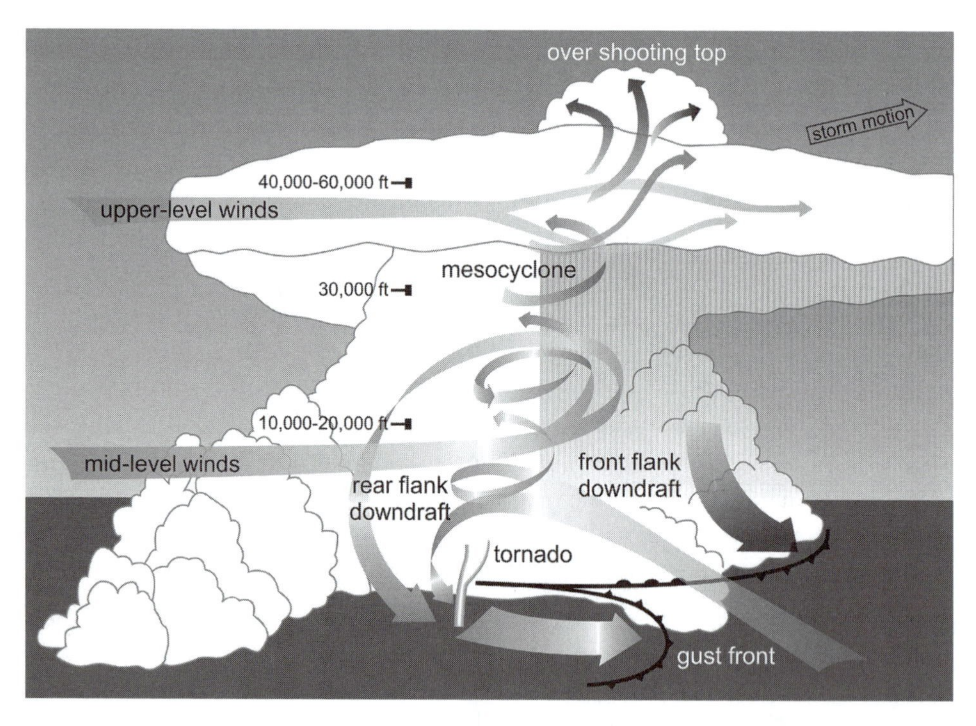

Supercell thunderstorms are caused by unstable conditions in the atmosphere that create very high temperature differences between the cloud and the surrounding air. Extremely high rotational speeds within the supercell are common and when downdrafts occur they can cause flash floods, tornadoes, or hailstorms. (NOAA)

severe thunderstorms as those that have winds in excess of 58 miles per hour, ¾-inch diameter hail, or frequent lightning. Rainfall rates in excess of 4 inches per hour are designated "downbursts" that can lead to flash flooding events in which streams overflow their banks with little warning. Flash flooding kills about twice as many U.S. citizens as does lightning.

Severe thunderstorms tend to be larger than air mass thunderstorms and can last as long as several hours. Sometimes large thunderstorms will align parallel to the passage of cold fronts and drylines. These linear zones can have significant surface wind converging into them and make maritime Tropical air rise violently. "Wind shear" is vital to the mechanics of most severe thunderstorms. If low-level winds into a thunderstorm are moist and very unstable, and if the winds in the middle and upper levels of the troposphere are much faster, then rising air starts to rotate with its ascension. This wind shear makes for supercell storms that have mesoscale cyclones with diameters of 10 km and capable of producing very high winds and tornadoes. Additionally, wind shear helps prolong the storm's duration by tilting the storm so that the updrafts and downdrafts do not interfere with each other. Squall lines of thunderstorms produced along fronts and drylines can exist for hours; downdrafts from one storm act like a miniature cold front to focus lifting and cause another storm nearby. Mesoscale convective complexes are large storms without a focus along fronts. These storms develop in warm, moist surface air in **regions** of weak winds aloft. Storms are created and the outflow from one becomes the inflow to the next. These convective complexes last for hours and can cover thousands of square kilometers. They move very slowly and train along the same paths thus making flooding rain the major hazard.

There is a well-known geography of thunderstorms. Mountains can lift air enough to become unstable (the orographic effect) and mountain ridges usually have more thunderstorms than their lowland surroundings. **Oceans**, at best, have moderately warm surfaces that do not engender extreme instability and, so, have modest thunderstorm occurrences. Thunderstorms occur year round in the warm, moist air of the Intertropical Convergence. Some rainforest locations receive thunderstorm precipitation almost 300 days per year. These locations experience increased cloudiness as instability builds during the day and then a brief thunderstorm. **Monsoon** climates experience large, long-lived thunderstorm deluges during the season of high sun angles. The middle latitudes tend to have most thunderstorms in the summer season because the winter brings much greater stability of air masses. Within the continental United States, thunderstorms are most numerous on the Florida peninsula and least numerous on the West Coast because of the stabilizing dual impacts of the influence of the Pacific subtropical high and the cold California current. Notable exceptions are the middle latitude Mediterranean climate regions where summers have stable air caused by subtropical highs and winter cold fronts bringing instability. Polar regions are noteworthy for the absence of such thunderstorms because polar air masses are usually stable. For instance, there has been only one thunderstorm recorded at Barrow, Alaska (71°N).

Toponymy

The sub-branch of cultural geography that studies the **location**, use, and origin of place names. A *toponym* is the name used to identify a specific location on the **landscape**. An examination of place names in a **region** can provide a great deal of information about the cultural landscape, both past and present, and may provide clues regarding **sequent occupance**. The etymology of a place name may reveal much about those who imposed it. For example, migrants sometimes name their place of settlement after some characteristic of the "home" they left, maintaining a sense of a familiar place, although the actual physical setting may be far different than the original. Hermann, Missouri, for example, was founded by German immigrants in the 1800s, and is named for a German hero who defeated the Roman Empire. Lawrence, Kansas, was founded in the 1850s by abolitionists from Massachusetts, who named the new settlement after a famous anti-slavery Massachusetts politician, Amos Adams Lawrence, reinforcing the status of Kansas as a free state. The influence of Spanish culture and settlement in southern California is evident in place names such as San Louis Obispo, San Diego, and Santa Barbara, along with dozens more. Across the United States, the ubiquitous presence of Native American place names for towns, counties, major cities, and states themselves indicates the lasting cultural impact of the region's original inhabitants. The specific language used in such naming indicates the home territory of individual groups—Tallahassee, Florida is derived from the language of the Muskogee people who lived in the region, while the name of Milwaukee, Wisconsin, comes from the Algonquin languages, spoken by several Native American groups who occupied the northern United States.

The origin of many place names is not as obvious as the examples mentioned above. Toponyms typically have two components, a specific portion and a generic element. In the case of Cartersville, Georgia, "Carter" is the specific part of the name, while "ville" is the generic segment. The use of certain generic components is sometimes unique to specific languages and/or ethnic groups, and linguistic geographers can often trace the origins of migrants through an examination of these patterns. For example, the generic suffixes of *boro, shire*, and *ford* used for town names (Foxboro, Wiltshire, Milford) are typical of English and Scottish settlements in New England. These generic components of place names have been transplanted to other areas of the United States, usually by migrants from the northeast. A suffix of *dorf* or *burg* in a city place name often indicates settlement by German immigrants. Studies of toponyms can uncover cultural relict **boundaries** in the landscape, because the toponyms used by a people often persist on the landscape long after the group has disappeared or moved on, or new boundaries have been formed. Of course, a toponym may provide descriptive information about the physical characteristics of a place as well. The qualities of Death Valley, California, and Pleasant Valley, Iowa, are reflected in their respective toponyms, at least according to those who named them.

Tragedy of the Commons

This phrase is the title of an article by the biologist and social commentator Garrett Hardin, first published in 1968. The article and the concepts regarding the conservation of resources it presents are still widely debated by scholars in the social sciences. The problem that Hardin describes, the proper management of resources held in common, represents a dilemma that has been addressed by philosophers and economists since ancient times. Hardin uses the metaphor of a "commons," a communal pasture, to illustrate how individual self-interest may be in conflict with community interests when utilizing a common resource.

In Hardin's example, a pasture is held in common by a group of farmers, where each farmer is allowed to graze as many cows as he/she wishes. Because there is no cost to any farmer to graze an animal, each will attempt to place the maximum number of animals on the land to gain the greatest amount of "profit" in the form of additional livestock. As each user adds additional cows, the quality of the pasture steadily declines due to overgrazing, but because none of the farmers actually owns the land, and there is no greater authority that can limit the pasture's exploitation, the land is degraded to the point that it can no longer support livestock. This occurs because the users of the resource (the pasture) have no motivation to conserve it, only to maximize their individual gain from it, and the damage from placing additional cows on the land is shared with the other users. According to Hardin, such abuse of a common property is in the rational economic interests of each individual farmer, because the advantage each receives, even after the pasture has been badly overgrazed, is still greater than any costs associated with using the "commons." The "tragedy" in Hardin's title refers to his view that such behavior is inevitable when humans attempt to utilize a "common" resource for which there are no associated costs.

Hardin's argument, in fact, is based on actual examples. The Boston Common, a central pasture located in the heart of the city in the 1600s, was closed to public grazing because of overuse early in its history. Much of the farmland in Great Britain was worked as common land before the enclosure movement of the 18th century, and some historians argue that enclosure was partly motivated by poor land management practices. The taking of millions of seals in Antarctica (a "common" land in that it was unregulated and unguarded) for fur and meat in the 19th century resulted in such a decimation of the animal **population** that the commercial sealing industry essentially collapsed by the early 20th century, and a similar over-exploitation of animals for fur occurred in Siberia during the 18th century. Indeed, much of the world today is still held in the form of "global commons," most notably the **oceans** and seabed beyond the territorial claims of countries, and the **atmosphere** outside the airspace claimed by sovereign states. Even outer space, at least in the zone near the **Earth**, may be considered an emerging "commons." A common space may be degraded not only through exploitation of the resources it contains, but also simply by transit—pollution of the international oceans and atmosphere by ships and aircraft is an example.

Hardin suggests that simply relying on individuals to preserve the commons and manage such areas wisely is unrealistic, because many users are selfish and are "free riders." In response to this dilemma, two schools of thought offer solutions to the "tragedy" that Hardin presents. One camp suggests that placing common resources into private hands presents the best method of conserving and maintaining the resource. Private ownership compels the user to manage the resource efficiently, because an economic cost is introduced with private control—poor use of the resource results in its degradation and loss of value. The second view postulates that control of common resources should be the responsibility of some larger authority, typically a local or national government, or in the case of the global commons, international regulatory agencies. These authorities also add value to the common areas and their resources through the levying of user fees, licenses, and taxes, or they may impose limits or quotas on how much a user may take from the commons. Examples of such regulation are international commercial fishing and whaling limits, entrance fees to national parks, and controls on the emission of air pollutants from automobiles in urban areas.

Transculturation

A concept that holds that the **cultural identity** of subjugated peoples during the era of **imperialism** was transformed, with the indigenous culture absorbing and modifying characteristics of the imposed European identity. In other words, colonized peoples or minorities construct cultural histories for themselves that at least partially incorporate the perspectives and logic of the dominant culture, resulting in a self-representation that is partially rooted in cultural constructs imposed from above. The term was first used in the 1940s and 1950s by Latin American intellectuals like the Cuban ethnographer Fernando Ortiz and the Uruguayan writer Angel Rama, who focused on the role indigenous culture played in shaping the character of national identity. A common manifestation of transculturation is **religious syncretism**, with the resultant faith sometimes gaining recognition as a new religious movement, as in the case of voodoo, Santeria, or the so-called "cargo cults" of the South Pacific. Transculturation is related to the process of modernization, but where the latter term implies an abandonment of cultural values and practices in favor of uniform, contemporary standards, transculturation suggests an adaptation based on the interaction and interpretation of social and historical evaluations from all the included cultural influences. The process of crafting a national identity in a multinational state that may eventually lead to the evolution of a **nation-state**, may be seen as a variation of transculturation. Such identity to be successfully constructed involves the adaptation and reconstruction at some level of values and mores of all member groups, who must by default constitute and recognize its legitimacy.

Some post-modern scholars of literature, such as Mary L. Pratt, have argued that transculturation is in reality a form of resistance to the **ethnocentrism** of the colonizing power, and that the process is most pronounced in so-called "contact zones." Such zones are fundamentally a geographic construct, because they represent the space in which the process of transculturation transpires, and in which elements of a cultural hybridization are incubated and ultimately expressed. In purely geographical terms, such spaces represent the reduction of distance in a **core and periphery** relationship linked to **cultural diffusion**. In a post-colonial context, **globalization** may be viewed as an expression of transculturation. This situation is more complex than the colonial example, in that globalization does not transpire via exclusive binary relationships, nor is there a single "hegemonic" culture, to borrow Pratt's term, that dominates, although some commentators would contend that Western culture itself serves as the hegemonic element in globalization.

Transhumance

A seasonal **migration** among nomadic herders between regions of different elevations. Transhumance is a type of **pastoralism**, but is specific to topography and should not be confused with more general nomadic herding or migration with animals. In the process of transhumance, the migration of humans and animals takes place as movement between pastures located at varying altitudes. Typically, during the warmer summer months, when **temperatures** are warmer and vegetation is available, herds are moved to higher locales. Herders, often with their entire extended families, move with their animals and take up residence at these places, living in either temporary shelters that may be relocated from pasture to pasture, or in more permanent buildings that are used at the same **location** from year to year. In late summer or early fall, again depending on the elevation of the pasture being utilized, the herd is moved downslope to a lower elevation, where temperatures at night are still tolerable to both human and beast, and sufficient plant material is available to maintain the herd. This process may continue over several weeks or even months, as pastureland at progressively lower altitudes is used in a step-by-step fashion, until the lowest land is reached during the coldest period of the year. The animals are kept at this location until warming temperatures in the spring allow for the use of higher grazing land, at which time the cycle begins anew. The most common animals involved in the process of transhumance are sheep and cattle, but other grazing animals may also be moved between elevations.

Many people across the globe practice transhumance, and pastoralists continue to employ this process even in technologically advanced economies—it is not strictly confined to developing regions of the world. In France and Spain sheep and goats are moved seasonally, and transhumance involving dairy cattle remains quite common in some European countries, especially Switzerland, Austria, and the Scandinavian countries. The historical record indicates that the practice of transhumance was common by the early Middle Ages and may have originated in alpine Europe as early as the Bronze Age. Some famous elements of the cultural geography of this region are connected to the use of mountain pastures, such as the use of the alphorn and yodeling, both of which almost certainly originated as means of communication between shepherds separated by wide distances. Only later did these forms of long-distance communication acquire artistic musical expression, with numerous melodies and songs emerging as part of the **folk culture** of Switzerland, Germany, Austria, and other alpine areas of Central Europe. Modern technology has altered some of the characteristics of transhumance in developed regions. For example, in some economically developed countries, those keeping flocks at high elevations during the summer are supplied via helicopter deliveries, whereas for centuries the traditional system relied on a supply system of mounted riders, who would periodically bring provisions to the herdsmen in the mountain pastures. Despite modern adaptations, the basic characteristics of transhumance remain a way of life for thousands of people across the globe.

U

Urban Decentralization

The diffusion of urban structures and functions outward from the urban core, or Central Business District (CBD). This process has occurred to some degree in metropolitan areas in the economically advanced countries because the advent of mass-transit systems in the 19th century, when trams, street cars, railways, and early subways allowed workers to live much greater distances from their places of work than had previously been the case. The development of such lines of transport allowed those with the economic wherewithal to relocate their places of residence further from the compact city center, along the various radiuses of the transportation network, resulting in a dispersion of city dwellers over a greater space. Residential clusters on the outskirts of the city were labeled "sub urban areas," or simply *suburbs*. As the customer base decentralized in urban areas, businesses also began to relocate to the suburbs, especially those that offered goods and services that were low cost and purchased frequently, such as grocery stores, barbers, clothing, etc. Land developers frequently worked in coordination with transit companies, planning residential communities that would have ready access to the transportation system, because many residents continued to work in the CBD and therefore commuted into and out of the city on a daily schedule. In North America, urban decentralization was spurred after the 1880s by rising standards of living, advances in transportation technology, and rapid **population** increases.

The period of greatest urban decentralization in the economically developed world occurred from the late 1920s to the 1970s. The affordability and convenience of the automobile, especially in the United States, resulted in an intensification of decentralization in metropolitan areas. After World War II, this movement became even more pronounced due to several factors, including a high rate of family formation, the availability of relatively inexpensive loans for home mortgages, and most importantly, the construction of the federal interstate highway system. Interstate highways were originally conceived as conduits for troops and supplies to major cities in the event of war, so the roadways were typically constructed directly through the heart of the city, adjacent to the CBD. By the 1970s, interstate bypasses, forming a loop around the city center, were common in many **locations**. This transportation geography led to the emergence of edge cities, so-called asylum suburbs, and related features of a decentralized urban **landscape**. Economic functions were also decentralized, and businesses that relocated toward the suburbs tended to cluster together in a single, concentrated space, the shopping mall, effectively displacing some of the benefits offered by the CBD. In the densely populated eastern seaboard of the United States, urban decentralization has resulted in an almost continuous band of urban development, the **megalopolis**. Some of the negative effects of urban decentralization include the economic decline of the inner city in many instances in the 1950s and 1960s; the appearance of so-called urban sprawl, characterized by vast tracts of land containing nearly identical homes; and greater social isolation of families and communities. Slowed somewhat by **gentrification** and other counter trends, urban decentralization continues to occur in many urban areas.

Urbanization

The process whereby an increasingly larger percentage of a given **population** lives in spatial clusters considered to be "towns" and "cities." The criteria that qualify a settlement as "urban" frequently vary from country to country, meaning that what is considered an "urban place" in one location may not be defined as such in another **nation-state**. In the United States, the U.S. Bureau of the Census considers any place having a concentrated population greater than 50,000 inhabitants to be an "urban area," and any population center of at least 2,500 residents outside of an urban area and having a community identity and name to be "urban." Urbanization is typically associated with a number of related processes: industrialization, economic development, declining average family size, rising levels of literacy and education, and others. In some cases a single urban place will come to dominate a country's urban geography in terms of both population size and economic function—this city serves as the **primate city**. In 2008 the United Nations estimated that for the first time in history, more than half of humanity lived in urban areas, and that the fastest rates of urbanization are occurring in the developing world. Urban places are not located on the **landscape** randomly. Many factors may affect the rate of growth of an urban area, including transportation linkages, **natural resources** located nearby, economic opportunities, and even the local physical geography can play a role in urbanization. Geographers have developed a number of approaches to explain the patterns of urbanization and to determine the characteristics that may influence urban growth and development. **Central Place Theory** holds that urban places are spatially arranged in a hierarchical framework, based on the sophistication of economic functions offered by each city or town.

Urbanization on a significant **scale** first took place in ancient Sumeria, along the course of the Tigris-Euphrates river system. The rich soils of these river valleys

Edge Cities

One of the newest concepts in the study of urban geography in the developed world is the edge city. Edge cities arise along the margin of major metropolitan areas, especially at the juncture of interstate highways, or they may be planned as independent suburban hubs, and in many cases are near large airports. They frequently are characterized by the presence of office parks, shopping malls, relocated corporate headquarters, and more green space than is found in the typical Central Business District. In addition, they are built for easy and quick access by automobile and did not appear on the urban **landscape** until the advent of the car as a major means of transport. The quintessential description of the emerging geography of the edge city is Joel Garreau's *Edge City: Life on the New Frontier*, the first book-length study devoted to the phenomenon. Garreau provides a typology of edge cities, as well as identifying the characteristics that uniquely distinguish this form of urban development. Edge cities have become a common element of the urban landscape in America, and to a lesser degree in other developed countries, indicating the dynamic nature of the modern cityscape in the information era.

allowed for the production of an agricultural surplus, which in turn led to a differentiation of labor, development of social classes, and the emergence of more sophisticated economic and political systems than had previously been the case. The morphology of many early cities was haphazard and chaotic, but some urban areas in the ancient world were surprisingly organized and well planned. The city of Mohenjo Daro, established around 2500 BCE in the Indus River Valley, was clearly a planned settlement with streets and buildings constructed following a grid system, covered sewers, and a municipal garbage collection system! But in general, cities for much of history were places of squalor, congestion, and disease until the late 19th century. Pestilence typically spread quickly in urban areas, due to high population densities and poor sanitation. Industrialization and the **agricultural revolution** that accompanied it initiated a surge in urbanization that continues in many parts of the world, as thousands of rural workers migrated to the cities seeking employment. By the mid-20th century, zoning legislation, **urban decentralization**, the development of **green belts**, and other factors had lowered population densities and improved the quality of life in many urban areas in developed countries, while large cities in the developing world continued to struggle with the problems of pollution, overburdened transportation networks, and shortages of adequate housing and basic services.

The process of urbanization may be driven by numerous forces. The population of urban areas may grow via a relatively high rate of natural increase, or by the arrival of migrants from other regions. The qualities of the physical landscape may coincide with the presence of a **break-of-bulk point**, or trade flows may lead to the emergence of an **entrepôt**, either of which may stimulate economic opportunities and urban growth. This in turn will lead to a higher demand for labor, attracting immigrants seeking employment opportunities. Subsequently, urbanization is compounded by the effects of **agglomeration**, which increase the scale of economic development and bring additional residents to the urban place. A classic example of this process is the growth of Chicago's population (along with other cities in the northern **region** of the United States) between 1910 and 1930, when hundreds of thousands of African Americans migrated from the American South to the industrial urban centers of the Midwest. The desire to escape discrimination in their home states drove many of these migrants northward, but the most important factor behind this movement was economic expansion in the north, resulting in the creation of tens of thousands of jobs. Chicago's black population expanded by nearly 200,000 in these 20 years, or an increase of almost five times the figure of 1910. This domestic **migration** of American blacks to northern cities followed decades of immigration to those locations by Europeans, who mostly came seeking the economic advantages such large, growing urban places offered.

Geographers have developed various models to describe the developmental patterns of urbanization. The **pre-industrial city model** was first articulated in the work of Gideon Sjoberg in the early 1960s. Sjoberg held that prior to the industrial age, social and political authority was concentrated in the center of the typical city, and that many cities in the developing world continue to follow this pattern. In the decades prior to Sjoberg's work, urban geographers had offered several theoretical perspectives on the spatial character of urban places, based on cities in the economically developed world. The first such effort was the concentric zone model, which was based on the development of Chicago in the 1920s. In this case, the urban space is divided into a series of rings around the Central Business District (CBD). Each ring is characterized by a specific type of economic use and activity. In 1939 Homer Hoyt presented a revised urban model, which became known as the **sector model**, or sometimes simply the Hoyt model. The sector model emphasizes the role of transportation linkages in the development of distinct urban functional zones, resulting in "sectors" of development that are associated with transport access and economic functionality. The multi-nuclei model, introduced after World War II, attempts to incorporate some elements from the previous models, while also accounting for the rise of extensive suburban development and other phenomena of urbanization that had appeared by the mid-1940s. All of these models are somewhat flawed, but nevertheless have contributed to our understanding of urbanization processes.

In the last decades of the 20th century, new trends in urbanization were identified, serving to highlight the complexity of the urbanization process. The emergence of so-called asylum suburbs is an example. In these locations residents live along

Rio de Janeiro is the second-largest city in Brazil, and one of the largest in South America. This photo indicates the intensive development of the urban environment in this coastal city, which serves a large hinterland located in the interior of the country. (Photo courtesy of Reuel R. Hanks)

the margins of urban space, enjoying the benefits of urban life from a distance and thereby avoiding many of the disadvantages of residing within the city proper, such as higher crime rates, congestion, higher taxes, higher levels of pollution, etc. In the 1970s and 1980s a trend of "counter-urbanization" was identified by urban geographers. Counter-urbanization occurred as urban populations voluntarily decentralized, with thousands of residents moving away from larger metropolitan areas, especially the emerging **megalopolis** along the eastern seaboard, into smaller communities lying outside the urban zone, or to smaller urban places located at some distance. More sophisticated and extensive transportation networks, including commuter rail linkages and expanded highways, provided the opportunity to live at a greater distance from the urban environment, but to retain employment there, as well as occasionally enjoying the entertainment and economic benefits the urban center typically offers. To offset the loss of residents to suburbs and rural areas, metropolitan authorities frequently use a strategy of **gentrification**, which may have the effect of attracting permanent residents, or revitalizing existing business districts to draw visitors for shopping, dining, and entertainment opportunities.

In many cities in the developing world, extremely rapid and unregulated urbanization has led to many problems, including congestion of transportation systems, high levels of pollution, poor and unsanitary housing conditions, and an inability to provide adequate services to the rapidly growing population. Mexico City is an example of a developing city that faces this spectrum of problems. Between 1940 and 1980 the population of the metropolitan area of Mexico City increased by almost 12 million, growing from just under 2 million to well over 14 million inhabitants. Most of this spectacular expansion was due to immigration from rural villages, as millions of poor Mexicans moved to the capital seeking greater employment opportunities, and settled in **squatter settlements**. The city's industrial base expanded to take advantage of the increasing supply of cheap labor, especially in the cases of petrochemicals, steel manufacture, and durable goods. Pollution from the dozens of large factories associated with this expansion plagued the city's residents for decades—from the 1960s to the 1990s Mexico City was frequently ranked as having the worst air quality in the world. The city's housing market simply could not keep up with demand, and many of the new arrivals could not afford housing in the urban area anyway. The result was a ring of poor neighborhoods composed of shacks around the city, known locally as *barrios*. Many of these districts have no sewer systems, with electricity and treated water in short supply. Many other cities in developing countries suffer from similar problems, due to rates of urban growth that challenge the most dedicated urban planners.

V

Von Thunen Model

An economic land use model developed to explain the variation in the pricing structure of agricultural commodities, based on zones of production. The model is still widely studied and critiqued as a tool of analysis in economic geography, and stands as one of the first examples of applied **locational analysis**. A subsequent body of theory in urban and economic geography emerged from revisions of von Thunen's original work, including Alonso's **bid-rent theory**, an application of von Thunen's agricultural-based ideas to the 20th-century urban environment. Johan Heinrerich von Thunen was a businessman and landowner living in northern Germany in the early 19th century. The fact that von Thunen was a resident of *northern* Germany is relevant, because unlike in Bavaria, the **landscape** in the north is relatively flat, a feature of the local geography that clearly played a part in the formation of his theory. As a businessman, von Thunen was interested in discovering new methods of maximizing his profits, and in particular he put considerable thought into the spatial ordering of production and transport of goods to the market. After a number of years of observation and the development of a mathematical model, he published his ideas in 1826 in his seminal work, *The Isolated State*.

Like all models, von Thunen's theory represents an idealized situation, and he begins with a series of assumptions that he recognizes do not precisely reflect reality. Nevertheless, these approximate the spatial and economic dynamics as von Thunen interpreted them in the pre-industrial economy he was observing. First, von Thunen assumed the existence of a single market town located at the center of a flat, uniform plain. There are no intervening settlements offering marketing opportunities between areas of agricultural production and this market—all production must be marketed and consumed in this single **location**. Second, the agricultural **region** surrounding the market town is homogeneous in terms of all factors affecting production. In other words, there is no spatial variation in soil quality, availability of water, input of labor, fertilizer, etc., and in theory, any crop may be grown for profit on any land around the market. Moreover, von Thunen assumed that only a single means of transportation was available to agricultural producers (an oxcart), and that all farmers living at a similar distance from the market had equal access to this type of transport. Farmers would carry their goods directly across the landscape to the market, as no roads or pathways exist. In order to market their crops, all farmers must deliver them in acceptable condition, meaning that any part of the harvest that is spoiled or stale cannot be marketed and has no value. Each farmer must pay for transport of his goods, and this cost is directly proportional to the distance from the market. Finally, everyone in the system is assumed to be motivated to maximize their profits and has the ability to switch to crops that are more profitable if that option is available.

Under these conditions, transportation costs become a major consideration in determining the margin of profit, and von Thunen argued that a series of concentric zones would emerge around the central market node. Each of these zones or rings would specialize in the production of specific commodities, based on the transportation cost of delivering them to market. The innermost zone would feature production of commodities that had relatively high market value, but which must be transported to market quickly due to their tendency to spoil quickly, and thus lose their economic value. Fresh vegetables and dairy farming for milk would be found in this zone, according to von Thunen. At some distance from the market center, however, increased transportation cost due to greater distance and losses due to spoiled products would force farmers located there to switch to producing an alternative commodity. This threshold distance marks the next zone moving outward from the market node. In von Thunen's original scheme, this second zone of production would be characterized by the production of firewood, which was the main source of fuel at the time. Firewood had a high demand and could be profitably transported greater distance than milk or vegetables because it could be carried in bulk and did not lose any value due to spoilage or damage. Subsequent zones grew various grains, and the final zone was a region of cattle ranching. Beyond the **boundaries** marking this outermost zone there was only unused land, because the transport costs to the marketplace were so great that no commodity could be produced profitably. The production zones thus created would appear as bands of rings circling the market node—economic geographers call these the von Thunen rings.

The model von Thunen proposed offers a means of determining the value of the land located in each ring of production. Von Thunen labeled this value "locational rent," but some economic geographers use the term "land rent." This value can be determined by the formula $R = V - (E + T)$, where R represents the "rent" or value of the land; V is the price the commodity produced on the land brings in the market; E equals the production costs on the farm; and T is the total transport cost. From this mathematical formulation, one can quickly ascertain that transport costs are the key geographical variable—as production is located at an increasing distance from the market, thereby increasing transport costs, the locational rent declines until a point is reached where the commodity being marketed can no longer be produced at a profit. This point marks the boundary of each von Thunen ring.

Beyond this margin a shift occurs to a commodity that has lower transportation costs to make production profitable. The model is not confined to the crops that von Thunen considered, of course; in theory, any group of various agricultural products will experience different prices per unit and varying transport costs based on the nature of the product, and these differences will result in a similar spatial arrangement around the central market location.

There are several weaknesses to von Thunen's model that detract from its usefulness. As he himself acknowledged, the assumptions he makes in the model are not reflected in reality. Even over a fairly limited area **soils** often tend to vary significantly in productivity, and this fact alone would undermine the formation of the neat, precise zones of production the model predicts. Of course, in the real world transport is not made cross-country, but via roadways or by rail, and some producers will have an advantage in transport costs simply by being located closer to these lines of transportation. Furthermore, it is unlikely that even on a plain unbroken by topographic barriers, only a single marketplace would service a region of any significant size. Rather, a hierarchy of market nodes would evolve based on the type of goods offered at each location (see **Central Place Theory**). Modern transport technology has greatly reduced some of the challenges of von Thunen's era, and this has altered the dynamic of transport costs between some types of goods. But despite its shortcomings, von Thunen's model made an enormous contribution to economic geography, and continues to influence the intellectual approach of modern scholars.

Vulcanism

Vulcanism is the term for the motion and landforming processes associated with molten rock. The term subsumes the more specific "volcanism" referring to shallow and external features and processes and "plutonic activity" when it is deep-seated.

As covered in the article about **Earth**, the interior of our planet is ferociously hot. This interior energy reaches all the way out to the surface, builds some **landscapes** and destroys others. Earth's crustal plates float on an upper mantle that is quite plastic and capable of deformation and flow. Concentrated flows far underground sometimes influence the surface on which we live. Other flows reach and spill out onto the surface. When this material is in Earth's interior it is known as magma, and when it flows over the surface it is called lava.

When magma solidifies under the surface the spate of forms created is known as intrusive vulcanism. Large magma intrusions under the surface are known as batholiths. They have sizes greater than one hundred square kilometers and extend downward as far as 50 km below the surface. They are composed of rocks with large crystals (e.g., granite) formed by slowly cooling magma. Batholiths are often associated with the presence of major mountain ranges in that the batholithic intrusion has pushed overlying rock to great altitudes. Well-known examples reside under the Sierra Nevada of California, the Extremadura region of Spain, the Darling Range of Australia, and the Transantarctic Mountains. Stocks and laccoliths are smaller intrusions and are usually grouped with batholiths in the category "plutons." The Black Hills of South Dakota, United States, the Tuscany region in Italy, and some of the High Himalayas are underlain by laccoliths. Minor intrusive forms include sills (horizontal) and dikes (vertical). A noteworthy example of the edge of a sill is the Palisades, which are a line of 100-m tall cliffs on the west bank of the Hudson River across from New York City. Dikes represent igneous intrusions that have been partially exhumed by erosion of the surrounding rocks. Dikes are typified by several 50-m tall walls radiating outward from the remains of a volcano at Shiprock in northern New Mexico, United States.

Volcanic activity leads to the extrusion of lavas over Earth's surface. This is common around volcanoes of all types, but is common over larger areas as the result of lava emerging from mantle plumes and midocean ridges in the absence of volcanoes. These areas have rocks with small crystals indicating the relatively quick cooling of Earth's exterior. Basalt is the rock type that takes up the most area. Great volumes of lava can be produced at various times resulting in thick

Fire

In Greek and other cultures of classical times, fire was thought to be one of the basic elements of the physical Earth. Although our knowledge of the physical **Earth** has become much more complete, fire still has a basic importance to our planet. Fire is natural and needs a fuel source in the presence of oxygen. If Earth's **atmosphere** contained more oxygen, fires would be much more difficult to control. Natural fires have several causes: lightning, spontaneous combustion of organic material, sparks from falling rocks, and volcanic eruptions. Fire has had a profound effect on the evolution of some plant and animal communities. **Climate** and vegetation type combine to make some **landscapes** more fire prone than others. Some life, like Sequoia trees (*Sequoiadendron giganteum*), has been tuned to resist damage by fire and, indeed, the Sequoia's dropped seed cones sprout after being heated by fire. Fire is dangerous to all living things, but humans learned how to—more or less—control fire upwards of 800,000 years ago. This was one of the most significant events in human history because it allowed for the cooking of food and the ready clearing of land. There is no doubt that humans are the most important starters of fire in today's world.

layers called flood basalts. Flood basalts can be several thousand meters thick and cover more than a half a million square km. The landscapes of the Deccan Plateau in India and the Columbia Plateau of the northwestern United States have been eroded out of these flood basalts.

Volcanoes are of four major types: (1) The most explosively eruptive are composite volcanoes made of alternating layers of pyroclastic materials and lava flows. These volcanoes have steep slopes (as great as 35°) and can be quite tall. Famous examples include the Cascade volcanoes of the U.S. Pacific Northwest, Mt. Vesuvius of Italy, and Mt. Fuji of Japan. (2) Shield volcanoes are named for the fanciful resemblance to the cross section of a warrior's shield. These volcanoes are less explosive and have much gentler slopes (commonly less than 5°) so the shield volcanoes are taller than wide and are comprised of much more material than composite volcanoes. Mauna Loa in Hawaii is the largest shield volcano on Earth and has a relief of over 17 km considering its base rises from a deep sea bottom. (3) Lava domes are composed of thick, viscous, high silica lava capable of explosive eruptions. They are considerably smaller than the first two volcano types and have heights less than 600 m; therefore they do not present threats over large areas. (4) Cinder cones are smaller yet and usually less than 500 m in height. They are made of ash and cinders (larger pieces) that hold slopes between 25 and 35 percent. Their smaller size bespeaks the relative brevity of their eruptive lives.

Volcanic eruptions can be of several types. Controlling factors appear to be the amount of silica (occurring as silica dioxide), the pressure of the magma's confinement, and the strength of surrounding rocks. Gases in magma tend to be explosively released if the magma has high silica content and an acidic chemistry. Cooler, basic lavas have lower silica contents and lower **temperatures** thus allowing gases to slowly escape making eruptions more placid.

Volcanoes are **natural hazards** to all manner of human endeavors. Agricultural activities have long hugged the fringes of volcanoes because of fertile andisols of volcanic origin. This symbiosis has been repeatedly broken. There have been many deaths recorded through history. Some populations at risk from the explosion of large composite volcanoes include: Naples, Italy; Tokyo, Japan; Mexico City, Mexico; and Seattle, United States. Many other areas run significant volcanic risk because they are in confined valleys near volcanoes or in narrow strips between volcanoes and coastlines.

It is common to think of the hazard from a volcano confined to the flow of lava. Although lava is a serious hazard, it pales in comparison to others because lava moves by gravity to low areas and does not spew evenly around a volcano in all directions. Lava flows at a variety of speeds depending on its viscosity. The most viscous flows are a few centimeters per day while the fastest flows are on steep upper slopes and may achieve 60 kph. Deaths by lava are rare in today's world but property damage can be immense if lava enters an urban area.

Earthquakes are associated with eruptions and some of these can be major causing widespread destruction. These short-periodicity motions of Earth's crust can result from the rearrangement of magma through injection or extraction of magma causing crustal rocks to flex and break. Eruptions are usually heralded by multiple earthquakes.

Volcanic gases and ash vented during explosions are nothing like the composition of normal air. Water vapor is the most prevalent volcanic gas, but toxic sulfur dioxide, fluorine, and hydrogen sulfide are also present. Jet aircraft turbines are easily fouled by ash and eruptions are scrupulously avoided. A frequent event in an explosion is a pyroclastic flow, which is a lethal mix of gases, ash, and rock fragments barreling down mountain slopes. Pyroclastic flows can travel at over 160 kph so that there is no escape. Famously, pyroclastic flows obliterated the Roman city of Pompeii in AD 79 and killed 20,000 people in Amero, Columbia in 1985.

Volcanic eruptions sometimes devastate low-lying areas by means of lahars, volcanic mudflows. In the 1980 eruption of Mt. St. Helens, Washington, the glaciers on the volcano's flanks catastrophically melted sending a steaming lahar through the Toutle River valley to the Columbia River over a hundred kilometers away.

Knowledge of **plate tectonics** provides scientists with fair knowledge as to the geography of volcanism around the planet. Large volcanoes and associated forms are concentrated along plate **boundaries** and in hot spots. Although scientists know much about volcanoes and keenly observe precursors such as mountain bulging and small earthquakes, there is no method by which to predict the time or severity of eruptions. Such efforts are ongoing in light of the many millions of persons living in volcanic hazard zones.

W

Weathering and Mass Wasting

The materials on the land surfaces of our planet are not everlasting and do not stay in the same place. Far from lasting forever, crustal materials are recycled on a grand scale (see **Plate tectonics**) and reworked on local **scales**. It is these local scales that are addressed in the concepts of weathering and mass wasting. Briefly, weathering is exposure of crustal materials to the atmospheric environment. This exposure eventually changes the materials via mechanical, chemical, and biological means. These materials are decomposed from large to small pieces and the pieces can be more readily transported than the original rock. Ultimately, all rock can be weathered into the tiny pieces called clay (see **Soils**) given enough time. The time is a function of rock type and **climate**. Mass wasting is the downslope movement of materials as caused by gravity. Weathering and mass wasting both happen on time scales ranging from seconds to thousands of years so that many of the actual processes are not frequently noted by humans.

Yet, the evidence is plentiful on the planetary surface. First, rock is not as uniformly "solid" as might be thought. There are many "nooks and crannies," especially in the rocks nearest the surface. These openings include faults, joints, cavities left by solution, and a range of other large and microscopic openings. Various rock types are associated with different types of openings. For instance, limestone tends to have regular horizontal and vertical jointing while this is absent in basalt that has resulted from lava flows. Basalt frequently has lava vesicles—small spaces from which gas has escaped during cooling into rock—while this is not present in limestone. No matter how they are formed, the openings are pathways into which the agents of weathering can enter to hasten the weathering processes. Rock is weathered from within and this vastly increases the area over which the weathering agents can act. Sometimes, the manner of rock openings exerts major control over **landscapes** that are produced. At Bryce Canyon National Park, located in Utah, there are fantastic forms caused by weathering along closely spaced vertical joints in limestone while in other places caverns are created by water dissolving materials along horizontal jointing planes.

Mechanical weathering entails processes that change rock from larger to smaller parts without chemical alteration. Over **Earth**, the mechanical process doing the most massive alteration is known as frost shattering or frost wedging. The greater the number of freeze-thaw cycles, the greater the amount of frost shattering. A unique characteristic of water is that it increases its volume by 9 percent as it freezes. This volume increase causes the water to exert considerable pressure on the walls of the confined openings in rocks. These pressures can exceed 20 million Pascals in winter and high-altitude situations. This is great enough to overcome the resistance of rocks to breakage.

Temperature changes also weather rocks. The diurnal temperature cycle heats and cools rocks and the various minerals within the rocks expand and contract at different rates. Millions of diurnal cycles are needed to mechanically reduce rock, although the intense temperatures from fire or lava can quickly do the same work. In general, temperature changes are accountable for limited weathering.

Salt wedging is a third form of mechanical weathering that occurs when salts form as water evaporates. The salt crystals can grow large enough to push against the rock and weaken its structure; this form is not nearly as potent as frost shattering.

Mechanical weathering alone is incapable of reducing rock to its smallest weathered remnants known as clays; chemical weathering takes care of that. The agents of chemical weathering are varied and work in combination Water is key in weathering because it is brought by the **hydrologic cycle** to all places on the continental surfaces.

Hydrolysis occurs when the hydrogen in the water reacts with minerals and becomes part of the crystalline structure of the rock. A common example is the alteration of the mineral feldspar, a major component of granite. The feldspar is changed into a type of clay known as kaolinite and the original rock is easily crumbled and removed.

As mentioned elsewhere, **Karst** solution dissolves all rock types, albeit at widely varying rates. This is because water readily combines with other common substances. The solution flows away from the original **location** of the rock that was dissolved. Oxidation is a profound expression of the action of water. Just as automobiles rust, so too do rocks. Oxidation chemically combines the oxygen atoms of water with metallic atoms in the rocks with the oxidation of iron-bearing silicate minerals being a widespread example. The resulting iron oxide and other oxides invariably take up more volume and are easier to remove than unaltered rocks.

Carbonation results from carbon dioxide dissolved in water such that the water becomes carbonic acid. Although the carbonic acid is not strong, it inexorably works on calcium-rich rocks to produce calcium bicarbonate that is much weaker and more readily removed than the original rock.

Life is capable of weathering crustal materials. Both animal and plant kingdoms are involved in this type of weathering. Although it can be locally important it cannot have the impacts of inorganic chemical and mechanical weathering, because

Angle of Repose

A term from **geomorphology** referring to the angle of loose **Earth** surface materials with the horizontal. In a simple way we can observe the angle by digging a hole in beach sand; steep-sided walls readily collapse and assume the angle of repose. The most prominent examples of the angle of repose are illustrated in the rock debris called talus (scree) that resides at the bases of cliffs after having broken off and fallen down the cliff; these steep rock piles can be many tens of meters tall and active ones are obvious by their lack of vegetation. The maximum sustainable angle of repose is also known as the *critical angle* at which the force of gravity is balanced with the force of resistance holding the material in place. Resistance is governed by the cohesiveness of the material, friction between the pieces of the material, its surface area, and its density. Resistance can be weakened by lubrication by water, **earthquakes**, and even digging by humans. Once gravity overcomes resistance the material must move downslope and assume a new angle of repose. The rate of movement ranges from inexorably slowly to well over 100 kph (60 mph). The maximum angle of repose varies markedly in differing materials and differing climatic environments. Dry sand assumes an angle of repose of up to 35 percent while some rocks repose at angles exceeding 45 percent.

the total mass of life pales in comparison to the mass of the outermost crust. Plant roots can mechanically increase the size of the openings in rocks. Plants can change the amounts of water in a soil profile and thus alter chemical weathering via soil moisture. The respiration of plants releases carbon dioxide, which combines with water to increase the acidity of water. Plants are able to shade the soil thus altering the temperatures and suppressing the rates of chemical reactions. After plants die and become incorporated in humus, humus keeps the soil moister by its impressive water-holding capacity. More remarkably, plants produce organic substances known as chelates capable of chemical alteration of rock by removal of positively charged ions. Some animals are able to ingest soil materials and this breaks apart some materials for further weathering. Burrowing animals move materials to other locations, including down and up within soil profiles, and instigate further weathering.

Rocks and **soils** readied by weathering are primed for transportation. Rocks, rock fragments, and soils on any slope are held against the force of gravity. The steepest angle of slope for a given type of material is called the angle of repose. Any material on a slope steeper than the angle of response is subject to downhill movement impelled by gravity. Such motion may be triggered because of lubrication by rainfall, **earthquakes**, the addition of materials onto the slope, the undercutting of the slope by a stream, or any number of other environmental events, including the actions of humans.

Table 5. Classification of Mass Wasting

Type	Material	Speed	Lubricating water
Rockfall	Rock	Fast	Dry
Landslide	Rock and soil	Fast	Moderately wet
Slump	Soil and Rock	Moderately slow	Moderately wet
Mudflow	Soil	Moderately fast	Wet
Earthflow	Soil	Moderately slow	Wet
Creep	Soil	Slow	Dry
Solifluction	Soil	Slow	Wet

There are several different types of mass wasting and these types are listed in Table 5. The types are differentiated by their speed and degree of lubrication as moderated by the type of material being mass wasted. Some of the types happen so quickly that they are deadly to anything in the way. For instance, rockfalls, landslides, and avalanches can have speeds in excess of 160 kph. They are usually quite localized. Soil creep is universal on soil-covered landscapes and averages a fraction of a centimeter per year. It is imperceptible except for traces such as leaning fence posts and cracked concrete walls.

Some of the terms in the table need elaboration. The familiar avalanche is a form of landslide with much snow accompanying the rock and soils. Earthflows are quite slow and generally over slopes with no connection to the drainage network; mudflows are much faster and confined to valley bottoms. Soil creep is a general form of exceedingly slow downslope movement in all latitudes, while solifluction is a specialized form of soil creep in the active layer of permafrost in tundra environments and is manifest as parallel arcuate lobes of soil material pointed in the downhill direction.

Wind Erosion and Deposition

Air is a form of fluid, albeit much less dense than water, and undergoes significant flow when heated and cooled. It readily expands, contracts, and moves horizontally and vertically. Despite the seeming ease through which we move through the air, it is composed of molecules and the molecules exert force when they are in motion. The force is enough to erode, transport, and deposit considerable amounts of materials. The **landscape** developments related to wind are corporately known as aeolian processes (derived from Aeolius, the mythical Greek keeper of the winds).

The force exerted by air depends on the air's velocity and the force exerted is proportional to the square of the velocity. This means that as velocity increases, the force rapidly increases. For instance, a doubling in wind velocity from 10 kph to 20 kph *quadruples* the force of the wind so that seemingly small wind variations can sometimes make significant differences in the work the wind performs. Along land surfaces, wind is able to modify those surfaces by eroding soil and weathered rock material already in small enough pieces to be moved. Known as deflation, this is especially evident in arid areas where there is little vegetation to hold the material. Unlike water, air is not limited to channels and is frequently capable of transporting materials uphill.

Aeolian transport has three modes depending on wind speeds and sizes of materials. The first mode is creep. Traction is when pieces are pushed along on the landscape. The eddying nature of surface wind is able to concentrate energy onto the surface and move materials. The second mode of transport, saltation, is the "skipping" of somewhat smaller pieces over the landscape. The materials are repeatedly picked up and dropped by whirling eddies that cause parabolic "skips" ranging from a few meters to a few tens of meters. The third mode of transport is suspension. This is generally reserved for materials less than 100 micrometers across, and it is through this mechanism that wind can transport materials over very long distances. For instance, satellite imagery observes dust from the Sahara being blown to the Caribbean basin and Mongolian materials being transported eastward far over the Pacific. When the aeolian material is concentrated it becomes a dust storm, which can be hazardous to plants, animals, and people. However, it has been found that aeolian materials deposited in the world's **oceans** provide a source of nutrients for the oceanic food chain.

As wind transports materials away its erosion creates several types of landforms. **Desert** pavement is an arid land surface created by the wind-caused disappearance of the finer materials leaving the larger size pieces behind; indeed, in places it appears as if someone has carefully laid out pavement. Blowouts are low places that are created when wind moves loose materials off of arid and semiarid surfaces. These basins are usually oblong. They are common on the U.S. Great Plains and Big Hollow west of Laramie, Wyoming approximates 18 km long by 6 km wide by 60 m deep. It is dwarfed by the Qattara Depression in Libya encompassing 15,000 sq km.

When considering arid landscapes like mesas, buttes, and canyons such as found plentifully in the American Southwest, one is tempted to ascribe many of the landforms to the action of wind erosion. Rather than being the major cause of these landforms, wind tends to put on the "finishing touches" by abrading surfaces through sandblasting. Even on the world's driest landscapes, it is actually water that accomplishes the bulk of the erosion.

All wind-blown material eventually settles and deposition is significant on parts of the landscape. Clearly visible are deposits of silts and sands. Silts are weathered pieces of rock that are quite angular in all dimensions and act quite a bit differently than the smaller pieces known as clay and larger pieces known as sand (see **Soils**). Land concentrations of silt are plentiful because of the Pleistocene glaciations that covered much upper middle latitude continental surface ground bedrock into those sizes. Barren of vegetation after the major melting commenced about 18,000 years ago, landscapes dried out and the silts were moved by the wind. The result is a widespread silt covering called löss (from the Swiss German meaning "loose"). Löss can also have origins out of arid regions and volcanic material. It probably covers a tenth of the world's land surfaces.

Löss is light brown in color and is capable of producing very fertile soils that are, unfortunately, very easily eroded via agricultural activity. Prominently, the Huang He (Yellow) River drainage basin of China has a thick löss covering causing the river and the adjoining sea to be yellow. Unlike sand, the angular pieces of löss can maintain vertical slopes. The bluffs along parts of the Mississippi River valley of the United States represent löss many tens of meters thick.

Hills of sand are common on **Earth** and are known as sand dunes. These are formed of materials 0.1 mm across up to 2.0 mm across and can be composed of weathered quartz, gypsum, calcium carbonate, or other materials. The world's largest dunes (in Peru) exceed 1,100 m from base to summit. It is tempting to believe that deserts are covered with ergs (from the Arabic for "dune field"). Sand is relatively plentiful in arid regions in that it is the result of incomplete chemical weathering. Aridity greatly slows weathering and larger sand particles are common rather than the clays in more humid environments. It should be noted that dunes occur in non-arid regions. For instance, major dune formations can be found along the shores of **oceans** and large lakes. Here, processes within the water present a plentiful supply of sand to the coast and then wind builds the dunes. The expansive Sand Hills of north central Nebraska are vegetation-covered relicts of the sand transported on the ending of the Pleistocene ice age.

There are three major dune types and other combinations and subtypes. The most prevalent type is the barchan, which results when wind-blown sand is slowed by a small obstruction. The sand builds up in an arc with the tips and steeper side faced away from the wind. This dune type is indicative of a small supply of sand with a consistent wind direction. Martian barchans have been observed by satellite, thus providing some **remote sensing** knowledge of conditions on Mars.

A second type of dune is the seif, from the Arabic word for "sword." The seif is also called a longitudinal dune. The form is that of a ridge with two slopes that average 33° inclination. The result is a long thin form that is about six times as wide as it is high. Seifs are formed in places with modest supplies of sand and seasonally alternating wind directions. Seifs are found in the Sahara and in Saudi Arabia and can be as many as 400 km long.

The third type of dune is the transverse dune. The transverse dune occurs on surfaces covered thickly by sand. They develop on a locally huge supply of sand with axes at right angles to the prevailing wind. They migrate slowly in the downwind direction. The transverse dune might be thought of as "what the desert is like," yet most of the planet's arid zones do not have a thick sand covering.

Winds and Pressure Systems

Wind is the horizontal motion of the air. It is important in the **Earth** system as a force capable of shaping and damage but also moves mass and energy over great distances thus significantly affecting **climate**. Wind is such an integral part of our geographic environment that humans have feared its extremes like tornadoes and **hurricanes** and have used its force to sail ships and grind grain and, now, to generate electricity.

Wind is caused by differential heating. There are many reasons why places are heated differentially (see **Temperature**) and these differences occur at various **scales**. Over small distances from a lawn to a sidewalk air will be forced to rearrange itself on a summer day because much more of the **solar energy** falling on the sidewalk is expressed as a rise of temperature than over the lawn where some of the solar energy is used to accomplish photosynthesis and evaporate soil moisture. The result of this differential heating is a modest flow of air from the lawn to the sidewalk. For an entire hemisphere, lower latitudes are heated more than higher latitudes because the lower latitudes have higher angles of sun in the sky. The vast difference in heating between equator and poles causes huge translocations of air known as the global wind and pressure belts that are explained below.

One might think that the heating of air causes higher pressure as it does in a confined pressure cooker. In the unconfined **atmosphere**, however, heated air molecules become excited, increase their speeds of movement, and average greater distances apart (lower density). Air density and air pressure are intimately related in that one tracks the other. As air heats, its pressure lowers and wind starts because of the differences in pressure.

Wind does not blow directly from high to low pressure. Instead, it is the product of three forces. Pressure gradient is the rate of pressure change over distance and is represented by lines of equal pressure called isobars; they are commonly denoted in units called millibars. Pressure gradient *force* is represented as a vector directed from high to low pressure at a right angle across the isobars. For wind to start moving there must be pressure gradient force and when isobars are relatively close together the wind will move quickly.

Coriolis force or Coriolis effect stems from the fact that Earth rotates under the wind resulting in the wind being bent (to the right in the Northern Hemisphere and the left in the Southern Hemisphere) of the pressure gradient direction. Coriolis effect is negligible in the tropics and at maximum in the polar regions. Additionally, the magnitude of Coriolis effect corresponds to the wind speed. Coriolis effect is crucial in making large atmospheric disturbances rotate.

Friction is the third force. Friction slows down the wind and alters its direction. Friction is imparted to the wind by contact with Earth's surface and is quite variable over the planet. The least friction is encountered over bodies of water and the greatest over mountains. The rougher the landscape, the greater the friction so that forests and cities are places with slower winds. Over a kilometer or so above Earth's surface air is virtually frictionless as distance from the friction-causing surface increases.

The result is the wind that actually blows near Earth's surface. Wind moves from the high to the low pressure side of the isobars. Winds aloft, over a kilometer from the surface, are frictionless and blow faster than the near-surface winds and parallel to the isobars. In this case, only the pressure gradient and Coriolis effect act on the air, and the wind is termed "geostrophic."

If near-surface isobars are arranged around a center of low or high pressure, air will rotate about the center. In the case of low pressure, the motion will be a counterclockwise spiral into the center in the Northern Hemisphere and a clockwise spiral into the center in the Southern Hemisphere; such is the case with **middle latitude cyclones**, hurricanes, and tornadoes. In the case of high pressure, the motion will be a clockwise spiral away from the center in the Northern Hemisphere and a counterclockwise spiral away from the center in the Southern Hemisphere.

In the vertical dimension, there is not as much flow in the atmosphere as in the horizontal. In fact, worldwide there is a vertical balance between pressure gradient force and gravity. Although air is transported upward and downward from imbalances in these two forces, horizontal imbalances are so much greater that horizontal motion transports thousands of times more air than vertical motion. Vertical imbalances, though, are crucial when considering the rise of air causing **precipitation** or the sinking of air causing cloud-free conditions.

Latitudinal imbalances of energy are key to understanding the world's wind and pressure belts sometimes called the global circulation. In the equatorial latitudes where there is an energy surplus, there is a broad zone of low pressure into which winds converge. This is known as the Intertropical Convergence or the doldrums. Winds are light and variable. The winds converging into the Intertropical Convergence are the trade winds that, on average, extend 5–20° in latitude away from the equator. These are the most dependable flows of air on the planet, and in the Northern Hemisphere blow from the northeast and in the Southern Hemisphere blow from the southeast. The trade winds flow out of the subtropical highs, which represent sinking, warming air forced by air that has risen in the Intertropical Convergence and spread north and south to sink at latitudes centered 25° away from the equator. The subtropical highs have light and variable winds and are the causes of the world's subtropical **deserts**. On the poleward side of the subtropical highs are the surface westerlies flowing into the polar front. The polar front is a stormy zone ranging over considerable latitude and represents the **boundary** between polar and tropical air masses. The polar zones are home to cold, stable areas of high pressure called the polar highs. Out of the polar highs and into the polar lows of both hemispheres flow the polar easterlies.

The entire global system of wind and pressure belts pulses with the **seasons**. For instance, the Intertropical Convergence moves northward in the Northern Hemisphere summer. Also, the northern Subtropical Highs enlarge and shift a bit poleward and the Polar Front weakens and shifts poleward. All this has the effect of making seasonal changes in temperature and precipitation. In the summer season of the Sahel **region** just south of the Sahara, the Intertropical Convergence comes over head for a few weeks providing some rain. When the Intertropical Convergence shifts equatorward, the Sahel becomes arid for the rest of the year.

World Systems Theory

A holistic, analytical approach to international relations developed primarily by the sociologist Immanuel Wallerstein. Although Wallerstein downplays the importance of geography in some of his writings, World Systems Theory (WST) is rooted in spatial concepts and may be viewed as a perspective on the **geography of economic development** at the global **scale**. The theory has much in common with Marxism, especially the notion that capitalism is an exploitative economic structure that enables certain classes (in this case, clusters of economic institutions) to manipulate the system to take advantage of their superior level of wealth, technology, and development, thereby collectively consigning lesser-developed clusters to a position of relative poverty. A second theory that foreshadowed the world systems concept is dependency theory, promulgated by Andre Gunder Frank and other scholars in the 1960s. Dependency theory holds that the global economic system restricts the ability of developing countries to advance, forcing them to remain "dependent" on the advanced economies for sophisticated technology and consumer goods, resulting in a **core and periphery** model of international economic relations. Frank himself adopted WST in a modified form. World Systems Theory adds to the binary arrangement of dependency theory a third dimension, the "semi-periphery," a zone of transition, in which countries are both victimized by the core group, and exploit those on the periphery. The tripartite nature of the system is crucial, because this prevents the formation of a unified opposition to the predominance of the core, because the members of the semi-periphery are unlikely to ally with either the core or the periphery. Two aspects of the theory are clear departures from earlier notions about the dynamics of global economic relationships. First, the system de-emphasizes the importance of the **nation-state** as an independent agent; rather, economic actors in each of the three divisions work across political **boundaries** to preserve and promote their interests. A prime example of this would be a multinational corporation. Second, the assumption that every country undergoes a similar progression in its economic development is rejected—proponents of WST argue that countries that are currently considered "developing" will not evolve economically in the same way that existing economic powers have.

Countries making up the core of the system share many characteristics. They have stable governments with strongly established state institutions, and often have significant military power. Furthermore, these states possess high standards of living, are highly industrialized, enjoy high levels of productivity in all **sectors of the economy**, can access large pools of capital generated by investment mechanisms, and have the most technologically advanced economies. According to Wallerstein, the initial set of economies that compose the core first emerged in the 1500s, as the maritime powers of Europe engaged in **imperialism**. Some proponents, like Frank, suggest that a world system was in place much earlier in history. Nonetheless, concentrated in Europe and later North America, within two centuries these countries established the world system in much the form that it exists today. Industrialization of the core during the 1800s further solidified its control of the world system. Core states typically compete for both resources and markets to maximize their economic advantage. Relations between core countries, and between the core, periphery and semi-periphery, may change as new states join this group and alter the dynamics of economic production. On occasion, one country has come to dominate the core. Most recently this has been the United States, which over the past 70 years has enjoyed the strongest currency, the largest economy as measured by total Gross National Product (GNP), and the most influential financial markets.

The periphery represents the collection of the world's poorest states. Most of these would fall within the more general category of "developing" countries, located for the most part in Africa, south Asia, and Latin America. Economies in the periphery are generally undiversified, and typically rely heavily on extractive activities such as mining, forestry, and agriculture. The level of technological development and application is also low when compared to the core economies,

Mercantilism

Mercantilism is an economic system imposed by colonizing powers on their overseas holdings during the age of **imperialism**, mostly in the 18th and 19th centuries. As European powers became industrialized, their need for both raw materials and markets increased dramatically. By establishing overseas colonies they satisfied both requirements: the colonies would provide new resource bases, and would also serve as emerging markets for manufactured goods from the home country. A great advantage as well was the fact that the colonial power set the prices of both raw materials and finished products, and prevented any competitors from entering the markets. The system kept the colonies economically underdeveloped, as the emergence of local industry was discouraged, if not outlawed. Great Britain, for example, exported large quantities of unprocessed cotton from India during the latter 19th century to supply textile mills in Yorkshire and other areas of the British Isles. The cotton was spun into cloth, clothing, and other manufactured goods, and then exported back to the Indian market in a classic mercantile relationship. The system was highly exploitive of the colonial populations, and created great resentment. In India, Mahatma Gandhi attempted to break the system by encouraging the development of a cotton "cottage industry," in which cloth was produced in local homes.

and both systems of governance and state institutions are unstable and inefficient. Countries located in this zone are often the target of exploitation by companies from the core due to either their natural resource base, or an abundant pool of cheap labor. According to proponents of WST, the countries of the periphery are kept in a state of developmental disadvantage compared to the advanced economies of the core by a host of mechanisms inherent to the system, including the stronger currencies of the core states, the ability of the core to dictate terms of loans, investment and trade (via agencies like the International Monetary Fund, World Bank, and World Trade Organization, as well as private banks and firms), and even the threat of military intervention. In theory it is possible for a country to emerge from the periphery into the semi-periphery, and to even eventually rise to the level of the core economies, but in fact this rarely occurs, because the system heavily favors the continued dominance of the core.

The grouping of countries that make up the semi-periphery represent either countries that have ascended from the periphery, or which have declined in status and influence and have dropped from a position among the core economies. An example of the former would be Argentina, a country confined to the periphery in the early 20th century, but which has experienced significant industrialization and economic advancement over the past several decades. The semi-periphery plays a vital part in maintaining the overall status of the system, because economies in this layer resist forming a united front with those in the periphery against the dominance of the core. In addition, some states in the semi-periphery acquire sufficient influence to exploit states located in the periphery, and therefore do not see it in their economic interests to undermine the commanding position of the core, which they seek to join.

World Systems Theory remains influential among many political geographers, who find the global scale of the analysis useful. Others have criticized the theory based on what they view as several weaknesses. One of the most common charges against WST is that it fails to properly take into account differences in culture and in historical circumstances. Cultural geographers and anthropologists attacking the theory on this basis hold that Wallerstein and others have ignored fundamental differences between cultures and have ignored the distinctive character of various historical civilizations, as well as differing social and economic value systems among countries. Other critics have pointed out that at the time Wallerstein and others were propounding the existence of a nearly ubiquitous world system committed to capitalism, dozens of countries, including the USSR and China, were pursuing economic advancement under a heavily socialized, state-controlled model, and they could hardly be considered as part of a global capitalist system. The rise of transnational corporations along with **globalization** of capital markets has also undermined some of the fundamental assumptions of WST, because such corporations frequently locate production facilities in the periphery, dramatically increasing industrial production there. Yet despite the critiques of its detractors, WST continues to be a widely discussed and debated theory of state relations and economic development among geographers and social scientists in general.

Z

Zoogeographic Regions

Traveling about **Earth**, it is manifest, even to the most casual of observers, that animal life has significant differences at **scales** of continents and large parts of continents. The concept of zoogeographic **regions** is explanatory of the animal geography of the planet. Rather than being controlled by the plant environment, a zoogeographic region is a large area based on forms of animals (taxonomy) and the evolutionary connections (phylogenetic relations) between them. The taxonomical hierarchy within the animal kingdom ranges from general to specific: phyla, class, order, family, genus, and species. Zoogeographic regions are focused on the nature of vertebrate families and species unique to that region. Vertebrates are those animals possessing backbones surrounding their nerve chords and include families of mammals, birds, fish, amphibians, and reptiles. That is, this regionalization is based on a minority of the animals present on the planet.

Rather than being thought of as exactly defined, zoogeographic regions are composites of the totality of animal life. The **boundaries** are dependent on major **climate** boundaries and/or distances over **oceans**. Considering other concepts such as the great amounts of time over which these families of animals have evolved, long-term changes in Earth's systems such as climate change and continental drift have played roles in the present day configuration of the zoogeographic regions.

Alfred Russel Wallace was, with Charles Darwin, one of the first two "co-founders" of the theory of evolution. Wallace was a naturalist and geographer and is the acknowledged scientific father of animal geography (zoogeography). In the 1850s he was on an expedition collecting animal specimens in the Malay Archipelago. He noticed that between the Malay islands and the nearby Celebes there was a significant divide in the taxonomy. To the west, the animals were similar to what was known in Asia and to the east the species more resembled those in Australia. This distinction was so marked over a few tens of kilometers that it became known as "Wallace's Line." This discovery inspired Wallace to study the rest of the world to identify major zoogeographic boundaries.

In his great 1876 work, *The Geographic Distribution of Animals*, Wallace identified six zoogeographic regions and component subregions for Earth. The regionalization was made from vertebrate animals with little else considered. Additional regions and boundary adjustments have been proposed by subsequent scholars, but the zoogeographic concept has remained intact for well over a century. It was not an attempt to micro-analyze the faunal world, but it provided a "big picture" of the planet that would help aid the scientific classification of species so very much debated in those times. In a sense, zoogeographic regions are the animal counterparts to plant-based **biomes**, but plants have secondary status in zoogeographic regions. When zoogeographic regions are considered along with biomes and the later-developed numeric climate classifications, there cannot help but be the sense of the grand organization of the Earth by energy and moisture availability.

Wallace's original map classifies oceanic islands as part of the six zoogeographic regions while not considering life *in* the oceans. In a sense this was proper because of the relative lack of knowledge about the oceans in the latter half of the 19th century. Later scholars have corrected this. A major addition to the concept has been the inclusion of ocean life zones. These are usually identified as the same zones used in describing the marine biomes of Earth.

Some studies have quantified boundaries via statistical analysis. Wallace's classification as given here ignores extinct families and families that were unknown in his day. Still, the zoogeographic concept was so strong that these names are still used in the literature.

Table 6 is a listing of the number of animal families and the number of endemic families of each zoogeographic region; an endemic species is one that has its natural distribution solely in a single region. There are several versions of world zoogeographic regions and they usually number six to nine. Here we use Wallace's original work.

The Neotropical Region is comprised of South America and tropical North America. Animal families are diverse and this reflects the long separation of South America from other regions and the variety of climates—from **desert** to tundra. This is the region with the largest number of endemic families and is particularly blessed with birds among the other vertebrate life.

The Nearctic Region is composed of non-tropical North America. The Nearctic has had considerable amounts of species influx from the Palearctic Region because of the land bridge between Siberia and Alaska in recent Earth history. The region is usually viewed as transitional between the nearby Neotropical and Nearctic regions. Although quite varied in climate, the Nearctic is not as diverse as its neighbor regions.

The Palearctic is the largest zoogeographic region and encompasses northern Eurasia and Africa north of the Sahara. The region is less diverse than those closer to the equator because of limitations of warmth and/or moisture over large expanses. Frequently, modern writers include the Nearctic and Palearctic as one region known as the Holarctic in deference to the faunal impacts of the land bridge.

Table 6. Terrestrial Zoographic Region as Delineated by Wallace

Zoogeographic region	Number of animal families	Number of endemic families
Neotropical	50	19
Nearctic	37	2
Palearctic	42	2
Ethiopian (including Madagascar)	52	18
Oriental	50	4
Australian (excluding New Zealand)	28	17

The Ethiopian Region is Africa south of the Sahara and the southern part of the Arabian Peninsula. The large expanses of ocean along the coastlines and the climatic barrier of the desert to the north have kept this realm isolated over long geologic periods although the isolation is a relative descriptor since it is far from complete. The region is tropical and subtropical so it has year-round warmth and, at least, seasonally available water. The result is a region of great diversity. Madagascar was a subregion on Wallace's original map, but subsequent investigations have shown the island to be unique enough in its vertebrate population to be accorded the status of a region.

The Oriental Region is composed of southern Asia and its associated islands south of the Himalayas and the other mountain chains that delineate the northern boundary of southern Asia. The Oriental Region is most similar to the Ethiopian Region, being composed of tropical and subtropical climates. Yet, the Oriental Region is usually thought of as being less diverse than the Ethiopian Region with the exception of families of reptiles and birds.

The Australian Region is usually thought of as being "most unique." There is not extreme diversity because of the arid and semiarid nature of most of Australia, but here can be seen a divergence in evolution from the other large landmasses of the world. Australia started its break from Gondwanaland about 120 million years ago. Animal life has taken its own course and, for instance, out of nine families of terrestrial mammals there are eight endemic to Australia. Of note are the monotremes, (egg laying) which are not found outside of the region, and the marsupials (the young live in pouches), only a few of which are found outside of Australia. Modern work has delineated the New Zealand Region as separate from Wallace's Australian Region, particularly because of the presence of many species of flightless birds.

Selected Bibliography

Alonso, William. *Location and Land Use: Toward a General Theory of Land Rent.* Cambridge, MA: Harvard University Press, 1964.

Arreola, Daniel. *Tejano South TA: A Mexican American Cultural Province* (Jack and Doris Smothers Series in Texas Life, History and Culture, No. 5). Austin, TX: University of Texas Press, 2002.

Beesley, Kenneth, Hugh Millward, Brian Ilbery, and Lisa Harrington, editors. *The New Countryside: Perspectives on Rural Change.* Brandon, MB: Brandon University Press, 2003.

Berry, Brian. "Approaches to regional analysis: A synthesis," *Annals of the Association of American Geographers*, Vol. 54, No. 1 (1964).

Black, W. "Transportation Geography," in *Geography in America*, edited by G. Gaile and C. Willmott. Columbus, OH: Merrill Publishers, 1989.

Bridge, G. "Grounding Globalization: The prospects and perils of linking economic processes to environmental outcomes," *Economic Geography*, Vol. 78, No. 3 (2002).

Brunn, Stanley D., Susan L. Cutter, and J. W. Harrington, Jr., editors. *Geography and Technology.* Dordrecht, The Netherlands: Kluwer Academic Publishers, 2004.

Butlin, Robin. *Historical Geography: Through the Gates of Space and Time.* New York: Edward Arnold, 1993.

Carville Earle, Kent Mathewson, and Martin S. Kenzer, editors. *Concepts in Human Geography.* Lanham, MD: Rowman & Littlefield, 1996.

Casey, Edward. *Getting Back into Place: Toward a Renewed Understanding of the Place-World*, 2nd edition. Bloomington, IN: Indiana University Press, 2009.

Castles, Stephen, and Mark J. Miller. *The Age of Migration, Fourth Edition: International Population Movements in the Modern World.* New York: The Guilford Press, 2009.

Chang, Kang-tsung. *Introduction to Geographic Information Systems.* Boston: McGraw Hill Publishers, 2008.

Clout, Hugh, editor. *Contemporary Rural Geographies*. London and New York: Routledge, 2007.

Coe, Neil, Philip Kelly, and Henry W. C. Yeung. *Economic Geography: A Contemporary Introduction*. London: Wiley-Blackwell, 2007.

Cohen, Saul B., editor. *The Columbia Gazetteer of the World*. New York: Columbia University Press, 1998.

Cohen, Saul B., editor. *Geopolitics. The Geography of International Relations*, 2nd edition. Lanham, MD: Rowman and Littlefield Publishers, Inc., 2009.

Cruikshank, James George. *Soil Geography*. New York: John Wiley & Sons, 1972.

Cutter, Susan, editor. *American Hazardscapes: The Regionalization of Hazards and Disasters*. Washington, DC: Joseph Henry Press, 2001.

Demko, George. *Why in the World: Adventures in Geography*. New York: Doubleday, 1992.

Evans, David, editor. *Geomorphology, Critical Concepts in Geography*, Vols. 1–7. London: Routledge, 2004.

Fonseca, James, and David Wong. "Changing patterns of population density in the United States," *Professional Geographer*, Vol. 52, No. 3 (2000).

Frey, William. "Immigration and internal migration 'flight' from U.S. metropolitian areas: Toward a new demographic Balkanization," *Urban Studies*, Vol. 32, Nos. 4–5 (1995).

Gaile, Gary L., and Court J. Willmott, editors. *Geography in America at the Dawn of the 21st Century*. Oxford, UK: Oxford University Press, 2003.

Garrison, Tom S. *Oceanography: An Invitation to Marine Science*, 2nd edition. Belmont, CA: Brooks/Cole, 2010.

Glassner, Martin, and Chuck Fahrer. *Political Geography*, 3rd edition. New York: John Wiley & Sons, Inc., 2004.

Gray, Colin S., and Geoffrey Sloan. *Geopolitics, Geography, and Strategy*. London and New York: Routledge, 1999.

Hagerstrand, Torsten. *Innovation Diffusion as a Spatial Process*. Chicago: University of Chicago Press, 1967.

Haggett, Peter. *Locational Analysis in Human Geography*. New York: St. Martin's Press, 1966.

Hanna, Stephen. "Finding a place in the world-economy: Core-periphery relations, the nation-state and the under-development of Garrett County, Maryland," *Political Geography*, Vol. 14, No. 5 (1995).

Hart, John Fraser. "The highest form of the geographer's art," *Annals of the Association of American Geographers*, Vol. 72, No. 1 (1982).

Hart, John Fraser. *The Rural Landscape*. Baltimore: Johns Hopkins University Press, 1998.

Hartshorne, Richard. *The Nature of Geography*. Lancaster, PA: Association of American Geographers, 1939.

Herschy, Reginald W., and Rhodes Fairbridge, editors. *Encyclopedia of Hydrology and Water Resources*. Dordrecht: The Netherlands, 1997.

Holt-Jensen, Arid. *Geography: History and Concepts*, 4th edition. Thousand Oaks, CA: Sage Publications, 2009.

Hooson, David, editor. *Geography and National Identity*. Oxford: Blackwell, 1994.

Hoyt, Homer. *The Structure and Growth of Residential Neighborhoods in American Cities*. Washington, DC: Federal Housing Administration, 1939.

Huggett, Richard John. *Geomorphology*, 3rd edition. London and New York: Routledge, 2011.

Hugill, Peter, and Kenneth Foote. "Re-Reading Cultural Geography," in *Re-Reading Cultural Geography*, edited by Kenneth Foote, Peter Hugill, Kent Mathewson, and Jonathan Smith. Austin, TX: University of Texas Press, 1994.

Huntington, Ellsworth. *Mainsprings of Civilization*. New York: John Wiley & Sons, 1945.

Kaplan, David, James O. Wheeler, and Steven Holloway. *Urban Geography*, 2nd edition. New York: John Wiley & Sons, Ltd., 2008.

Kuby, Michael, John Harner, and Patricia Gober. *Human Geography in Action*, 5th edition. New York: John Wiley & Sons, Ltd., 2009.

Lillesand, Thomas, Ralph W. Kiefer, and Jonathon Chapman. *Remote Sensing and Image Interpretation*, 6th edition. New York: John Wiley & Sons, 2007.

Livingstone, David. *The Geographical Tradition*. Malden, MA: Blackwell Publishing, 1993.

MacKinder, Halford J. "The geographical pivot of history," *The Geographical Journal*, Vol. 23, No. 4 (1904).

Malthus, Thomas. *Essay on the Principle of Population* (Oxford's World Classics). New York: Oxford University Press, 2008.

Marsh, George Perkins. *Man and Nature*. Edited by David Lowenthal. Seattle: University of Washington, 2003.

Martin, Geoffrey J., and Preston E. James. *All Possible Worlds: A History of Geographical Ideas*, 3rd edition. New York: John Wiley & Sons, Ltd., 1993.

Mathez, Edmond A. *Climate Change: The Science of Global Warming and Our Energy Future*. New York: Columbia University Press, 2009.

McKnight, Tom, and Darrell Hess. *Merriam-Webster's Geographical Dictionary*, 3rd edition. Springfield, MA: Merriam-Webster, Incorporated, 2007.

McKnight, Tom, and Darrell Hess. *Physical Geography: A Landscape Appreciation*, 10th edition. Upper Saddle River, NJ: Pearson Prentice Hall, 2011.

Meinig, Donald. *The Interpretation of Ordinary Landscapes: Geographical Essays*. New York: Oxford University Press, 1979.

Meinig, Donald. *The Shaping of America: A Geographical Perspective on 500 Years of History*, Vol. 1. Binghamton, NY: Vail-Ballou Press, 1986.

Mikesell, Marvin. *Geographers Abroad: Essays on the Problems and Prospects of Research in Foreign Areas*. Chicago: University of Chicago Research Papers, 1973.

Muehrcke, P. *Map Use: Reading, Analysis and Interpretation*. Madison, WI: IP Publications, 1983.

Mumford, Lewis. *The City in History*. San Diego: Harcourt Brace Jovanovich, Publishers, 1961.

Murphy, Alexander, Terry Jordan-Bychkov, and Bella Bychkova Jordan. *The European Culture Area: A Systematic Geography*. Lanham, MD: Rowman & Littlefield Publishers, Inc., 2009.

Murphy, Alexander, Terry Jordan-Bychkov, and Bella Bychkova Jordan. *National Geographic Desk Reference*. Washington, DC: National Geographic Society, 1999.

O'Brien, Larry. *Introducing Quantitative Geography, Measurement, Methods and Generalised Linear Models*. London: Routledge, 1992.

Oliver, John E., editor. *Encyclopedia of World Geography*. Dordrecht, The Netherlands: Springer, 2005.

Olwig, Kenneth. "Rediscovering the substantive meaning of landscape," *Annals of the Association of American Geographers*, Vol. 86, No. 4 (1996).

Orme, Anthony R., editor. *The Physical Geography of North America*. Oxford, UK: Oxford University Press, 2002.

Pacione, Michael, editor. *Applied Geography: Principle and Practice*. London: Routledge, 1999.

Pattison, William. "Four Traditions of Geography," *Journal of Geography*, Vol. 63, No. 5 (1994), pp. 211–216.

Peet, Richard. *Modern Geographical Thought*. London: Wiley-Blackwell, 1998.

Rodrigue Jean-Paul, Claude Comtois, and Brian Slack. *The Geography of Transport Systems*, 2nd edition. London: Routledge, 2009.

Rohli, Robert V., and Anthony J. Vega. *Climatology*. Sudbury, MA: Jones and Bartlett Publishers, 2008.

Rowntree, Les et al. *Globalization and Diversity: Geography of a Changing World*, 3rd edition. Boston: Prentice Hall, 2008.

Sauer, Carl O. *Agricultural Origins and Dispersals*. New York: American Geographical Society, 1952.

Sauer, Carl O. *The Morphology of Landscape*. University of California Publications in Geography, 2 (1925), pp. 19–54.

Shortridge, Barbara, and James R. Shortridge, editors. *The Taste of American Place: A Reader on Regional and Ethnic Foods*. Lanham, MD: Rowman & Littlefield Publishers, Inc., 1998.

Sjoberg, Gideon. *Preindustrial City*. New York: Free Press, 1960.

Smith, Neil. "Gentrification and the rent gap," *Annals of the Association of American Geographers*, Vol. 77, No. 3 (1987).

Smith, W. D. "Friedrich Ratzel and the Origins of Lebensraum," *German Studies Review*, 3 (1980).

Sopher, David. *Geography of Religions*. Englewood Cliffs, NJ: Prentice Hall, Inc., 1967.

Spykman, Nicholas J. *America's Strategy in World Politics: The United States and the Balance of Power*. New York: Harcourt Brace, 1942.

Straussfogel, Debra. "Redefining development as humane and sustainable," *Annals of the Association of American Geographers*, Vol. 87, No. 2 (1997).

Stump, Roger. *The Geography of Religion*. Lanham, MD: Rowman & Littlefield Publishers, Inc., 2008.

Timothy, Dallen, "Political boundaries and tourism: Borders as Tourist attractions," *Tourism Management*, Vol. 16, No. 7 (1995).

Timothy, Dallen, and Daniel Olsen, editors. *Tourism, Religion and Spiritual Journeys*. London and New York: Routledge, 2006.

Tuan, Yi-Fu. *Space and Place*. Minneapolis: University of Minnesota Press, 1977.

Tuan, Yi-Fu. *Topophilia. A Study of Environmental Perception, Attitudes and Values*. Englewood Cliffs, NJ: Prentice Hall, Inc., 1974.

Turner, Billie, and Paul Robbins. "Land-change science and political ecology: Similarities, differences, and implications for sustainability science," *Annual Review of Environment and Resources*, Vol. 33 (2008).

Vogler, John. *The Global Commons: Environmental and Technological Governance*, 2nd edition. New York: John Wiley & Sons, Ltd., 2000.

Wallerstein, Immanuel. *World Systems Analysis: An Introduction*. Durham, NC: Duke University Press, 2004.

West, R., editor. *Pioneers of Modern Geography*. Geoscience and Man 28. Baton Rouge, LA: Louisiana State University Press, 1990.

Wilbanks, T. "Sustainable Development in Geographic Perspective," *Annals of the Association of American Geographers*, Vol. 84, No. 4 (1994).

Wilford, John Noble. *The Mapmakers*, 2nd edition. New York: Random House, Inc., 2001.

Zelinsky, Wilbur. *The Cultural Geography of the United States*. Englewood Cliffs, NJ: Prentice Hall, Inc., 1973.

Index

Note: Page numbers in bold indicate main entries in the encyclopedia. Page numbers followed by "t" indicates a table and "i" indicates an illustration.

About the Author

Reuel R. Hanks is Professor of Geography at Oklahoma State University and serves as the editor of the *Journal of Central Asian Studies*. He received his doctoral degree from the University of Kansas and has taught geography at the University of Missouri, the University of Kansas, Kennesaw State University, and Oklahoma State University. He has been a Visiting Professor at Tashkent State Economics University, Samarkand State Institute for Foreign Languages, KIMEP (Almaty, Kazakhstan), and Eurasian National University (Astana, Kazakhstan). Dr. Hanks was a Fulbright Scholar in Tashkent, Uzbekistan in 1995 and has published more than 20 articles and book chapters on geography and Central Asia. He is the author of three books: *Uzbekistan,* an annotated bibliography in the World Bibliographical Series, *Central Asia: A Global Studies Handbook*, and *Global Security Watch: Central Asia*. He resides in Stillwater, Oklahoma with his wife, Oydin Uzakova, and daughter Kamila.